Technology for the United States Navy and Marine Corps, 2000-2035

Becoming a 21st-Century Force

VOLUME 2 Technology

Panel on Technology
Committee on Technology for Future Naval Forces
Naval Studies Board
Commission on Physical Sciences, Mathematics, and Applications
National Research Council

NATIONAL ACADEMY PRESS
Washington, D.C. 1997

This work was performed under Department of the Navy Contract N00014-96-D-0169/0001 issued by the Office of Naval Research under contract authority NR 201-124. However, the content does not necessarily reflect the position or the policy of the Department of the Navy or the government, and no official endorsement should be inferred.

Copies available from:

Naval Studies Board
National Research Council
2101 Constitution Avenue, N.W.
Washington, D.C. 20418

Printed in the United States of America

PANEL ON TECHNOLOGY

JOSEPH B. REAGAN, Saratoga, California, *Chair*
HERBERT RABIN, University of Maryland, *Vice Chair*
SUSAN D. ALLEN, Florida State University
RONALD CLARK, Lockheed Martin Corporation
ANTHONY J. DeMARIA, DeMaria ElectroOptics Systems, Inc.
DANIEL N. HELD, Northrop Grumman Corporation
RAY L. LEADABRAND, Leadabrand and Associates, Inc.
DAVID W. McCALL, Far Hills, New Jersey
WILLIAM J. PHILLIPS, Northstar Associates, Inc.
DENNIS L. POLLA, University of Minnesota
MARA G. PRENTISS, Jefferson Laboratory, Harvard University
JOHN W. ROUSE, JR., Southern Research Institute
ALBERT I. SCHINDLER, Rockville, Maryland
STEVEN J. TEMPLE, Raytheon Company
EDWARD W. THOMPSON, Hughes Research Laboratory
ROBERT M. WESTERVELT, Harvard University

Invited Participants
ARISTOS CHRISTOU, University of Maryland
FRANK A. HORRIGAN, Raytheon Electronic Systems
JOHN W.R. POPE, JR., Tri-Tech Microproducts
TIMOTHY D. ROARK, TRW
HOWARD STEVENS, Vector Research (as of January 1, 1997)

Navy Liaison Representatives
PAUL G. BLATCH, Office of the Chief of Naval Operations, N911E
FRED WOLPERT, Office of the Chief of Naval Operations, N911E1

Consultants
LEE M. HUNT
SIDNEY G. REED, JR.
JAMES G. WILSON

Staff
RONALD D. TAYLOR, Director, Naval Studies Board
PETER W. ROONEY, Program Officer
SUSAN G. CAMPBELL, Administrative Assistant
MARY G. GORDON, Information Officer
CHRISTOPHER A. HANNA, Project Assistant

COMMITTEE ON TECHNOLOGY FOR FUTURE NAVAL FORCES

DAVID R. HEEBNER, Science Applications International Corporation (retired), *Study Director*
ALBERT J. BACIOCCO, JR., The Baciocco Group, Inc.
ALAN BERMAN, Applied Research Laboratory, Pennsylvania State University
NORMAN E. BETAQUE, Logistics Management Institute
GERALD A. CANN, Raytheon Company
GEORGE F. CARRIER, Harvard University
SEYMOUR J. DEITCHMAN, Institute for Defense Analyses (retired)
ALEXANDER FLAX, Potomac, Maryland
WILLIAM J. MORAN, Redwood City, California
ROBERT J. MURRAY, Center for Naval Analyses
ROBERT B. OAKLEY, National Defense University
JOSEPH B. REAGAN, Saratoga, California
VINCENT VITTO, Lincoln Laboratory, Massachusetts Institute of Technology

Navy Liaison Representatives
RADM JOHN W. CRAINE, JR., USN, Office of the Chief of Naval Operations, N81 (as of July 4, 1996)
VADM THOMAS B. FARGO, USN, Office of the Chief of Naval Operations, N81 (through July 3, 1996)
RADM RICHARD A. RIDDELL, USN, Office of the Chief of Naval Operations, N91
CDR DOUGLASS BIESEL, USN, Office of the Chief of Naval Operations, N812C1
PAUL G. BLATCH, Office of the Chief of Naval Operations, N911E

Marine Corps Liaison Representative
LtGen PAUL K. VAN RIPER, USMC, Marine Corps Combat Development Command

Consultants
LEE M. HUNT
SIDNEY G. REED, JR.
JAMES G. WILSON

Staff
RONALD D. TAYLOR, Director, Naval Studies Board
PETER W. ROONEY, Program Officer
SUSAN G. CAMPBELL, Administrative Assistant
MARY G. GORDON, Information Officer
CHRISTOPHER A. HANNA, Project Assistant

NAVAL STUDIES BOARD

Marine Corps Liaison Representative
LtGen PAUL K. VAN RIPER, USMC, Marine Corps Combat Development Command

RONALD D. TAYLOR, Director
PETER W. ROONEY, Program Officer
SUSAN G. CAMPBELL, Administrative Assistant
MARY G. GORDON, Information Officer
CHRISTOPHER A. HANNA, Project Assistant

Preface

This report is part of the nine-volume series entitled *Technology for the United States Navy and Marine Corps, 2000-2035: Becoming a 21st-Century Force*. The series is the product of an 18-month study requested by the Chief of Naval Operations (CNO). To carry out this study, eight technical panels were organized under the Committee on Technology for Future Naval Forces to examine all of the specific technical areas called out in the terms of reference.

On November 28, 1995, the Chief of Naval Operations requested that the National Research Council initiate (through its Naval Studies Board) a thorough examination of the impact of advancing technology on the form and capability of the naval forces to the year 2035. The terms of reference of the study specifically asked for an identification of "present and emerging technologies that relate to the full breadth of Navy and Marine Corps mission capabilities," with specific attention to "(1) information warfare, electronic warfare, and the use of surveillance assets; (2) mine warfare and submarine warfare; (3) Navy and Marine Corps weaponry in the context of effectiveness on target; [and] (4) issues in caring for and maximizing effectiveness of Navy and Marine Corps human resources." Ten specific technical areas were identified to which attention should be broadly directed. The CNO's letter of request with the full terms of reference is given in Appendix A of this report.

The Panel on Technology was constituted to address item 1 of the terms of reference:

> Recognizing the need to obtain maximum leverage from Navy and Marine Corps capital assets within existing and planned budgets, the review should place

> emphasis on surveying present and emerging technical opportunities to advance Navy and Marine Corps capabilities within these constraints. The review should include key military and civilian technologies that can affect Navy and Marine Corps future operations. This technical assessment should evaluate which science and technology research must be maintained in naval research laboratories as core requirements versus what research commercial industry can be relied upon to develop.

Panel membership included broad expertise in managing large-scale technology development programs, as well as specific expertise in condensed matter physics; electronics and electrical engineering; photonics, optics, and electro-optics; materials science, including micro- and nanofabrication; chemistry and chemical engineering; computer and information science; and space science and engineering. The panel held 12 meetings over the course of a year during which it received input from scientists, engineers, and decisionmakers from government, industry, and academia.

Acknowledgments

The Panel on Technology would like to acknowledge the critical help of the following experts who made important contributions to this report: M. Al-Sheikhly, S. Ankem, R. Briber, W. Chappas, I. Lloyd, J. Quinn, R. Ramesh, G. Rubloff, and M. Wuttig, all from the University of Maryland; R.C. Cammarata, from Johns Hopkins University; L.-Q. Chen, from Pennsylvania State University; C.K. Cowan, P.G. Mulgaonkar, and V. Shastri, from SRI International; R. Crowe, from the Defense Advanced Research Projects Agency (DARPA); M.J. Daily and K. Reiser, from Hughes Research Laboratory; D.E. Dietrich, from Mississippi State University; J. Dorr, from Duke and Associates; R.C. Herndon, G. Jones, T.N. Krishnamurti, and J.J. O'Brien, from Florida State University; M.E. Fine and S. Vaynman, from Northwestern University; O.T. Inal, from the New Mexico Institute of Mining and Metallurgy; D.V. Burke, Jr., M. O'Brien, M. Prestero, P. Rosenstrach, and G.T. Schmidt, from the Charles Stark Draper Laboratory; R. Spindel, from the Applied Physics Laboratory at the University of Washington; and C. Tschan, from the U.S. Air Force's 50th Weather Squadron.

In addition to acknowledging the valuable help of the outside experts noted above, the panel's chair would like to acknowledge the following panel members and invited participants for their extraordinary efforts during the course of this study: S.D. Allen, from Florida State University; A. Christou, from the University of Maryland; R. Clark, from Lockheed Martin Corporation; F.A. Horrigan, from Raytheon Electronic Systems; W.J. Phillips, from Northstar Associates, Inc.; J.W. Rouse, Jr., from Southern Research Institute; and E.W. Thompson, from Hughes Research Laboratory.

Contents

Executive Summary

The future national security environment will present the naval forces with operational challenges that can best be met through the development of military capabilities that effectively leverage rapidly advancing technologies in many areas. The panel envisions a world where the naval forces will perform missions in the future similar to those they have historically undertaken. These missions will continue to include sea control, deterrence, power projection, sea lift, and so on. The missions will be accomplished through the use of platforms (ships, submarines, aircraft, and spacecraft), weapons (guns, missiles, bombs, torpedoes, and information), manpower, materiel, tactics, and processes (acquisition, logistics, and so on.). Accordingly, the Panel on Technology attempted to identify those technologies that will be of greatest importance to the future operations of the naval forces and to project trends in their development out to the year 2035.

The primary objective of the panel was to determine which are the most critical technologies for the Department of the Navy to pursue to ensure U.S. dominance in future naval operations and to determine the future trends in these technologies and their impact on Navy and Marine Corps superiority. A vision of future naval operations ensued from this effort. These technologies form the base from which products, platforms, weapons, and capabilities are built. By combining multiple technologies with their future attributes, new systems and subsystems can be envisioned.

The panel also attempted to identify those technologies that are unique to the naval forces and whose development the Department of the Navy clearly must fund, as well as commercially dominated technologies that the panel believes the Navy and Marine Corps must learn to adapt as quickly as possible to naval

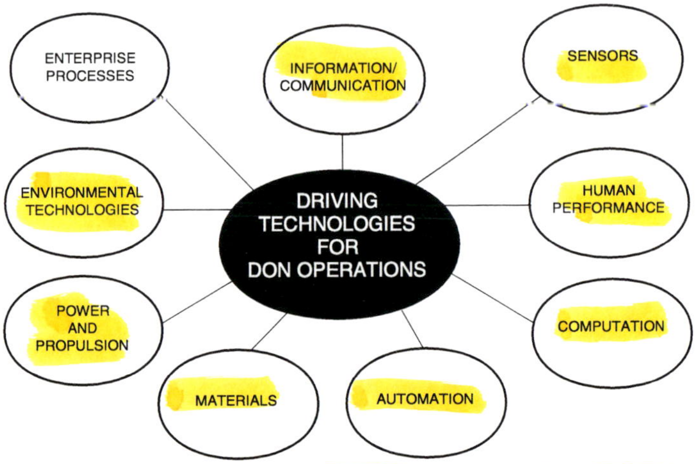

FIGURE ES.1 The driving technology areas for future Department of the Navy operations.

applications. Since the development of many of the critical technologies is becoming global in nature, some consideration is given to foreign capabilities and trends as a way to assess potential adversaries' capabilities.

Finally, the panel assessed the current state of the science and technology (S&T) establishment and processes within the Department of the Navy and makes recommendations that would improve the efficiency and effectiveness of this vital area.

MAJOR TECHNOLOGY APPLICATION AREAS

The panel identified more than 100 important technologies that will form the technology base for the future naval forces and grouped them into the following nine major application areas (Figure ES.1):

- Computation,
- Information and communications,
- Sensors,
- Automation,
- Human performance,
- Materials,
- Power and propulsion,
- Environments, and
- Enterprise processes.

Each technology is described in the body of the report, its relevance to the Navy and Marine Corps identified, the current status established, and the future trends forecast. The developments necessary to achieve the benefits of the forecast, some estimates of the time scale for these developments, and the status of foreign capabilities are also assessed.

To compile a comprehensive list of technologies that are highly relevant to future naval operations, the panel not only relied on its own expertise but also surveyed numerous technologies used by the U.S. Air Force and U.S. Army and those under development by the Defense Advanced Research Projects Agency and other government agencies. Future trends in these technologies, as expressed in such studies as *STAR 21*[1] and *New World Vistas*,[2] were evaluated. The panel also evaluated new technologies in the commercial marketplace.

As expected, the panel had extensive discussions with the science and technology community of the Department of the Navy, including the Office of Naval Research (ONR), the Naval Research Laboratory (NRL), the Naval Command, Control and Ocean Surveillance Center, the Office of the Chief of Naval Operations, and the Marine Corps. A visit to the USS *Yorktown*, the U.S. Navy's "smart" ship, proved particularly informative. The smart ship is a project sponsored by the Chief of Naval Operations to reengineer a combatant Aegis cruiser with new operational processes and procedures supported by new technologies in the short period of 12 months. The smart ship has achieved reductions in manpower and maintained operational capabilities through the use of advanced technologies. The success of the smart ship project rests on its ability to empower ship commanders, reduce cycle time, and take timely advantage of the latest technologies.

From the more than 100 technology areas important to future naval operations, the panel has extracted several specific technologies that will likely have the biggest impact on changing the way the Department of the Navy conducts operations in the future and that, because of their importance, deserve careful attention in the future. These include:

- Micro- and nanoscale technologies
  —Microelectromechanical systems
  —Nanoscale electronic circuits
  —Systems-on-a chip;
- Teraflop affordable computers and petaflop high-performance computers;

---

[1]Board on Army Science and Technology. 1993. *STAR 21: Strategic Technologies for the Army of the Twenty-First Century,* National Academy Press, Washington, D.C.

[2]Air Force Scientific Advisory Board. 1995. *New World Vistas, Air and Space Power for the 21st Century,* United States Air Force, December.

- Genomics—the marriage of biotechnology, genetics, and electronics;
- Smart materials involving nanophase materials engineering;
- Ubiquitous wideband communications and connectivity;
- Global distributed collaboration;
- Multisensory virtual reality environments;
- Information warfare, defensive and offensive;
- Autonomous agents; and
- Signature management and warfare.

TECHNOLOGY INVESTMENT TO SUPPORT FUTURE CAPABILITY

For each of the more than 100 technologies identified in this report, the Panel on Technology attempted to determine whether the Department of the Navy must invest heavily in the future because of unique applications—such as torpedoes, stealth for ships and submarines, and unmanned underwater vehicles—or whether the commercial sector could be expected to drive future developments, such as those in computers, software, and materials, in which case the Department of the Navy must be a smart and active user that adapts the commercial technologies to naval forces' use.

The world of technology is advancing at a blistering pace. For example, the speed of computers, memory size, and associated elements have doubled every 18 months over the past 40 years. Many advanced technologies are now available globally and with the rapid expansion in telecommunications will continue to be available to future adversaries. To maintain technological superiority and dominance of the battlefield, the naval forces must reengineer themselves continually to exploit these rapid advances. In contrast to the past, many of these advances will come from the commercial sector. The rapid growth in information and telecommunication technologies is a good example. The Department of the Navy must be an agile user and adapt these commercial technologies rapidly for military use.

It is a daunting task to forecast technology trends some 40 years into the future. Everyone tends to be guided by experiences of the past. The panel knows that it will surely make mistakes, mostly of omission. For example, 40 years ago even the best technology minds missed the phenomenal growth and impact of computers on society. Merely identifying the future capability of a technology or how several technologies might be used together to create a new capability does not necessarily mean that these events will happen. Vision, belief in the forecasts, strategic planning based on that belief, investments in development and applications, and rapid adoption are all additional ingredients necessary to turn a technology dream into real systems for the future.

In future warfare, the naval forces will be operating predominantly in cooperative engagements as part of joint and multinational task forces. This mode of operation brings the requirement for compatibility of communications, databases, and operating systems. The Navy and Marine Corps must adapt to this new environment through adherence to structured methodologies, international standards, open systems, and rapidly reconfigurable systems.

Investments in science and technology have served the naval forces well over the past 50 years. Military technology requirements will continue to drive the state of the art in the future. The panel strongly believes that the Department of the Navy must continue to support the development of those fundamental sciences and technologies that are relevant to naval operations. This includes the nurturing of discoveries and inventions in universities through prototype applications in industrial and government laboratories into full-scale proof-of-concept demonstrations. The panel sees no alternative to this course of action. No other government entity will ensure the advances and delivery of technologies of vital interest to naval operations. This is especially true of areas that are unique to the Navy, such as oceanography.

The panel is concerned that the research and development (R&D) emphasis within the Department of the Navy has shifted from longer-term research to shorter-term applied development. Accompanying this shift is an increase in the amount of process, auditing, and scrutiny that is distracting scientists and engineers in universities and industrial and government laboratories from the real work of research and development. This is not to say that scientists and engineers should be isolated from understanding the future needs of the naval forces. Those future needs should be made abundantly clear to members of the science community supported by the Department of the Navy; their efforts must be directed toward those objectives and they should be held accountable. Fundamental research is necessary for the future, and the Department of the Navy must recognize that making progress involves risk and failures. New ideas and concepts need receptive support and stable funding. Fundamental research should be buffered from the same criteria that are applied to higher-level development efforts.

Finally, the Department of the Navy should capitalize on the emerging technological opportunities described in this report, even in the face of severe budget constraints and reduced manpower. To do so demands prioritization of goals, significant changes in the way the Department of the Navy conducts business, tradeoffs between modernization and infrastructure, and the use of the technologies identified in this report, not only to ensure the dominant warfighting capability of naval forces, but also to reduce the size and cost of the infrastructure. It is clear to the panel that technologies, particularly information technologies, must be applied to a reengineered infrastructure to significantly reduce cost and to gain the needed resources to modernize weapon systems and platforms as well as retain within the naval forces the warfighters of the future.

CONCLUSIONS AND RECOMMENDATIONS

The Panel on Technology makes the following recommendations to the Chief of Naval Operations:

- Information technology will dominate future warfare and must be elevated in priority. Rapid access to appropriate knowledge at all levels will optimize warfighting and crisis response capabilities. Commercial technologies in knowledge extraction, data management, and data presentation, together with unique military technologies in data fusion and automatic target recognition to deal with the increased complexity and tempo of warfare, must be pursued. Department of the Navy information systems should be protected against increased software and electromagnetic warfare attacks and other vulnerabilities. The Department of the Navy should develop offensive information and electronic warfare technologies to find, identify, and attack adversary systems and to strengthen naval systems.
- Computer technology will be a major enabler of future naval operations. Computers will enable enhanced situational awareness, realistic modeling and simulation, faster warfighting decisions, more effective weapons, lower-cost platforms, and more efficient and effective use of people. The Department of the Navy should exploit the continual evolution of commercial computer technologies into robust computational systems.
- The Department of the Navy should undertake early exploitation of the new innovations in commercial communication satellites and fiber optics to acquire the necessary increased bandwidth and diverse routing for future networking needs.
- Naval operations are increasingly dependent on enhanced sensor data to provide situational awareness, target designation, weapon guidance, condition-based maintenance, platform automation, personnel health and safety monitoring, and logistic management. The Department of the Navy should provide continuing support of sensor technology for areas critical to future naval operations. Special attention should be given to applications of microelectromechanical systems technology because it offers the advantage of low-cost, high-capability systems-on-a-chip that will enable future cooperative sensor networks.
- Automation increases manpower effectiveness and warfighting capability by performing routine functions, conducting superhuman and hazardous operations, and minimizing casualties. The Department of the Navy should field a vigorous program in the technologies for ship automation that will realize these benefits. Unmanned aerial vehicles and unmanned underwater vehicles will play a major role in future naval warfare as surveillance, communication, targeting, and weapon-guidance platforms. The Department of the Navy should support technology developments to increase mission duration and operational capability, enhance sensor payloads, and increase survivability.
- Economic and social conditions will force the Navy and Marine Corps to

conduct future missions with fewer people and lower overall manpower costs. To accomplish this, the Department of the Navy should exploit the technology advances in communications, information, health care, biotechnology and genetics, and cognitive processes to enhance human performance through expanded education and training, improved personal health and safety, and enhanced quality of life throughout the Navy and Marine Corps.

- Materials are used in every aspect of naval operations. In the future, entirely new and enhanced existing materials will be designed and manufactured using a computational approach in which the physical and mechanical properties of materials are understood on an atomic scale. The nanophase materials engineered in this way will be tailored to meet specific requirements and to be reliable and robust at lower life-cycle cost. The Department of the Navy should strongly support the development of this new materials design and processing approach.
- Direct electric drive for ships and submarines offers unique advantages for the future naval forces in the areas of reduced volume, modular flexible propulsion units, lower acoustic signatures, enhanced survivability, and the enabling of new capabilities. The power and propulsion technologies of efficient gas turbine propulsion units, modular rare-earth permanent magnetic motors, and power control modules have matured to the point that the Department of the Navy should place a high priority on the development of new all-electric ships with the associated drive, power-conditioning, and distribution systems. In the future, incorporation of superconductivity into motors, energy storage, and power distribution will further increase capabilities.
- Battle-space awareness, communications, target identification, navigation, weapon guidance, and tactical planning all require real-time understanding and forecasting of the atmospheric, space, and sea environments of operation. Global weather models with improved satellite data on winds, temperature, solar inputs, and so on will permit the generation of accurate weather forecasts. Space weather forecasting of solar disturbances, scintillation phenomena, and other disturbances will be modeled based on real-time satellite data. The Department of the Navy must support the development of this modeling capability. The Department of the Navy should continue to support the measurement and modeling of not only the deep-ocean environment, but also, especially, the littoral waters to improve mine and submarine detection.
- Large-scale processes within the Department of the Navy, such as platform acquisition, logistics management, resource planning, mission planning, and personnel management, are major cost drivers of naval operations. Information technologies are becoming available that can revolutionize the execution of these enterprise processes with a resultant substantial reduction in manpower, cycle time, risk, and cost. Simulation-based acquisition is an example of a revolutionary approach for the acquisition of complex platforms and systems. The Department of the Navy should strongly embrace and support these information technologies for enterprise-wide processes.

- Science and technology will continue to be the essential underpinning for maintaining superior Navy and Marine Corps warfighting capabilities. The Department of the Navy should follow a three-pronged strategy: (1) exploit rapidly evolving commercial technologies, such as computer, information, and communication technology, and biotechnology; (2) maintain technical leadership in noncommercial areas of naval importance, such as weapons, sensors, oceanography, and naval platforms; and (3) continue to support vigorously those areas of fundamental, long-term basic research, primarily conducted at universities, from which new understanding and new naval technologies evolve.

1

Overview of Technology Opportunities

The Panel on Technology identified and focused on the fundamental technologies of importance to the future operations of the Department of the Navy. The panel attempted to project the future trends in these technologies out to the year 2035 and assessed what impact they will have on future Navy and Marine Corps mission capabilities. Other panels in the study focused on future systems that will incorporate these technologies.

It is a daunting task to forecast technology trends some 40 years into the future. We tend to be guided by experiences of the past. The panel will surely make mistakes, mostly of omission. For example, 40 years ago even the best technology minds missed the phenomenal growth and impact of computers on society today. Merely identifying the future capability of a technology or how several technologies might be used together to create a new capability does not necessarily mean that these events will happen. Vision, belief in the forecasts, strategic planning based on that belief, investments in development and applications, and rapid adoption by the Navy are all additional ingredients necessary to turn a technology dream into real systems for the future.

DRIVING TECHNOLOGY APPLICATION AREAS

The foundation of future naval superiority will be information. Computing capability, application software, and telecommunications are the underpinnings of this foundation. The panel therefore places a high priority on these technologies which are growing and changing rapidly, driven predominantly by commercial business objectives. An explosive growth in commercial telecommunication and in Earth-imaging satellites is under way. These global commercial services will have a profound impact on future naval operations. They will be available to cooperative forces as well as adversaries around the world. The Department of the Navy must adopt these services but adapt them to fit its particular needs while providing the necessary security.

Computation

Computers and related software that form computational capability are among the most important technologies that will control the makeup and performance of future naval forces. As information is key to military success, computation is key to information, its acquisition, its extraction, and its use. In Chapter 2, the panel examines the status and trends in computer hardware, software, and key applications, such as computational fluid dynamics (CFD).

Silicon microelectronics has been the engine of the tremendous computer technology growth over the past 40 years. The linewidths used in making computer chips have decreased exponentially, and the number of transistors per chip has increased correspondingly over this time period. This has resulted in ever-increasing computational power. Although the panel sees a number of practical problems that must be solved, there appears to be no fundamental physical limit to continuing this growth out to the year 2035, the time horizon of this study.

The panel analyzed and projected computer performance in two distinct families: high-performance computing and functional/affordable computing. High-performance computing represents the state of the art in computational speed, storage capacity, and transmission rates. The Cray T90 is an example of this type of computer. The panel studied the progression of such computers over the past 40 years and used these data to extrapolate into the future. The processing speed of high-performance computers has consistently increased at an annual rate of 60 percent, doubling every 18 months, or 100 times per decade. Many computer and materials scientists believe that the number of transistors per chip will continue to increase over the next 40 years despite the growing cost of fabrication facilities and likely changes in building-block materials, architectures, and lithography techniques.

The quest for ever-greater computing power will be driven by the demands of cyberspace, entertainment, and other commercial businesses. If the past trends continue, the operating speeds of high-performance computers in 2035 are expected to be in excess of 1 petaflop, i.e., 10^{15} floating point operations per second, and memory size will be 10 terabytes. It is estimated that the human brain has a processing power of 1 petaflop, and so computers in 2035 may well become on-body personal assistants that provide a high degree of intelligent support.

Functional or affordable computing refers to computers that are mass produced and sold at costs that make them accessible to many people. A current example is the Intel Pentium Pro personal computer. The growth in capability of affordable computers has been somewhat slower than that of high-performance computers, with the processing speed doubling every 24 months. If this trend continues, a sailor or marine in 2035 will have access to a teraflop (10^{12} floating point operations per second) desktop computer at an affordable price.

Military computers in general applications have tended to lag in time behind the capabilities available in the civilian world. The exceptions to this are high-performance computers where the Department of Defense has been a major funding source for such as applications as nuclear weapons simulation. The challenge for the future Department of the Navy is to adopt the state of the art in commercially available computers as rapidly as possible into naval forces applications.

Chapter 2 also deals with the status and developments in the software languages that allow computers to perform applications. The evolution in software has been from assembly language to more compact, higher-order languages, which has resulted in an improvement in productivity, as measured in average cost per function point, of about 5 percent per year. The size of software application tasks has, however, continued to grow at a rate more like that for the computer-performance gains, i.e., at 60 percent per year, so that software runs a risk of being the Achilles heel of the information revolution. The error rate, in defects per line of code, also increases with the size of programs, compounding the productivity problem. The panel believes that a possible solution to the software dilemma is to create domains of expertise within naval operations, such as avionics, missile guidance, command and control, and so on, and to practice within these domains a higher degree of reuse of software modules and develop software logic synthesis tools, automatic programming tools, and other methods that will boost productivity significantly. Object-oriented programming, an approach that demands well-defined building blocks and that encourages reuse, should be a major factor in future software development.

Chapter 2 also treats the expected progress in digital and analog signal processing. Significant progress in this area is critically tied to progress in high-performance computing. The benefits of pattern recognition in automatic target detection have long been recognized, but empirical evidence over the past 30 years has shown this to be a very difficult problem. Neural networks, which have been shown to have good pattern recognition ability, do not seriously compete with human performance. Many believe that the human brain simultaneously attempts all possible solutions to a problem and is able to reach a conclusion by sheer computational power. As high-performance computational power in 2035 approaches that of the human brain, the performance of a wide range of pattern recognition technologies will substantially improve.

Computational fluid dynamics represents the use of modern computational techniques to model fluid flow. CFD is used extensively in aerodynamic design of modern aircraft and missiles and is coming into greater use in ship and submarine design. The modeling of turbulent flow is, however, a fundamental limitation in the technique, and continued support from the Department of the Navy to develop better models to deal with strongly separated flows, flows in strong vortices, and flows in powered-lift vertical landing is needed. Achievement of laminar rather than turbulent flow over the entire surface of a vehicle could, in principle, reduce parasitic drag by as much as an order of magnitude while also lowering the target signature.

Information and Communications

Information and communications will dominate future naval warfare and must be elevated in priority within the Department of the Navy. Information technologies include the underlying computation capability, connectivity, and networking and the interaction of humans to reach decisions based on captured distributed knowledge. In Chapter 3, the panel addresses the technologies involved in distributed collaboration, software engineering, information presentation, human-centered systems, intelligent systems, planning and decision aids, communications, and defensive and offensive information warfare.

The panel views communications technologies as the backbone for connectivity of the naval forces. With the explosion in commercial communications via satellite and fiber, the panel envisions that in the time frame of this study, naval forces will have access to transparent wideband (capable of handling video data) global communications on demand. This capability (Figure 1.1), when coupled with the use of common databases and intelligent support systems, will provide naval forces, whether at sea or on land and anywhere in the world, with "any data all the time" access to information. Users will have full control over the quantity and quality of information that is needed to accomplish their mission. The concept is analogous to the capability made possible by the rapidly growing Internet. Much of this capability will come from the commercial sector. The

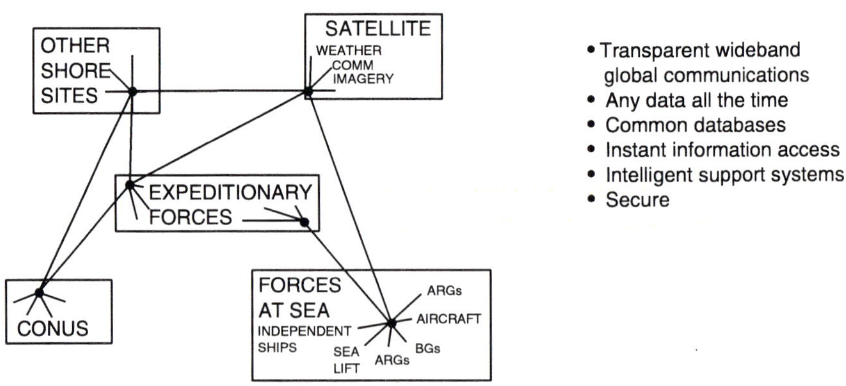

FIGURE 1.1 Global distributed collaboration is key to future naval operations.

challenge for the Department of the Navy is to adapt to this commercial resource while maintaining secure operations and graceful degradation at times of conflict.

The Department of the Navy should develop strategies and practices for defending hardware and software systems of the future against attack by viruses, software agents, and externally generated radio-frequency signals. As information systems become more important, as computer chips become more dense and contain more functionality, as the Department of the Navy moves more to commercial sources of software, hardware, and communications, and as the need for greater interoperability increases, naval systems become more susceptible to deliberate attack. The emergence of low-cost, small-size, ultrawide-bandwidth (UWB) radio-frequency impulse transmitter systems that can deliver intense, disrupting pulses into electronic circuits creates a new threat that must be countered.

Since the United States currently has a commanding lead in the development of advanced information systems, the opportunity exists to exploit this lead to develop offensive information weapons of the future that would have the same or greater impact than bullets or missiles. The panel encourages the Department of the Navy to aggressively pursue these technologies and applications within the constraints of international agreements.

Sensors

The Department of the Navy is a user of a wide variety of sensor systems including electromagnetic, acoustic, mechanical, chemical, biological, nuclear, environmental, and temporal. These sensors, as described in Chapter 4, are strongly constrained by the physics of the phenomena to be sensed, and hence the size, shape, material, and configuration are matched to the sensitivity required. Some of the technology trends that are common to all classes of sensors are the shift to solid-state electronics, a movement toward atomic-level devices (quantum wires and dots), digital implementation, distributed system implementation (networking, data fusion, and societies of microsensor), multidimensional signatures (multispectral, hyperspectral, and data fusion), and multifunctional sensor systems (common apertures for radar, communications, electronic warfare, and so on).

The critical technologies common to all sensor classes are semiconductors, superconductors, digital computers, and algorithms. The growth in these critical underlying technologies will determine the sensor capabilities available to future naval forces. The panel anticipates a significant change in the use of sensors as we know them today. Figure 1.2 illustrates the expected evolution in sensors over the time frame of this study. Today, a conventional sensor system is composed of a sensing element and one or more processing elements. Each element is discrete, relatively expensive, and subject to component failures. In the near future, the panel sees this situation moving to smart monolithic sensor systems,

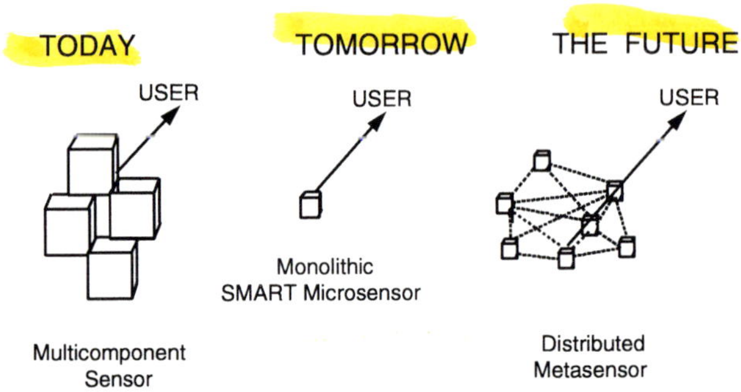

FIGURE 1.2 The evolution of sensors for Department of the Navy operations.

enabled greatly by new technologies such as MEMS. These sensors-on-a-chip will be inexpensive, mass producible, and highly reliable and will contain far more intelligence and decisionmaking capability than today's sensors. In the 2035 time frame, the panel expects distributed societies of these smart monolithic sensors to form metasensors that cooperate and provide much more information about the operational environment and an adversary's capabilities.

The DOD has been and will continue to be a major driver of progress in advanced sensor systems. The Department of the Navy has specific sensor needs that will not be addressed by any other agency. These include sensors for sensing and imaging in littoral environments, synthetic aperture sonars (SAS), and communications in the ocean. The Department of the Navy should be the major investor in these sensor technologies.

Automation

In the future, affordable automation systems and equipment, as described in Chapter 5, will supplement or supplant many current operator functions in shipboard routine and casualty operations, such as firefighting and trauma care, to reduce ship size and manpower costs and to minimize the number of personnel exposed to hostile action.

Teleoperated and autonomous platforms including UUVs and UAVs will provide the Department of the Navy with new operational capabilities in undersea warfare, reconnaissance and surveillance, and control of the air battle space above the fleet and the expeditionary Marine Corps forces on land. UUVs will be operated both by submarines and surface combatants and will provide economical force multiplication, increased and extended underwater battle-space awareness, and reduction in exposure of humans to hostile action. UAVs will view the battle space with EO/IR and SAR and other types of sensors, participate in cooperative engagements against difficult targets, such as low observable sea-skimming missiles, collect enemy radar and communication signals, hunt mines and submarines, jam enemy electronic systems, guide precision weapons, and potentially act as precision weapons themselves.

Autonomous navigation and guidance and automatic target recognition are among the fundamental technologies critical to these UUV, UAV, and other robotics systems. Expendable robots that can be used for reconnaissance and surveillance as well as weapon delivery platforms will also be pervasive.

Human Performance

During the next three decades, revolutionary developments are expected in several technology areas that will greatly enhance human performance in ways that will significantly benefit the Department of the Navy. In Chapter 6, the panel suggests that the following technologies are of particular significance to improvements in human performance: (1) communications, including global access to people and information sources, distance learning, and interactive multimedia information transfer; (2) information processing, including man-machine interface, computer-processing speed and memory, intelligent software, modeling and simulation, and embedded information (e.g., smart cards); (3) health care, including telemedicine, drug delivery systems, surgical techniques, and disease prevention and treatment; (4) biotechnology and genetics, including gene therapy, human genome mapping, DNA identification and categorization, susceptibility testing, disease diagnosis, and vaccine and drug development; and (5) cognitive processes, including knowledge of brain functions, optimal learning modes, memory-expanding drugs, and fatigue management. These areas will experience extraordinary leaps in knowledge and applicability in the foreseeable future.

These technological advances in human performance will have a profound impact on many aspects of life in the naval services, including education and training, operator performance, health and safety, and quality of life. The Department of the Navy has placed a high priority on reducing crew sizes both aboard ships and on shore. By adapting these technology improvements, the Department of the Navy can achieve a higher level of manpower readiness than ever before and potentially accomplish its mission with fewer, better-prepared people.

Materials

In Chapter 7, the panel envisions a systems approach to materials development in the future, whereby the physical and mechanical properties are understood on a fundamental atomic scale. Materials will be designed based on this atomistic approach, which will also be extended to process and fabrication simulation. The panel believes this computational materials design approach will enable breakthroughs in ferrous alloys, titanium matrix composites, polymer composites, high-temperature ceramics, wideband gap semiconductors, and smart materials based on ferroelectrics, ferromagnetic materials, and ferroelastic materials. Unique nanophase materials for new applications and smart materials based on shape memory alloys, ferroelectrics, and ferromagnets that can be adapted for mechanical and electrical applications and on wideband gap semiconductors will be the technology drivers of the future.

Power and Propulsion

Warfare in the 21st century likely will bring the use of advanced technologies and combat systems with ranges, lethality, and detection capabilities surpassing anything known in contemporary warfare. In the battle zone surrounding and including the battle group and expeditionary forces, there will be requirements for electrical power at levels from tens of watts (for surveillance and communications) to kilowatts (for radar and a variety of electric motors) and hundreds of kilowatts (for field-based power demand) and multi-megawatts (for advanced weaponry and countermeasures systems). Mobility of all systems will be essential for survival and success.

In Chapter 8, electric power technology is evaluated in two areas: (1) power and distribution on the electric ship and (2) energy storage and recovery for ultimate power delivery. The electric ship area includes nuclear-electric, turboshaft engine, piston-type, magneto-hydrodynamic, and hydrocarbon-fueled generators for continuous and pulsed power and conditioning of that power for distribution aboard ships and submarines. The technologies available today and expected in the year 2035 are evaluated against six general figures of merit. In the energy storage and recovery area, primary batteries, secondary (rechargeable) batteries, fuel cells, and flywheels are evaluated against similar figures of merit.

The panel strongly supports the concept of the electric ship. The reduction in volume and signature and the design flexibility gained by the use of modular electric motors, including high-temperature superconducting motors in the future, and direct electric drive are powerful motivators. With the availability of all-electric power and propulsion systems, the possibility of electric-driven weapons and launch systems becomes attractive.

The U.S. Navy has developed a strategic plan for developing and implementing electric ships of the future. The focus is on a scalable and flexible architecture that can be applied across ship classes, making the implementation and acquisition affordable. The plan has insertion milestones predicated on commercial and Navy-developed availability of critical technologies. The panel endorses this plan and believes that the Department of the Navy should proceed as rapidly as possible toward implementation.

Environments

Two very distinct and different classes of environments important to naval operations in the future are described in Chapter 9.

The first class is the traditional physical environment in which naval forces operate, including long- and short-term weather forecasting and the modeling of ocean and littoral waters in order to accurately sense and locate submarines and mines.

The second and newer class deals with protection of the environments in which naval forces operate. Public awareness of the sea as a critical component of Earth's environment is growing. New restrictions on ocean dumping have already been placed on ships at sea and in harbors. Such restrictions will only increase in the future. The Department of the Navy should plan for and implement logistics system technologies that minimize the production of waste. Commercial technologies to not only limit sources of pollution but also to practice efficient and environmentally benign waste disposal in the future must be adopted. An issue of importance is the growing public concern about the possible harmful effects of active sonars and underwater explosives on marine life. To avoid serious curtailment of naval operations in the future, the Department of the Navy should develop the necessary data to understand the effects of current and proposed underwater acoustic sources on marine life.

In the area of weather forecasting, the panel believes that in the near term (3 to 5 years), high-resolution modeling using the improvements in shore-based high-performance computers has the potential of providing reliable 3- to 5-day forecasts of mesoconvective rainfall, cyclone tracks and intensity, ocean waves, and coastal mesoscale phenomena. In the mid term (5 to 10 years), improved models that include ocean/atmosphere interactions could provide high-resolution weather forecasts for battle groups and expeditionary forces. In the long term (10 to 20 years), ships will be provided with local high-performance computing capability, along with having massive, stored databases for autonomous high-resolution recognition of current weather as well as 30-day weather forecasts displayed multidimensionally and multisensorially.

The near-shore waters are driven by the sun, tides, winds, land runoff of fresh water and sediments, and offshore currents. The actual shape of the bottom and its capacity to reconfigure are critical sources of data. Because naval operations in littoral waters are of growing importance, the Department of the Navy will require large-scale models that can accurately simulate the physical state of a littoral zone under various scenarios of weather. The panel envisions that in 10 years models could be developed that are time-dependent and three-dimensional with outputs that would be presented in multidimensional graphical visualization systems to identify patterns of currents, temperature fronts, and so on. The modeling of acoustic phenomena in littoral waters to identify mines and submarines is challenging: The temperature and salinity of the water change with the season; sea surface waves change quickly and frequently; the behavior of bottom reflectivity and absorption changes within waters. Real-time sensor measurements will be required along with sophisticated models to deal with these challenges.

Enterprise Processes

The Department of the Navy is a very large business enterprise. Multiple large-scale processes such as platform and weapon acquisitions, logistic management, resource planning, personnel management, and training are manpower- and cost-intensive elements of operating this enterprise. Powerful information technologies, described in Chapter 10, are becoming available that can be applied to these enterprise processes to significantly reduce the cycle time on weapons and platforms development, to improve logistics system efficiencies, and to reduce manpower and cost. The panel believes that these technologies will have a major impact on the entire spectrum of Department of the Navy and DOD activities in the future.

Simulation-based acquisition (SBA) is a revolutionary process for the acquisition of complex platforms and systems. The information technologies in SBA integrate all acquisition activities, starting with requirements definition and continuing on through production, fielding and deployment, and operational support in order to optimize the design, manufacturing, business, and support processes. As envisioned, at the initiation of a proposed acquisition, government and industry teams will create virtual prototypes of candidate systems such as a new warship, a new airplane, and so on. These virtual prototypes will be fully linked to an integrated product database for that platform that includes its attributes, and they will be used during the acquisition, development, and operational lifetime of the product.

The SBA computer environment integrates models and simulations, engineering and physics, operations, and doctrine to evaluate overall product value and cost, guide the development process, and establish support operations of the product during its life cycle. SBA also features an interactive, immersive synthetic environment, integrated collaboration tools, legacy analysis and modeling tools that are "wrappered" to achieve compatible interface standards, smart agents to aid information collection and integration, advanced information sharing and distribution, and advanced product and process optimization tools.

SBA will give decisionmakers the ability to examine design and operational issues by interacting with the virtual prototype. High-fidelity simulations will allow acquisition authorities to visualize and analyze the virtual prototype in distinct phases of the life cycle. They will be able to conduct studies that include mission value, affordability, performance tradeoffs, manning, maintenance, and so on. SBA will have a major impact on the Department of the Navy and DOD because it will reduce the time, resources, and risks in the acquisition process and will simultaneously increase the quality of the systems being acquired. The ability to operate SBA in real time will be directly tied to progress in computing technologies. Analysis codes, such as CFD to determine drag or finite element to determine structural loading, may take hours to run today but will be reduced to minutes or seconds in the future.

Similar software and computer technologies are being developed and will be applied to the manufacturing, logistics, resource planning, dynamic mission planning, simulation of war, and other domains.

Other Technology Application Areas

There are four important areas in which many of the above technologies have application for the Department of the Navy. In Chapter 11 the panel emphasizes the importance of these four areas—space applications, signature management, chemical and biological warfare (CBW), and combat identification applications—to future dominance of the battlefield.

For military use, the four major applications of space are navigation of platforms and weapons, communications, environmental monitoring, and surveillance. It is clear to the panel that for the Navy and Marine Corps space offers innumerable advantages for rapid and secure global communications, monitoring of the air and sea environment, locating and targeting adversary naval and air forces, and carrying out potentially soft and hard kills of targets. Space will continue to provide the high ground for line-of-sight access to the ocean areas of the globe and is integral to the mission responsibility of the Department of the Navy. The panel recognizes that the Department of the Navy, while a major user of space, is not a developer of major space systems. However, the panel believes that to remain a "smart buyer and user" of national space assets, the Department of the Navy must maintain a knowledgeable cadre of professional personnel versed in the latest space technology, equipment, and systems. The Navy space cadre must participate actively with the other major organizations constituting the national programs through joint training and job rotation. This cadre must have opportunities for professional growth and recognition. They must also have access to the highest decisionmaking levels of the Department of the Navy so that space requirements, alternatives, and options can be considered and acted on in the best overall interests of the department.

Signature management is the application of many of the technologies described in this report to permit the purposeful reduction of the observability of a platform, a vehicle, a facility, personnel, and so on to accomplish one or more of the following objectives: (1) hide its existence from the enemy, (2) confuse the enemy's ability to locate it, (3) confuse the enemy's ability to identify it, or (4) reduce its vulnerability to attack. Over the past decade, the DOD has made remarkable use of signature management to enhance the probability of success on the battlefield. The F-117 stealth fighter clearly demonstrated the advantage of surprise in Desert Storm in terms of inflicting initial devastating destruction upon the adversary while remaining virtually invulnerable. U.S. Navy submarines through their shape, material coatings, and operational practices have been the silent service for decades. The B-2 stealth bomber and the Sea Shadow stealth ship developments, while not yet deployed in warfare, are other examples.

In the past decade or so, use of particular materials to control the absorption and reflectance of radar and optical signals and the shaping of objects have been the principal methods of controlling and reducing the signature of platforms to make them stealthy. Electromagnetic analysis codes have evolved at a rate similar to that of computers to improve the fidelity with which the signature of an object can be determined. Progress should continue to parallel the computer technology trends described above, provided that the Navy invests in the development of software code. In many cases materials for these applications have been complex and expensive. To make signature management available in a wider variety of applications, investment also must be made in appropriate material technologies with the goal of significantly reducing their cost and enhancing their ease of application and maintainability.

With the advent of MEMS technology, the availability of systems-on-a-chip, and advances in smart materials, computing power, and intelligent software systems, the possibility exists in the future that the signatures of important assets such as aircraft, UAVs, and UUVs could be actively controlled. This capability could lead to signature warfare—the ability to present in real time a different signature to an adversary as the warfighting scenario evolves. Signature warfare could change the outcome of a battle by changing an adversary's perception and visualization of the battle space. Because of the importance of signature management to future warfare, the panel strongly encourages the Department of the Navy to invest in and exploit the technologies described in this report for applications to reduce and control the signature of important assets.

Chemical and biological warfare has become a credible threat, both strategically and tactically. As strategic weapons of mass destruction biological weapons are predicted to be effective over several hundred square kilometers and capable of causing hundreds of thousands to millions of deaths. Chemical weapons are more tactical in nature and are estimated to be effective over a few square kilometers in causing hundreds to thousands of deaths.

Self-protection from CBW agents involves detection and identification of the agents and evasion and/or provision of a safe haven until the threat has abated. Based on rapid advances in MEMS technology, sophisticated CBW detection devices are being developed by DARPA and others that show promise of being sensitive, small, rugged, and low in cost and hence could be deployed on UAVs to provide advanced warning. The panel encourages continued support of this exciting technology as a keystone means for staying ahead of the CBW aggressor.

The other aspect of the system for defense against CBW is the safe haven to protect military personnel from CBW agents. The Department of the Navy has been active in this area through the development of the CPS, which provides filters for CBW agents in protected areas of a ship. CPS can be designed into the air-conditioning system of new ships or backfitted into critical compartments on existing ships to provide protection. The panel supports the Navy technology development efforts in this area and encourages as rapid an implementation as possible.

In nearly every military operation, the ability to distinguish friendly assets from enemy assets is extremely important. The need to reduce "friendly fire" accidents is growing, particularly as the tempo of warfare increases. Accurate identification becomes especially important when U.S. assets in the form of fighter aircraft fall into the hands of future aggressors through earlier foreign military sales. In the past this capability for identification was referred to as identification, friend, or foe (IFF) but is now called combat identification (CID). The need for a dependable, multiservice, multipurpose CID that will be effective in air-to-air, air-to-ground, ground-to-air, and ground-to-ground engagements is clear. The panel encourages the Department of the Navy to pursue new technologies in the form of low probability of intercept (LPI) and other covert communications to develop this badly needed capability.

KEY BENEFITS

The driving technology areas identified in this report, while important in their own right, have the greatest synergy for the Department of the Navy when they are combined to provide new and more powerful capabilities. The panel believes that the three key benefits to be realized through combining several of the driving technology areas are as follows:

- Efficient and effective use of people,
- Superior weapons and platforms, and
- Dominant naval operations.

The first key benefit is illustrated in Figure 1.3. By combining computational power, information, communications, sensors, and the technologies that enhance human performance with more efficient enterprise processes, the Department of the Navy will realize much more efficient and effective use of people. The Navy and Marine Corps will be able to operate with fewer but better-informed and better-equipped sailors and marines. They will be more professional and have higher morale through a greater sense of job responsibility and satisfaction, which, in turn, will lead to higher retention rates. The increased use of automation aboard ships and the growing numbers of UUVs and UAVs will minimize human casualties in battle. The great advances expected in the

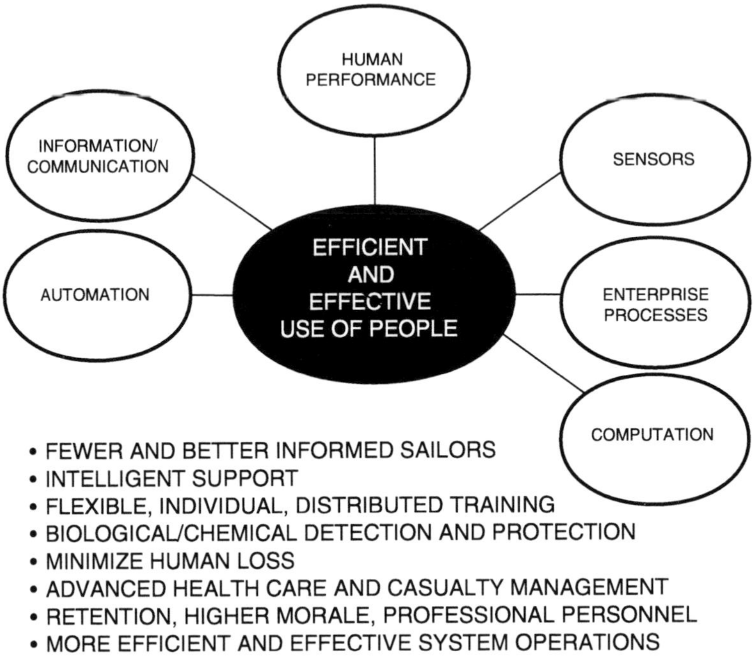

FIGURE 1.3 A key benefit of using the technologies shown is the efficient and effective use of people.

genomics area (the coming together of biotechnology, genetic engineering, and microelectronics technologies) will protect them against threats and cure them faster when needed.

By streamlining design and manufacture through the use of powerful new information technologies, such as simulation-based acquisition, and by involving the logisticians and other support areas from the inception, superior weapons and platforms will be developed in the future. This key benefit is illustrated in Figure 1.4. Interactive virtual platforms will be created in the computer before the physical units are ever built. Users and support personnel will interact with the design in real time, feedback will be rapid, and cycle time for producing new products will be greatly reduced. High-fidelity testing on the computer will greatly reduce the necessity of subsequent testing in wind tunnels, on ranges, and at sea. The platforms and weapons designed in this manner will be more robust and survivable through a greater use of passive and active signature management, smart materials, smart sensors, and improved navigation and guidance.

Through the application of these powerful technologies the Department of the Navy can realize dominance in future naval operations (Figure 1.5). Advanced sensors, long-duration UUVs and UAVs, high-performance on-board

FIGURE 1.4 The use of the technologies shown will produce superior weapons and platforms.

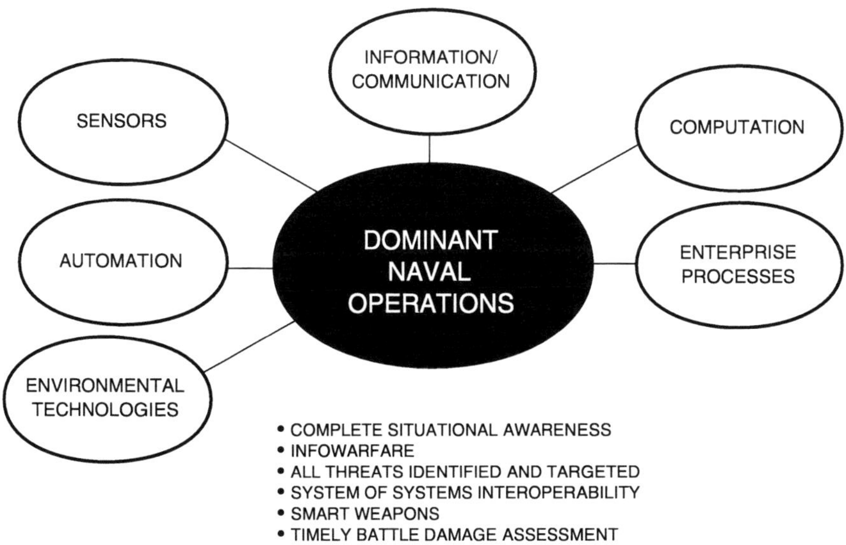

FIGURE 1.5 In the future, dominant naval operations will result from the use of these technologies.

computers, transparent communications, and automated intelligence analysis will provide Navy and Marine Corps personnel with high-resolution situational awareness. A great deal of information will be available on the enemy—location, strengths, weaknesses, and intentions. Many of the potential threats will be identified (as will friendly forces and assets) and targeted. Cooperative engagements against difficult targets will be routine. Smart standoff weapons with high-precision guidance will destroy enemy targets with great efficiency. Battle damage assessment will be prompt and accurate, further adding to the weapon efficiency.

EXCITING NEW TECHNOLOGIES

The panel examined more than 100 technology areas important to future naval operations. Many of these areas will experience evolutionary growth over the time frame of interest. A few will experience revolutionary technical breakthroughs. Others will be subject to explosive commercial growth. This progress will enable naval capabilities and operations that are not possible today. The panel highlights below those technologies that are likely to have the biggest impact on changing the way the Department of the Navy conducts operations in the future. Because of their importance, these technologies deserve careful attention in the months and years ahead:

- Micro- and nanoscale technologies
  —Microelectromechanical systems
  —Nanoscale electronic circuits
  —Systems-on-a chip;
- Teraflop affordable computers and petaflop high-performance computers;
- Genomics—the marriage of biotechnology, genetics, and electronics;
- Smart materials involving nanophase materials engineering;
- Ubiquitous wideband communications and connectivity;
- Global distributed collaboration;
- Multisensory virtual reality environments;
- Information warfare, defensive and offensive;
- Autonomous agents; and
- Signature management and warfare.

SCIENCE AND TECHNOLOGY SUPPORT

Investments in science and technology have served the Department of the Navy well over the past 50 years. Military technology requirements will continue to drive the state of the art in the future. The panel believes strongly that the Department of the Navy should continue to support the development of those fundamental sciences and technologies that are relevant to naval operations. This includes the nurturing of discoveries and inventions in universities through prototype applications in industrial and government laboratories into full-scale proof-of-concept demonstrations.

One of the elements of the national technology infrastructure that undergirds all of the technologies and technology applications discussed in this report is the system of standards-setting bodies and institutions that conduct metrology research. These include the National Institute of Standards and Technology within the Department of Commerce, other national laboratories, and a host of voluntary standards-setting organizations and industry associations. Standards and advanced metrology techniques for testing, inspection, verification, and validation are required for all aspects of technology development. For example, the ability to share and accept test data among different players in the R&D process is critically dependent on the acceptance of common standards and metrology tools. Whenever necessary, the Department of the Navy must be prepared to participate in and support the nation's standards and metrology infrastructure as required for its own technology development programs.

FOCUSED RESEARCH AND DEVELOPMENT

No modern, technology-intensive enterprise can prosper without sustained research and development support focused on the enterprise's main objectives. This truism has been recognized for the armed forces since World War II, but the nation may be losing sight of it today as budget concerns occupy the nation's attention. The environment in which future naval forces will exist and in which they will have to function effectively will be characterized by continuing budget stringency, barring the emergence of some future mortal threat to the United States and its allies. Regardless of the level of resources that will be allocated to support the creation of the entering wedges of capability that this full study on technology for future naval forces foresees as essential to future naval force viability, and however they are found, the R&D part of those resources will have to be invested as effectively as possible, and in a timely manner.

In addition to effective technical management, a key step in effective use of resources for R&D will be to focus the R&D effort on those elements that are unique to military and naval forces, and for the rest to capitalize to the greatest extent possible on R&D and technology emerging from the civilian, commercial sector. The technology areas examined during the course of this study were reviewed to see where relevant R&D is currently performed and is likely to continue. The review showed extensive scientific and technology development effort in the civilian sector that can be of value and use to the naval forces in the following technology areas or clusters:

- Information technology,
- Technologies for human performance,

- Computational technologies,
- Automation,
- Materials,
- Power and propulsion technologies,
- Environmental technologies, and
- Technologies for enterprise processes.

Although particular areas of science and related military and naval applications will always require Department of the Navy investment and attention, military R&D in the above areas can concentrate heavily on adapting the civilian and commercial technologies and their products to naval force use.

This orientation must be adopted with caution, however, because in many areas commercial industry is also deferring long-term R&D in favor of short-term programs offering a quick payoff in highly competitive markets. The Department of the Navy must thus remain vigilant to ensure that its needs will indeed be met in these areas by the civilian world. In no sense, therefore, should comments on priority in this regard be taken as a suggestion that basic, long-term research be foregone by the Department of the Navy in all these areas without first ascertaining that research needed for naval force purposes will in fact be performed by the commercial sector. The Navy Department must also be ready to recognize and adapt wholly new advances that can change how military tasks are performed, equipment is brought into being, and kinds of equipment created. The naval forces must remain open to new and vital knowledge. The issue is to apply appropriate judgment to allocation of scarce research resources.

With due attention to these caveats, it appears now that science and technology for military and naval force use will have to be especially sustained by the military R&D community (where possible and beneficial, in cooperation with the civilian community) in the following areas because, in the absence of large civilian markets, no one else is likely to support it (the inclusions in parentheses give examples of the kinds of capabilities and devices that would be included in each):

- Sensor technologies (electronically steered and low-probability-of-intercept [LPI] radar, IR and advanced infrared search and track [IRST], multispectral imaging, embedded microsensors and "smart" skins and structures, lasers, SQUIDs);
- The sensor technologies would be joined in application with specialized information technologies (secure data access; stealth and counterstealth; ASW; chemical, biological, and nuclear weapons detection; automatic target recognition) to contribute to the military parts of the information-in-warfare system. (Fundamental research into the theoretical basis of naval warfare underlying modeling and simulation must obviously be supported by the naval forces, as well.)
- Military-oriented materials (energetic materials, including explosives and

rocket propellants, high-temperature materials for engine turbine blades and combustors, and composites, among others);

- The materials together with power and propulsion technologies (rocket engines, warheads, and advanced aircraft, ship, and submarine power plants) would contribute to the creation of advanced weapon systems and, in the form of long-life and high-power-density power sources, to reducing equipment loads and logistic resupply requirements.

In many of these areas, the naval forces will have to join with the other military departments to share the applied R&D and advanced development loads so that the total resources are spent as efficiently as possible. R&D expenditures by the Navy Department in these areas, and in the adaptation of civilian technology to naval force purposes, must be focused in two areas: development of unique naval force capabilities needed to support ongoing force improvement and creation of future capability; and development, by work-sharing arrangements in the joint environment, of capabilities that all the Services will be able to use. Deciding the allocation of resources between these two areas of effort will obviously be the responsibility of the Department of the Navy working with the Joint Chiefs of Staff, the other military departments, and the Office of the Secretary of Defense. Some of the jointly agreed on R&D will help the naval forces, just as some of the Navy Department R&D will help meet needs of the other Services.

Finally, it must be emphasized that some major system advances take place in major steps after ongoing research and advanced development have created new opportunities. This has been especially apparent in the aviation area, where ongoing R&D in propulsion, aerodynamics, and structures leads periodically to a major advance in capability embodied in a new class of aircraft. For this to happen, the R&D must be supported in a sustained, long-term program in which each step is built on the last, such that at significant points a new system can be built on the advances achieved to that time. An example is the Integrated High Performance Turbine Engine Technology (IHPTET) program, jointly sponsored by the Office of the Secretary of Defense (OSD), the Military Departments, and industry. This program, together with its predecessor Service programs, has led to major advances in turbine and compressor materials, advanced combustors and engine controls, and overall engine designs. These advances have led in turn to major improvements in thrust, thrust-weight ratio, and fuel economy, leading to the superior U.S. military aircraft engine performance we see today, and to significant advances in civilian aviation as well.

The areas of surface ship and submarine design and construction, antisubmarine warfare, and oceanography also need a similar model of integrated, sustained R&D support, with clearly defined goals and schedules, industry-government collaboration, and stable funding, to achieve the potential seen for them in this study.

2

Computation

INTRODUCTION: WHAT IS COMPUTATION?

Computation is one of the most important technologies that will control the makeup and performance of future naval forces. As information is key to military success, so computation is key to information—to its gathering, its extraction, and its use.

In the past two decades, digital computers have become ubiquitous—from our computerized cars and personal computers, to Wall Street, Hollywood, and the Internet. Our society (and the military) clearly has been drawn irreversibly into what has come to be known as the information revolution. It has changed how we work, how we communicate, and how we play. Its impact on naval forces over the 40 years of this study will be enormous, as is evident already from the recent appearance of such military terms as information warfare and the digital battlefield.

By computation we mean digital computation—the crunching of bits by some form of computer for some purpose mathematical, textual, or graphical. Of course analog computation certainly plays a role in a number of important sensor systems—for example, the lenses of optical systems and the dish antennas of radars function as analog computers performing naturally millions to billions of operations for which digital beamforming alternatives so far are only rarely attempted. But these applications are usually limited and specialized to the physics of the specific situation, whereas digital processing offers almost unlimited flexibility of application through its software programmability.

Computation has a wide range of digital hardware support—from special-purpose application-specific integrated circuits (ASICs), to digital signal processing (DSP) chips, reduced instruction set computers (RISCs), complex instruction set computers (CISCs), central processing unit (CPU) chips, desktop computers, workstations, signal processors, mainframes, massively parallel processors (MPPs), and all kinds of multiprocessor supercomputers. The capability of this hardware has been growing exponentially for two to three decades and shows no signs of slowing down. This explosive growth has created the information revolution and will propel it into the future.

As the hardware has grown in capability, new tasks have been imagined and an ever-increasing myriad of algorithms has evolved for accomplishing those mathematical, textual, and graphical tasks. To provide the software for programming these algorithms, dozens of computer languages and software programming tools have been developed over the years, and new concepts (e.g., Java) continue to appear. This ability to accomplish new functions, previously not feasible, has led to many significant advances, including virtual reality, simulation-based design, Kalman filtering, better cars, and so on. Unfortunately, as the availability of computational power has increased, the size of software packages has also grown, with error densities and development costs growing even faster, until software has become one of the most critical issues of our time—perhaps the Achilles' heel of the whole information revolution. Many large computation-dependent systems today find their reliability and availability determined almost completely by software bugs, with few limitations supplied by the hardware.

In short, computation and its exponential growth offer enormous opportunities and many challenges for the future of society and the military. Whatever the challenges, there is no doubt that computation, in a multitude of forms, will be a dominant factor in all aspects of future naval forces.

RELEVANCE: WHAT WILL COMPUTATION DO FOR THE NAVY AND MARINE CORPS?

Computation and computers are already an integral part of all Department of the Navy activities. High-performance computers and workstations support finite element thermal and mechanical analysis and electrical and logic design and simulation of components and systems, as well as large-scale, linear and nonlinear analysis of critical physical phenomena, such as CFD, which is used extensively in aerodynamic design of modern aircraft, missiles, and gas turbines and is coming into greater use in ship and submarine design. Today's complex military systems can no longer be designed and manufactured without computer assistance. Desktop computers support the fusion of information and data from multiple sources; provide the capabilities for identifying, extracting, and displaying the meaningful information contained in the masses of data available; and supply the tools for planning and implementing naval force and logistics functions.

Long-range sensors, such as radar and sonar, are thoroughly dependent on embedded signal processors to transform the raw data into detailed information

Box 2.1 Naval Applications of Computation

- High-performance computing
  - — Computational physics, e.g., CFD
  - — Weapons analysis
  - — Simulation-based design
  - — Virtual reality
- Functional and affordable computing
  - — Data fusion
  - — Information mining
  - — Planning and logistics
- Embedded computing
  - — Sensors
  - — Adaptive signal processing,
  - — Automatic targeting recognition
  - — Aim point selection
  - — Controls
  - — Guidance

about the targets and the environment being observed and to supply adaptive real-time modifications of the sensors' parameters to optimize their performance. In other scenarios, embedded computers in missiles support the tracking sensors and provide the automatic target recognition, aim point selection, and detailed weapons guidance commands required to accurately counter hostile threats. Applications of computation are indicated in Box 2.1.

TECHNOLOGY STATUS AND TRENDS

Microelectronics

Mechanical assistance for computation has been evolving continuously, albeit slowly, for the past several thousand years—the abacus, the slide rule, Babbage's Difference Engine, and so on. During and just after World War II, the simultaneous needs of sophisticated nuclear weapons design and the challenges of cryptography (code breaking) led to the introduction of the first electronic computers—the Mark I built by IBM for Harvard University in 1944 and the Electronic Numerical Integrator and Computer (ENIAC) built at the University of Pennsylvania beginning in 1946. These early electronic computers employed rooms full of vacuum tubes for the implementation of the digital binary logic and as a result were large, slow, and power hungry. By the 1950s, the concept of the stored program had been introduced and the architectural concepts of the modern computer were beginning to take shape. It needed only the inventions of the transistor in 1947 at AT&T's Bell Laboratories and the integrated circuit (IC) in the late 1950s to point the computer down the exponentially improving path it now follows into the information revolution. Without the integrated circuit, and the sophisticated semiconductor microelectronic technology that has developed around it, today's small, powerful, and relatively inexpensive computers—from DSP chips to supercomputers—would be unthinkable.

After a couple of decades of fundamental investigations of semiconductor materials and devices and the introduction of the IC in the 1960s, microelectronics based on silicon emerged as the front-running technology. Since the 1970s, silicon technology has been improving at an incredible rate. Most parameters that characterize microelectronics, such as increase in clock speed, decrease in size and cost, components per chip manufacturable with high yield, and the like, have been growing exponentially for the last two or three decades. These component-level technology growth patterns combine with computer architectural and algorithm and software advances to produce the explosive growth of computational power we are witnessing today.

The envelope of the technology growth curves is remarkably exponential, suggesting that extrapolation of these trends into the future should supply valid guidance as to the level of performance to be expected at various times. What these extrapolations do not indicate is which technology or specific devices will actually be developed to achieve the predicted envelope performance—these specifics are very difficult to forecast, but the levels of performance predicted are much more reliable. As long as no fundamental limits are in sight (e.g., the speed of light or the laws of thermodynamics), and this seems to be the case at least for the next 20 or more years, it is reasonable to assume that the observed exponential growth envelopes will continue into the future. As each implementation reaches its inherent limits, new concepts are introduced, improve rapidly as they stand on the shoulders of all that came before, achieve the increased level of performance, and saturate in their turn, and so on. As major changes in materials, architectures, processes, and fundamental physical operation occur, the slope of the growth curve may flatten for a time but is likely to resume along the exponential envelope. Obviously reaching out 40 years to 2035, as this study is attempting to do, involves a great deal of uncertainty but from the present point of view, the only other alternative seems to be pure conjecture, with even less basis and more uncertainty. The use and validity of envelope technology growth curves for technology forecasting are discussed again in Chapter 4 on sensors, again in the context of microelectronics.

Figure 2.1 illustrates the history and projected future growth of two of the key parameters that characterize modern microelectronics manufacturing—fabrication linewidth and transistors per chip. Obviously, decreasing the linewidths by which the individual devices are photolithographically defined permits more devices to be included on a given size chip, but the transistors-per-chip growth is

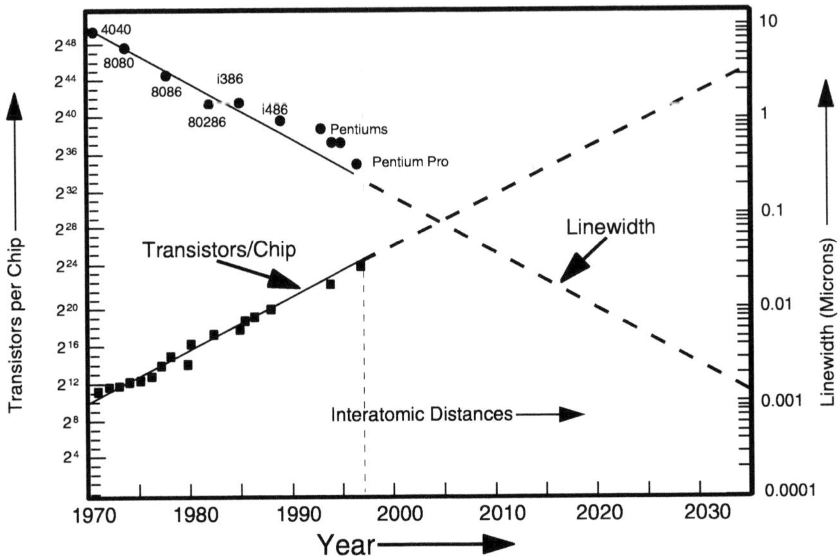

FIGURE 2.1 Microelectronic technology growth—fabrication linewidth and transistors per chip. SOURCE: Computation from data in Conrad, Michael, 1986, "The Lure of Molecular Computing," *IEEE Spectrum,* 23(10):55-60, October. See also information in Figures 4.2 and 4.7 in Chapter 4 of this report.

actually increasing faster than this because there is a concurrent but slower exponential increase in the size of defect-free chips. The two effects combine to produce the observed device-per-chip growth.

In 1965, Gordon Moore, founder of Intel, first publicized the exponential increase in transistors per chip, which at that time was doubling every 12 months. In practice, as Figure 2.1 indicates, the growth rate slowed somewhat, and Moore's Law, that is, the exponential growth of transistors per chip, has held consistently for almost three decades (i.e., from 1970 to the present) roughly doubling every 18 months. There is no obvious physical limit on devices per chip—e.g., table-sized chips are not practical, but are not excluded by any fundamental law of nature.

On the other hand, fabrication linewidth as shown in Figure 2.1 will encounter a fundamental limit near the end of our 40-year study window, namely interatomic dimensions. Dominated today by optical lithography, which has been refined well past the apparent limits of the technology as envisioned a decade ago, the current manufacturing state of the art is at the 0.25-μm level, with x-ray and electron beam technology ready to carry the linewidths smaller. Optical lithography is clearly heading toward additional practical obstacles in the next decade, as the wavelength of the light used moves further into the ultraviolet

(UV) where material absorption makes it harder and harder to find transparent optical materials for lenses. But these, too, may be overcome with new concepts, just as phase approaches have enabled the present practice to perform much better than the apparent wavelength limit.

Manipulation of nanosized clusters of atoms and even single atoms has been demonstrated. As the dimensions grow smaller, more quantum effects are encountered and the device technology will inevitably change from the structures of today's transistor concepts to advanced devices like single-electron transistors (SETs), which, although still macroscopic from an atomic point of view, operate on the basis of detecting the motion of single electrons. Room-temperature SETs have recently been reported, with transverse dimensions of 7 nm (0.007 μm) to 10 nm, which is in the neighborhood of the fabrication linewidths predicted for the 2020 to 2025 time period, although the fabrication processes used do not yet lend themselves to reliable large-scale manufacturing.

As none of the other key microelectronic parameters is expected to reach fundamental limits before 2035 (there are certainly numerous practical limits to be overcome), it would appear that we can have some confidence in the expectation that digital microelectronics will continue to grow throughout the 40-year time period of interest. Actually, the speed of light introduces obstacles to the implementation of very high speed logic (e.g., 300-GHz clock speeds are anticipated by 2035), but this does not seem to offer the same kind of impenetrable barrier that atomic dimensions present for linewidth. There are options to deal with the high-speed logic problem, e.g., superconducting rapid single-flux quantum (RSFQ) logic, optical interconnects, parallelism, and so on.

Chapter 4 of this report, dealing with sensors, provides more detailed discussions of these issues and gives a more complete discussion of silicon and other semiconductor technologies relevant to present and future micro- and nanoelectronics.

Computer Performance

With the vacuum tubes replaced by transistors, computers have exploited the ever-smaller, faster, and cheaper development of microelectronic integrated circuits to provide a similar exponential growth in available computational power. Figure 2.2 illustrates this growth in a schematic way, i.e., without specific performance points indicated but with a smooth curve suggesting the general average level of performance that characterized the time at which each of the indicated computers was introduced. Two distinct families of processors are indicated: (1) the high-performance computing processors, i.e., supercomputers, that have been largely government-sponsored, are expensive, and at any given time represent the highest level of performance available, and (2) the functional and affordable computing processors, i.e., personal computers (PCs), that generally have been

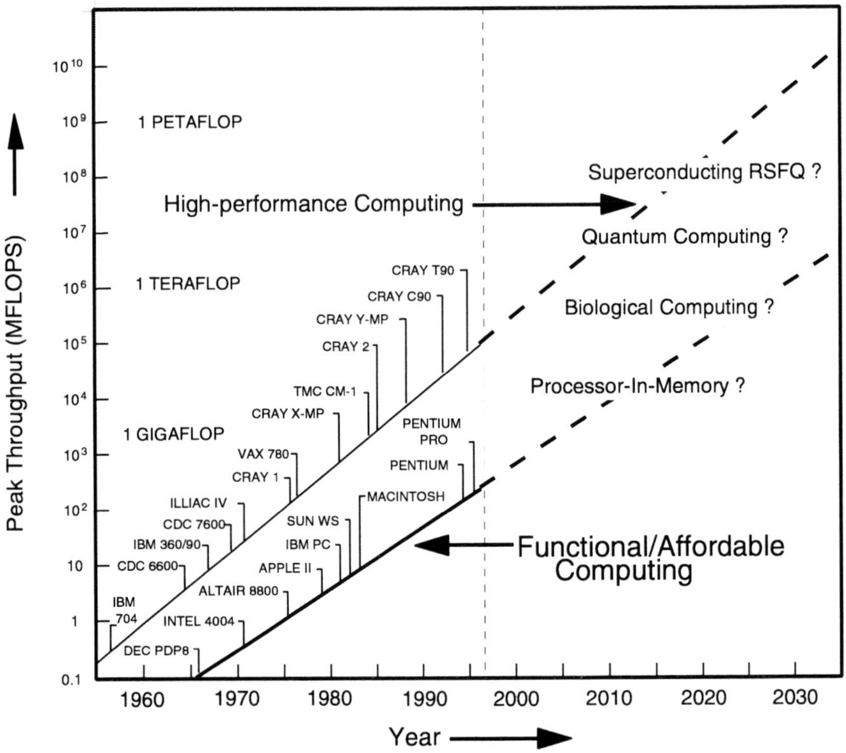

FIGURE 2.2 A history of computer throughput. SOURCE: Adapted from Brenner, Alfred E., 1996, "The Computing Revolution and the Physics Community," *Physics Today,* 49(10):24-30, Figure 1, October.

commercially developed, have lesser performance capability, and are relatively inexpensive.

The costs for high-performance computers have continued to rise with time as the supercomputers have grown more powerful—from the $4 million IBM 704 in 1956 to the Cray T-family of supercomputers that sell in the range of $10 million to $12 million each today. On the other hand, the costs for the functional/affordable computers have fallen dramatically as the processors have soared exponentially in performance—a DEC PDP-8 sold for several tens of thousand dollars in 1965, but a 200-MHz Pentium Pro PC can be purchased today almost anywhere for only a few thousand dollars, with a performance more than 1,000 times that of the PDP-8. It is difficult to believe that the cost trends exhibited by the affordable computers are not going to saturate soon, for costs on some quite powerful PCs have already dropped into the $300 to $500 range because of the fierce competition that drives this commercial market.

In the immediate future, there are plans to install a new generation of teraflop (10^{12} floating point operations per second) supercomputers at three national laboratories: a 1.8-teraflop Intel Pentium Pro-based multiprocessor at Sandia National Laboratories in late 1997, a 3.1-teraflop Cray/Silicon Graphics multiprocessor at Los Alamos National Laboratory in late 1998 or early 1999, and a 3-teraflop IBM multiprocessor at Lawrence Livermore National Laboratory sometime in 1998. Each of these falls somewhat above the extrapolated average performance curve for the time frame for which it is planned, but when the new supercomputers really come on-line is yet to be determined. In any case, high-performance computers will be limited in growth in the foreseeable future by technical issues, but economic constraints may also be a factor. People are already at work exploring the issues of achieving petaflop (10^{15} floating point operations per second) performance, which the extrapolation in Figure 2.2 suggests will become available near 2025. It seems possible that this level of high-performance computing will be achieved within the time frame of this study.

On the affordable PC side, Intel has announced plans for its P6 microprocessor line that will run at speeds of up to 400 MHz—twice the speed of the current high-end product, the 200-MHz Pentium Pro. Introduction of the P6 family over the next few years will conform closely to the extrapolated curve shown in Figure 2.2 for functional and affordable computing.

Lying between the two families of processors illustrated in Figure 2.2 are a number of powerful workstations that are not explicitly indicated. The Alpha chip, which powers the DEC workstation, clocks well above the PC state of the art at about 500 MHz and is a very capable microprocessor. There is no difficulty in envisioning this kind of chip migrating into PCs, and so the projected growth curves for functional/affordable computing will easily be met for the next decade or so. But, as is discussed further in Chapter 4 (Sensors), clock speeds will continue to increase—from decreasing linewidths (a 0.18-mm complementary metal oxide semiconductor [CMOS] circuit has been clocked at 1.14 GHz at Lincoln Laboratory, Massachusetts Institute of Technology) or perhaps from a change of technology, say from pure silicon to silicon germanide (SiGe), gallium arsenide (GaAs), indium phosphide (InP), silicon carbide (SiC), or some other system currently unimagined. Gigaflop (10^9 floating point operations per second) PCs by 2002 and teraflop capacity on the desktop by 2030 are realistic possibilities.

Multiprocessor Architectures

A computer's throughput performance depends not only on the speed of its hardware but also on its architecture—that is, the nature of and the relationships between its constituent parts. With clock speeds peaked at whatever the current technology can supply, additional performance can be extracted by exploiting parallelism—distributing the computations over multiple hardware components.

At the level of a single CPU, this is achieved by the parallel execution of instructions, through the incorporation of multiple adders, multipliers, and so on on a single processor chip. In this manner, rather than using on the order of 10 clock cycles to accomplish a single instruction, as was the case for the earlier microprocessors (e.g., the Intel 8086 or the Motorola 68000), today's microprocessors, through parallel execution, accomplish multiple instructions per clock cycle.

To achieve even more computational throughput, it is natural to combine the resources of multiple CPUs to produce distributed arrays of processors—that is, multiprocessors. Supercomputers have long employed this principle of parallel hardware units. The three supercomputers scheduled for installation at Sandia, Los Alamos, and Lawrence Livermore national laboratories in the next 2 years are all multiprocessor architectures, one of which envisions combining 4096 IBM/Motorola/Apple PowerPC processors and another of which links 9072 Intel Pentium Pro processors.

Cooperative behavior of multiple processors to provide efficient computation is not easy to accomplish. Depending on the tasks undertaken and the instantaneous state of the hardware, there can be many conflicts for access to the resources—the processors, the busses, the memory, the input/output (I/O) to the outside world, and so on. Many options have been explored and no single best multiprocessor architecture has yet been defined. And so architectures continue to evolve through engineering trades, improvements introduced to correct problems encountered in earlier generations, and the introduction of new component technologies and software concepts that alter the balance of the earlier tradeoffs. It seems likely that the architectures will continue to evolve throughout the window of this study, with completely new concepts appearing from time to time.

In terms of how instructions and data interact, multiprocessor architectures are broadly classified into two categories—single instruction multiple data (SIMD) and multiple instruction multiple data (MIMD). SIMD machines perform the same instructions on multiple sets of data simultaneously and thus are ideally suited for applications such as signal processing, image processing, or the investigation of physical phenomena involving multidimensional, multivariable differential or integral equations. With these special applications come special configurations of the hardware. The early Cray supercomputers, for example, contained pipelined vector processing units that repeated the same operations over and over as the data streamed through. They were extremely efficient once set up but took a significant amount of time to be reconfigured for different tasks. A decade ago, with large amounts of DOD support, one-dimensional vector processors were superseded by two-dimensional MPPs. These highly specialized processors consisted of two-dimensional arrays of relatively simple CPUs (sometimes a 1-bit processor), each with its own large memory stack, which all executed the same instruction synchronously with the others, but on different data. Ideal for image processing and the solution of some physics problems, MPPs have been deployed already as signal processors in many problems. But the SIMD machines, although undeniably powerful for certain applications, have proven to be inflexible and somewhat difficult to program. They are not suited to general-purpose computation, and most, if not all, of the companies producing this class of processor are now out of business.

MIMD machines do not synchronize instructions but allow each processor to operate independently. Although somewhat less efficient on problems for which the MPP is matched, the MIMD class is completely flexible and, although still a challenge, easier to program than a SIMD. All future high-performance computers are likely to be MIMD multiprocessors.

On the battlefield, it will become more and more critical to ensure that the computational resources always work when required, make no errors, and when and if they do fail, do so in such a manner as to not endanger the people or the projects involved. In other words, the computers must be highly available, fault tolerant, and fail-safe. As the information revolution progresses, these features will cease being options and will simply be mandatory—whatever the apparent cost of implementation. It is already clear that we are so computerized and networked today that a single failed computer has the potential to bring an office, a factory, a city, a military force, and even a whole country to its knees—the costs of failure are going to overwhelm the cost of implementation. Hardware redundancy and careful design of the software and operating systems are the keys to achieving this robustness for both commercial off-the-shelf (COTS) and Navy Department computer systems.

Hardware Technology

Microprocessors

Complex Instruction Set Computing

Original microprocessor CPU designs were developed and incorporated into the hardware mechanisms for implementing all potentially useful elementary instructions as the assembly languages. Because the combinations of instructions that a programmer could employ were very large and unconstrained, designers added hardware mechanisms to deal with all potential on-board hardware conflicts. The result was a complex chip that generally needed multiple clock cycles to accomplish any given instruction. The processor, before implementing the next instruction, had to check to see if the last one was still using a resource (e.g., register, adder or multiplier, and the like) and then resolve any conflicts. These early architectures have come to be known as CISCs, in contrast to the more recent innovation of RISCs.

Reduced Instruction Set Computing

As programmers made the transition from assembly language to a strong reliance on high-level computer languages and compilers, it was realized that, in practice, very few instructions were used frequently. By limiting the instruction set supported directly by the hardware, the instruction decode logic could be reduced to 20 to 30 percent of the chip real estate versus 75 to 80 percent for CISC chips. A reduced instruction set with limited pipelining and instruction decode logic was first seen on the Cray-2 series of computers and achieved outstanding performance, exploiting the resulting small chip sizes and the fast clock speeds that the small chips permitted. Not yet known as RISC, the Crays were thought of as streamlined vector processors, optimized to carry out efficiently those arithmetic operations needed for the solution of high-powered mathematical and physics problems of nuclear weapons design, for which they were largely employed. Modern DSP chips follow this same path—they were efficient for repeated mathematical operations but not terribly flexible with respect to general programming.

In the 1980s, it was realized that most useful compiled software programs required only a small subset of the available CISC instructions. Also, once the code was written, the potential on-board hardware conflicts were defined and knowable in advance, before the code was run. Thus the final code could be passed through an optimizer software module that configured the sequence of instructions to avoid all on-board conflicts. The result was the RISC chips, which are simple and fast. Adding multiple arithmetic units on-chip and pipelining the instructions permitted the early RISC processors to perform an instruction per clock cycle. In the late 1980s, this provided an instant jump in microprocessor performance by a factor of almost 10, without any increase in clock speed. Examples of such RISC processors include the Sun Sparc, the HP PA-RISC 7000, the Motorola 88K, and the MIPS R3000.

With time, improvements were introduced to capitalize on the increased number of devices on a single chip and to run faster, as well as to overcome deficiencies in the earlier implementations or limitations of the associated components. To minimize main memory access (dynamic random access memory [DRAM] access times remained at 70 ns for a long time), large caches of up to 4 megabytes have been provided on-chip. Multiple arithmetic units with vector capability have been combined with multiple (up to four), long (up to six or eight stages) instruction pipelines and on-chip branch prediction logic added to deal with data-dependent branches, which are not predictable in advance. The resulting chips are known as superscalar RISC and include such examples as the DEC Alpha, the Sun Ultra Sparc, the HP PA-RISC 8000, the Motorola 600, and the MIPS R10000. These processors operate at speeds of up to 200 MHz with the Alpha reaching 300 to 500 MHz. To achieve these features, the superscalar RISCs are beginning to resemble CISC, in that a larger fraction of the real estate

TABLE 2.1 Performance of Typical Microprocessors

Processor	MHz	SPECint95	SPECfp95
Power PC 604	225	8.7	7.5
HP PA-8000	180	10.8	18.3
MIPS R10000	200	9.2	17.6
DEC Alpha	500	12.8	18.3
Sun Ultra Sparc 2	200	6.8	11.5
Pentium Pro	200	9.0	6.2

SOURCE: Data from Raytheon Electronic Systems.

is necessarily devoted to instruction decode, cache management, and branch prediction (currently up to 75 percent of the chip area).

Table 2.1 summarizes the current state of the art for microprocessors, most of which are superscalar RISC architectures. The Intel Pentium chips resemble superscalar processors in terms of on-chip functionality except that they have relatively large, CISC-like instruction sets. The performance is measured separately for integer operations (SPECint95) and for floating point operations (SPECfp95).

Interconnects and Packaging

Interconnections between the processors and memories in a multiprocessor are key to performance. Data and information must flow quickly to where it is needed if computation is to proceed smoothly at the level desired. Many schemes for interconnecting processors have been considered, including two-, three-, and higher-dimensional arrays of busses and a variety of processor-to-processor configurations with two-, three-, and higher-dimensional connections to nearest neighbors, n-cube topologies, and others. Perhaps the most common today is the simple two-dimensional rectangular north-south-east-west array with each processor communicating only with its four adjacent neighbors. Obviously, minimizing the number of internal data transfers required is highly desirable as it brings up questions of communication routing, resource conflicts, and latency, which all can adversely affect overall computational performance. As the hardware speeds up, it becomes more and more important to avoid having any unplanned delays, for even a brief pause in a 100-GHz data stream can cause an enormous pileup of data and rapidly exceed temporary storage capability.

For large distances (i.e., centimeters to meters or kilometers) and high data rates, it is clear that fiber-optic telecommunication is the preferred approach for digital data transfer. Multigigahertz rates have been demonstrated, and higher are no doubt possible. Many modern radar systems with high data-rate requirements already use COTS fiber-optic links to interconnect signal processors and control computers. This is clearly the right technology for these applications, although costs are still high for the highest-performance links. The cost of optical links will inevitably decrease under the impetus of the commercial telecommunication explosion.

But it is not only at the processor-to-processor level that optical communication technology will find a role in computation. Even in communicating digital signals board-to-board and chip-to-chip, optical technology offers solutions to future problems. A laboratory curiosity today, free-space optical interconnects between boards and chips will soon find their place in computer technology. Although the optical componentry is still lacking in some respects—in diode threshold power and reliability, in particular—as the clocks speeds increase, the optical interconnects will become more acceptable than electrical. Figure 2.3 illustrates the tradeoff between the use of a 3-μm-wide aluminum conventional electrical interconnect and a free-space optical interconnect using a 1-mW threshold laser operating at 9 percent efficiency. The curve is the break-even line length as a function of frequency for which the electrical and the optical interconnections require the same power to transfer the data. The results vary slightly for narrower Al lines and higher-power lasers, but the results are similar. Even at 50 MHz, for distances beyond 5 or 6 mm, optical interconnects provide a lower-

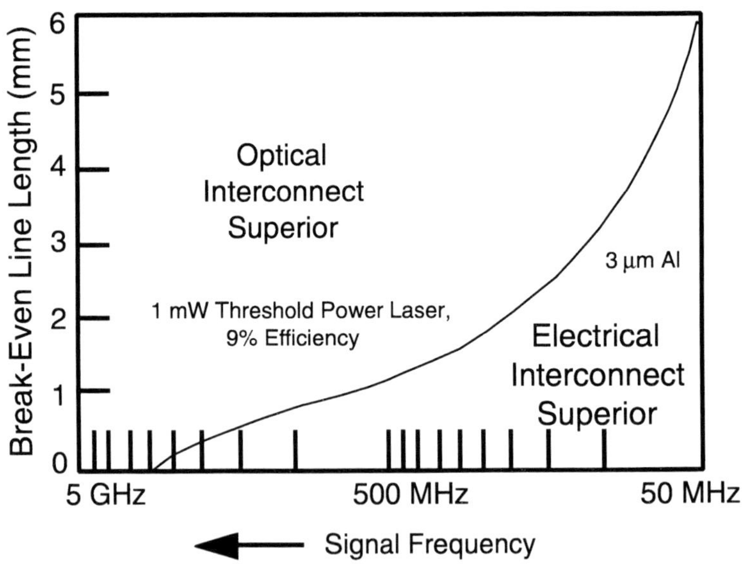

FIGURE 2.3 Break-even transmission line length—optical vs. electrical interconnect. SOURCE: Adapted from Cinato, P., and K.C. Young, Jr., 1993, "Optical Interconnections Within Multichip Modules," *Optical Engineering*, 32(4):852, April.

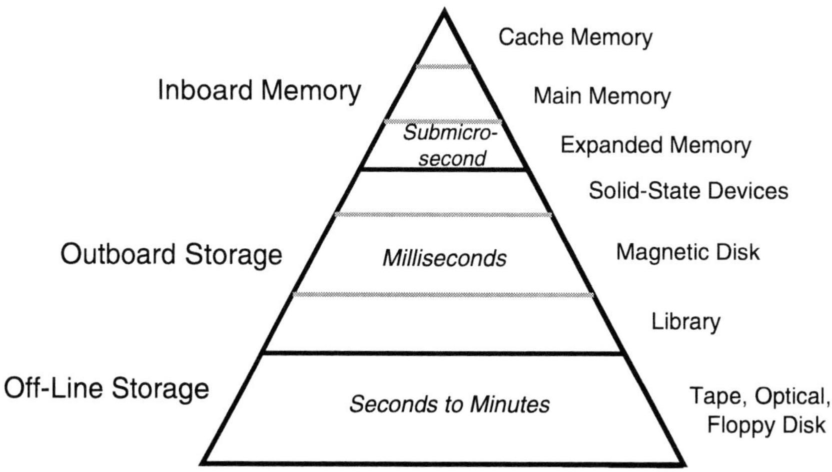

FIGURE 2.4 Computer data storage hierarchy. SOURCE: Adapted from Wood, C., and P. Hodges, 1993, "DASD Trends: Cost, Performance and Form Factor," *Proceedings of the IEEE*, 81(4):573-585, April.

power solution. For rates of 500 MHz and above, the break-even distance drops below 1 mm, which means that very soon, optical interconnects will be viable for chip-to-chip interconnections inside mine countermeasure packages. In addition to these power advantages, optical interconnections have other virtues, such as that (1) they are free from capacitive loading and electromagnetic interference (EMI), which afflicts conventional electrical approaches, and (2) they are inherently parallel and permit a high level of multiplexing because of their high bandwidth, thereby offering excellent solutions for resolving chip I/O issues.

For the next 40 years, optical interconnections will play a major role in facilitating the implementation of very high speed (multigigahertz clocks) CPUs and multiprocessors for both high-performance and functional and affordable computing applications.

Memory

Memory, its storage and its access, is another key to the success of computation. Computers, from PCs to multiprocessor supercomputers, store data in a hierarchical form, with very-fast-access, small on-chip cache memories at the top, and very large, slow, off-line mass data storage at the bottom. Figure 2.4 illustrates this hierarchy. The performance and the price per bit stored generally rise as you move from the bottom to the top. At the top are semiconductor memories, with access times in the nanosecond range. At the bottom are removable media, tapes and disks stored off-line, having access times measured in seconds to minutes. In the middle are the off-board hard disks and solid-state devices with access times measured in milliseconds.

As discussed above, many CPUs, particularly the superscalar RISCs, incorporate significant amounts of cache memory directly on the processor chip, while increasing amounts of discrete random access memory (RAM) characterize all modern computers. It is not unusual to find 32, 64, or even 128 megabytes of RAM in a PC these days. In keeping with the industry trends, we can expect this class of memory to continue to grow in size and to reduce access time, while the price is simultaneously decreasing. Fortunately, for these classes of working memory applications, volatility is quite acceptable, for it has proven extremely difficult to achieve nonvolatility while retaining access time performance in semiconductor memories. Although memory speed and capacity have been growing, they have not been growing as rapidly as processor speed and capacity. It is possible that memory limitations could impede the full utilization of the processing power that is expected to be available in the future.

The middle ground, i.e., outboard storage, has been, and continues to be, dominated by the magnetic hard disk. Access times are determined by the mechanical motions of the read/write arm and the rotation of the disk itself and are typically measured in several to tens of milliseconds. In the 1980s it was believed that magnetic recording had reached the technological limits for feasible data density (bits per unit area). However, the introduction of the optical disk in 1982, which at that time exceeded the density of the state-of-the-art magnetic devices by an order of magnitude, stimulated a new look at the subject, and continuous increases in recording densities have followed. By 1990, the optical disk advantage had vanished, and today, magnetic storage densities are increasing at a rate of 60 percent per year. The most advanced disks today have densities exceeding 150 megabits/square inch, and at least a gigabit per square inch is expected by 2000.

Solid-state competition for the hard disk, i.e., semiconductor memory packaged for lower price and lower performance than on-board RAM, does provide the fastest memory for outboard applications but is volatile. To date, nonvolatile solid-state memory, e.g., charge-coupled devices (CCDs), magnetic bubbles, and holographic storage, has not been able to match the price and performance of the hard disk, often falling short by orders of magnitude. At the bottom of the pyramid, optical disks and tapes have come into their own, competing successfully with magnetic tape for high-volume, low-cost removable storage. Optical disks and tapes have the advantage of much greater storage lifetimes.

Optical Data Storage

CD Technology. The practicality of optical storage of digital data is obvious today. For music and movies, digital optical recordings such as compact disks

(CDs) and laser disks continue to increase in popularity, threatening to eliminate their analog magnetic tape competition, in spite of being more expensive. In the computer world, all new PCs come equipped with integral CD read-only memory (ROM) drives, and most new software is readily available only in CD form.

It is often suggested that the advantage of optical over magnetic digital data storage lies in an inherent ability of optics, highly focused laser beams, and the like to produce higher-bit densities than does magnetic recording. This, however, is not the case today. The very real advantages of optical over magnetic recording lie elsewhere.

When first introduced, in the early 1980s, the optical media area density of 155 kilobits/mm^2 greatly exceeded the 10 or 11 kb/mm^2 of the magnetic recording technology of the time. The latter was thought to be approaching fundamental limits. But, as is frequently the case, the magnetic incumbent, with much to protect, responded vigorously and by the early 1990s, the advantage of optical over magnetic recording had vanished through inventions such as the magneto-resistive head. Today, practical area densities of magnetic hard disks can exceed those of comparable optical CDs by as much as an order of magnitude.

The optical CD drive is very similar in performance to the magnetic hard drive, in that it provides fast random access to any data recorded on the disk at rates and access times that approach those of the magnetic drives. The critical differences are in the optical recording media, which are simple, inexpensive, and durable, and the use of optical techniques that permit the read/write heads to stand well back from the recording surface with relaxed tolerances (disk surface motions are accurately tracked by sophisticated optoelectronic feedback loops, keeping the laser read/write beams continuously in focus on the surface where the data bits are recorded). The result is a removable optical disk with the performance of a magnetic hard disk, far exceeding the capabilities of removable magnetic floppy disks of similar size. Removable random-access data storage with long life without degradation is where optical storage excels.

On the other hand, although magnetic disks are infinitely rewritable, rewritable optical disks have proven more difficult to master. Writable and rewritable optical disk systems are available in limited quantities and are still in development and, as a result, are often expensive. Rewritable optical recording requires a different form of recording from the pits and bumps of read-only CDs, and two such candidates have been exploited so far—magneto-optic recording and phase-change recording. The magneto-optic systems can be directly rewritten without erasing, whereas the phase-change systems require a controlled anneal to restore the medium to writable form. Recently a CD recordable drive for less than $500 has become available. Progress in these areas has been rapid, and without doubt such versatile optical disk systems will soon be available at affordable prices.

Optical Tape. Beyond removable random-access high-capacity storage, optical storage technology offers a promising extension in the form of the optical tape recorder for massive archival data storage where the high capacity and longevity of the optical medium offer significant advantages over magnetic storage. Using slightly modified read-write concepts, the optical tape stores its data in closely spaced parallel tracks along the length of the tape, which can be read out in parallel onto CCD detector arrays. With oversampling, i.e., more detectors than data tracks, electronic rather than physical tracking can be easily implemented. Because tape is so thin, enormous amounts of data can be stored in a very small volume. A standard 3480 magnetic tape cartridge today can hold 10 gigabytes of data. A comparable cartridge loaded with optical tape can handle almost 30 gigabytes. Reduction in tape thickness to 0.025 mm and the use of blue (415 nm) rather than red (633 nm) laser diodes will permit up to 200 gigabytes of data to be compressed into the same cartridge by the year 2000. The parallel readout enables continuous data readout rates of about 3 megabytes/s today and can be expected to reach 10 megabytes/s by 2000. As is characteristic of all tape storage concepts, random-access times are on the order of seconds (for tens of meters of tape running at tens of meters per second).

The Future. The area densities of two-dimensional optical storage devices will continue to improve through increases in the numerical apertures of the lenses, i.e., smaller spots, and the introduction of shorter-wavelength laser diodes (i.e., blue). Figure 2.5 illustrates the projected growth for both optical and magnetic disks and tapes over the next few years.

Three-dimensional Holographic Memories

Present disk and tape systems store data in two-dimensional formats on a surface. In a holographic memory, bits are stored in two-dimensional patterns, known as pages, using standard holographic interference techniques. An information and a reference beam are combined in a photorefractive material that modulates the index of refraction of the material in response to the interference pattern's distribution of electric fields. To pack more information into the same volume, additional pages are recorded on top of each other by varying the angle of incidence of the reference beam. Illuminating the storage medium by a reference beam reads out only the page that was recorded with that particular reference beam by projecting an image of the page on a two-dimensional photodetector array.

Although the number of superimposed holograms is limited by the dynamic range of the photorefractive material, in lithium niobate as many as 10,000 superimposed holograms have been demonstrated. Read-out rates are limited by the requirements for rapid, accurate generation of randomly accessed reference beams. The photorefractive materials identified so far tend to be expensive to grow in bulk form. Breakthroughs are being sought in the form of multilayer thin-film composite structures. While promising much, particularly for archival

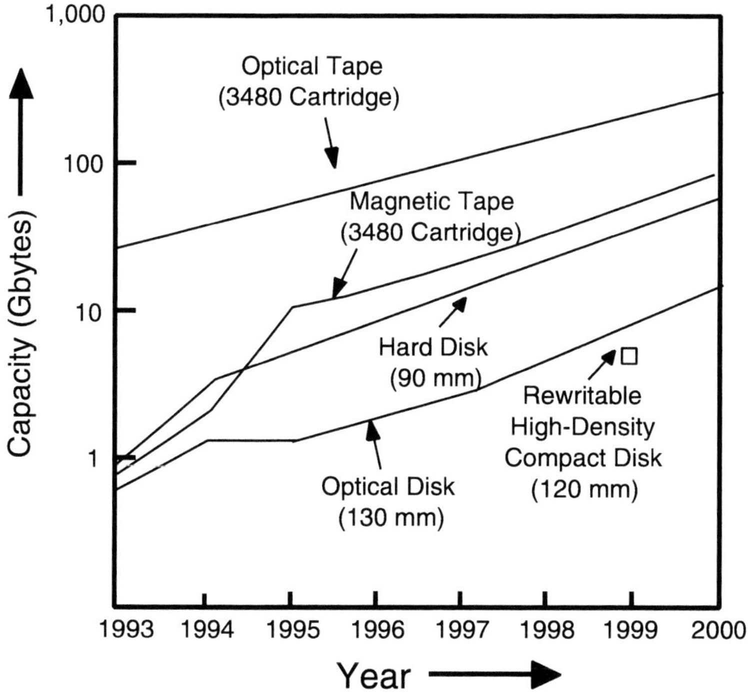

FIGURE 2.5 Comparison of optical and magnetic data storage. SOURCE: Adapted from Asthana, P., and B. Finkelstein. 1995. "Super Dense Optical Storage," *IEEE Spectrum*, 32(8):25-31, August.

applications, holographic optical storage's success seems to depend heavily on the occurrence of a significant advance in materials. It is too early to predict with any confidence what will happen to holographic memories.

Software

Computer Languages

As the hardware has grown in capability, new tasks have been imagined—and an ever-increasing array of algorithms have evolved to accomplish these tasks. To provide the software for programming computers to implement these algorithms, dozens of computer languages have been developed. Beginning with the low-level assembly language, which is different for each processor implementation and exercises direct control of the elementary steps of the computer hardware, a series of higher-order languages have evolved. Through a brilliant bootstrap technique called the software compiler, the programmer has been re-

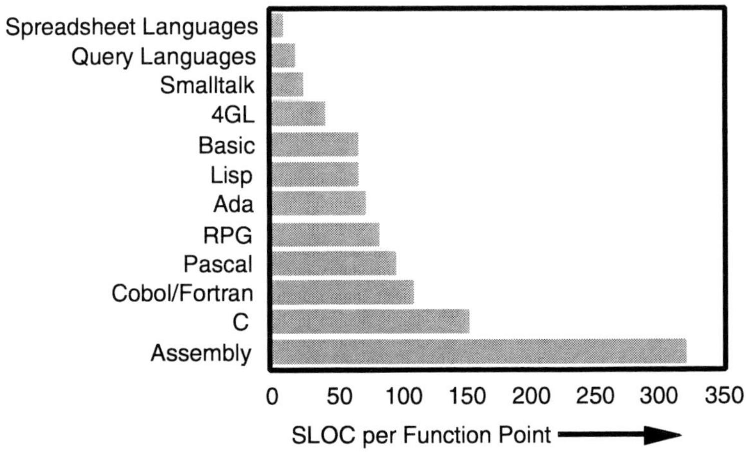

FIGURE 2.6 Power of programming languages in terms of source lines of code (SLOC) per function point. SOURCE: Adapted from Lewis, T. (ed.), 1996, "The Next 10,000 Years: Part II," *COMPUTER*, 29(5):81, Figure 5(b), May.

lieved from the necessity of dealing with every minute detail, and greatly improved programming efficiency has resulted.

Figure 2.6 compares the programming power of different programming languages in terms of the average number of source lines of code (SLOC) in that particular language required to implement a function point (FP), i.e., a specific block of code. The function point, a useful metric introduced in the 1970s by Allan Albrecht of IBM, measures both the content of computer programs and the productivity of the programmer by counting the number of steps (input, output, memory access, and general processing) that define the block of code. As can easily be seen, there is a large advantage in compactness of code for all the high-order languages as compared with assembly language. Somewhat surprisingly, C (and C++, it must be assumed), which is the most popular programming language today, is the least compact high-order language as compared with assembly language. Actually the low-level characteristics of C are what makes it attractive to many programmers as it offers a deep level of control. Ada, on the other hand, the DOD's preferred but fading language, is more compact than C by a factor of two and has proven in practice to be an excellent language for the disciplined development of large military programs. Yet C remains enormously more popular, explained, perhaps, by the observation that Ada has garnered very little COTS support in the United States (more in Europe, ironically) and few, if any, universities offer courses in Ada whereas C and C++ enjoy all these advantages.

The evolution up the chart to more and more compact languages has been roughly, but not precisely, in temporal order. With the growth of more efficient high-order languages, programming productivity, as measured in average cost

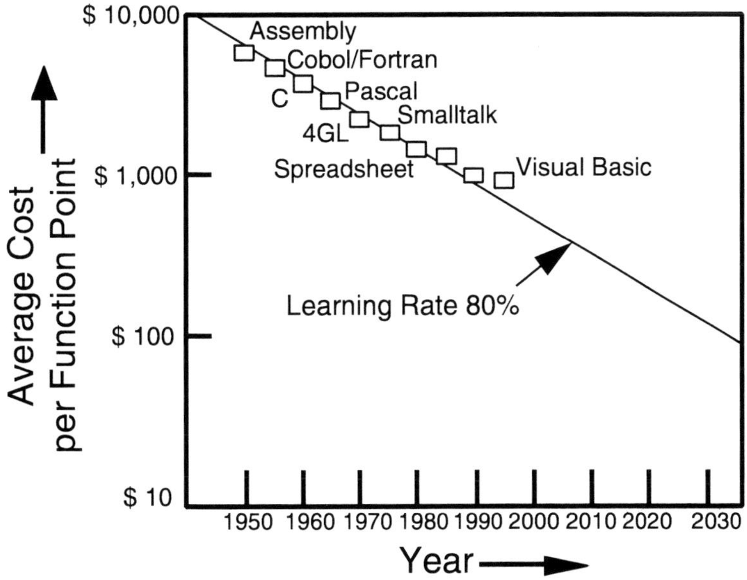

FIGURE 2.7 Average cost per function point. SOURCE: Adapted from Lewis, T. (ed.), 1996, "The Next 10,000 Years: Part II," *COMPUTER*, 29(5):81, Figure 5(b), May.

per function point, has been improving exponentially, as with the hardware parameters that drive the information revolution. Figure 2.7 shows costs per function point as a function of time. The exponential decrease indicated by the straight-line extrapolation represents a learning rate of about 80 percent over 5 years or roughly 5 percent per year, compounded.

Whether or not the slight deviation from the curve, evident over the past decade, is indicative of some kind of fundamental limit, is impossible to determine at this time. New concepts for even more productive languages may yet arise, and the final goal for 2035 of $100 per FP does not seem completely unreasonable.

Software Growth

As computational resources have grown, the size of the tasks undertaken have grown correspondingly. Unpleasantly, this exponential growth in the magnitude of software systems has serious consequences for cost and reliability. The error rate (i.e., errors per SLOC) does not remain constant but increases with the size of the programs. Figure 2.8 shows the growth in total code size, as measured by the size of the memory necessary to contain the code, for a number of National Aeronautics and Space Administration (NASA) space and military aircraft pro-

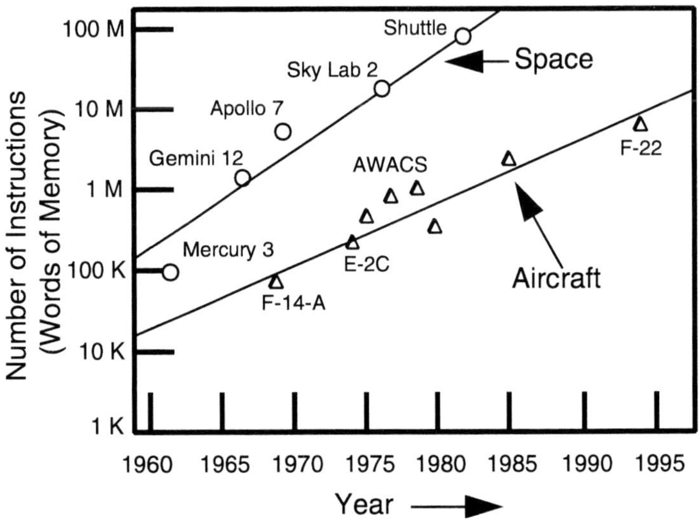

FIGURE 2.8 Growth in size of memory for application software. SOURCE: Adapted from Buckley, CAPT B.W., USN, 1996, "Software Production Efficiency Lags Computer Performance Growth," a viewgraph included in the presentation "Future Technologies for Naval Aviation," Naval Research Laboratory, Washington, D.C., July 22.

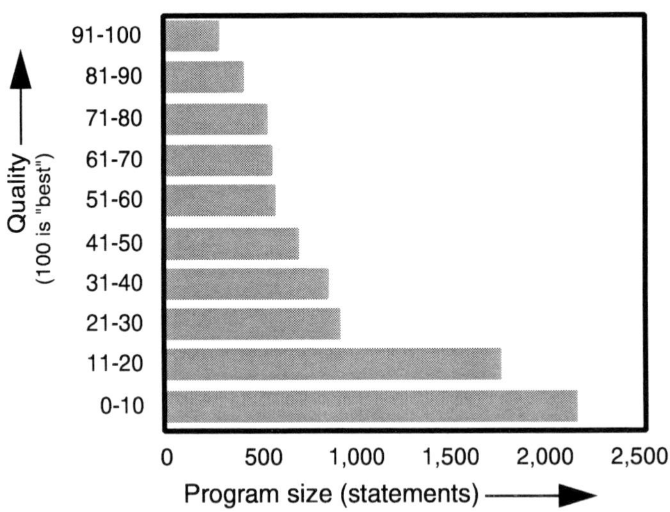

FIGURE 2.9 Software quality falls off as programs become larger. SOURCE: Adapted from Lewis, T. (ed.), 1996, "The Next 10,000 Years: Part II," *COMPUTER*, 29(5):85, Figure 8, May.

grams. Extrapolating out to 2035, software code storage volumes considerably larger than 10 billion words of memory are suggested.

In strong contrast to most of the other growth patterns that characterize the information revolution, this trend should be reversed. Software quality definitely falls off as the programs get larger. Figure 2.9 illustrates this phenomenon. It suggests that a solution is to build large programs from small, reusable modules that can be independently and separately constructed and debugged. The growing size of software coupled with the strong negative correlation of software quality with size has focused attention on the software development process. Although software provides the source of the capability and much of the value in most large commercial and military systems, it also seems to be the source of most major overruns and schedule slips in the development stage, and it increasingly represents the major contributor to system failures in the field.

In recognition of the critical nature of software and in response to the above problems, the DOD initiated, about a decade ago, a software engineering initiative (SEI) to monitor and certify the software development processes in use by DOD contractors. Initial surveys revealed that most software developers had no structured and documented practices at all. Software development was thought to be more of an art than an engineering discipline. But as programs grew to millions of lines of code, the physical impossibility of generating a handful of artists became evident. With time and with SEI certification a formal requirement of the contract, contractors established structured software development practices and introduced the routine use of computer tools that have greatly reduced but not eliminated the risks in software development today; as the appetite of the information revolution inevitably grows and program sizes reach tens of hundreds of millions of lines of code, these problems will continue to plague us.

In the area of computer-aided design (CAD), software lags well behind hardware design, surprising in view of the early introduction of software tools for compiling, assembling, and so on. Given the desired input and output from a digital circuit (DC), logic synthesis programs routinely create credible logic implementations, which are correct by construction, by automatically performing the appropriate Boolean equations. Hardware designs need no longer be constructed by hand through analysis and intuition as software designs still are. Equivalent logic synthesis tools for software design should also be possible, although the underlying mathematics will not be a simple Boolean algebra. However, the software tools available today are frequently little more than bookkeeping tools that do help to maintain consistent forms and practice but do not help much in achieving a good design. A good software design is still largely up to the talents of the individual programmer. One possible solution to the software development cost, schedule, and quality problem is to develop software logic synthesis capabilities. This should be an ongoing topic of R&D for the future.

Object-oriented Programming

For most of the almost 50 years that people have been programming computers, the approach to constructing the software has been based on a viewpoint that emphasizes the function associated with the design. Emphasizing the flow of mathematical operations and the algorithms, this approach has been highly productive. For small systems it works well, giving excellent performance, and has produced proven methodologies for constructing hard real-time software systems, as many sensor, control, and weapons systems require. In addition, most operating systems developed to date, including the popular UNIX and WindowsNT operating systems, have been based on a functional decomposition model. During the last 5 years, UNIX attained major visibility in workstation applications, but recently interest has shifted rapidly in favor of WindowsNT because of its natural support for networking.

As the software program sizes continued to expand over the past decade, attention turned toward achieving software reusability, so that systems could be constructed from generic and reliable, small standard building-block modules. But reuse of modules constructed on functional decomposition principles turned out to be difficult because of a general fuzziness in the area of module-to-module interface definitions. A more promising approach was sought and found in the object-oriented programming (OOP) approach.

OOP takes quite a different point of view from that of functional decomposition, but one that is much better attuned to reusability. OOP emphasizes the objects that are acted on by the program, such as the data types, data structures, and interfaces, rather than the actions themselves. OOP looks at the nouns, whereas functional decomposition views the verbs in the ordinary language description of what the program is supposed to do. As such, these two approaches are duals of each other—the blobs in one representation of the program become the lines in the other. The OOP approach has its root in the proven methodology developed in the context of database management systems. Because of its explicit attention to interface definitions, OOP modules look like well-defined building-block modules and provide greater reuse possibilities because the data can be mixed and matched in many ways. On the other hand, OOP probably suffers performance penalties over functional designs because of data movement within structures and some difficulties in enforcing hard real-time constraints. But increasingly higher-performance processors and larger on-board caches will easily compensate for these deficiencies in the future.

Supported by several high-order languages (e.g., Smalltalk and C++), OOP promises to be a major factor in future software developments. A recent development, stimulated by the Internet, has been the creation by Sun Microsystems of Java. An interpretative, rather than compiled, object-based language, Java has the concept of objects, classes, and methods fully integrated within the language. Java has no operating-system calls. All such functions are accessed through a universal I/O set of classes. Exception handling is accessed through a set of exception classes, graphics capability is accessed through a set of graphics classes, networking is supported through a set of network classes, and so on. The language allows the users to define their own classes and methods and provides unparalleled flexibility. Without the need for operating-system calls, Java programs become independent of the specifics of the platform and the operating system. The penalties associated with using an interpretative mode will fade in importance as processor speed and performance continue to improve. Other Internet providers are exploring similar concepts, and both Netscape and Active X offer products, with similar features to Java, that give them platform and operating-system independence. Java and other similar approaches offer new programming models that deserve further serious exploration.

Software Vulnerabilities

In many ways the software itself is its own biggest liability, for everything can be done by software, including crashing the system and producing unexpected or erroneous results. The software can do this through design faults or with outside help through the malicious introduction of viruses, bugs, and bombs. For more detail, see Chapter 3, Information and Communications.

Reliability

As the role of software grows, the impact of software defects on the overall system reliability becomes a critical issue. In contrast to hardware, software does not wear out with time and usage and become failure prone. In spite of good intentions, defects are built in through faulty logic, omission, coding errors, and the like. What is statistical are the number and kinds of defects that have been created that escape the developer's best attempts at testing and the associated user's mix of programming requests. To properly understand these factors, continuous and appropriate measurements must be made throughout the coding and testing and integration phases. We would like to know the rate of creation and the distribution of defects that are introduced, the efficiency of the testing procedures in uncovering them, the rate at which repair introduces new errors, and so on. Many statistical models are available that have been created to address these scenarios. These should be routinely applied to the software development process to improve and control practices and the quality of the final products.

Correctness

Evaluating software through traditional methods of exhaustive testing can never prove that the software is correct, that is, that it will do only what was intended and nothing else, independent of the inputs. On the other hand, formal specifications, expressed generally in some executable high-order specification language, allow mathematical theorem-proving techniques and tools to be applied in practical ways that permit genuine proofs to be convincingly constructed. In other words, instead of building the software in the traditional manner and then testing it to discover what has emerged, we apply structured processes to its development to ensure that it is correct by construction. More generally within the industry, alternate methods of software development, such as IBM's so-called clean room technique, have been suggested that also rely on formal techniques and that are claimed to produce more defect-free code than does the traditional approach. For these several reasons, it is very appropriate to take a serious look at the role of formal methods in the efficient generation of reliable, safe, and robust software.

Information Warfare Attacks: Viruses and Bugs

The vulnerability of software-dominated systems to viruses and bugs, whether deliberately or inadvertently introduced, is an everyday experience for most computer users. The network server goes down and every computer connected to it freezes, or an electronic-mail message arrives and with it comes a virus that starts to do unpleasant things, or a proven piece of COTS software suddenly quits, dumping the file you just spent the morning on because of a bug. As the information revolution progresses, our whole society, the naval forces included, will become increasingly dependent on computational software and computers and thus increasingly vulnerable to deliberate interference with the software. Although virus-detection add-on software exists, it would be preferable if somehow the defense mechanisms were build into the software from the start as part of the overall design. How to design robust software, maximally resistant to external interference, is a theoretical question of great interest and should be the subject of research for the future.

Applications

The applications of computation are practically limitless, ranging from explicit mathematical calculations for the solution of equations describing physical phenomena and engineering analysis or the interpretation of sensor data, to the production and management of text files for documents and communication, and the generation and display of graphics, from simple line drawings to full-color, high-resolution, natural-looking synthetic scenes for virtual reality. The panel has chosen to classify these applications into three categories in terms of the computer hardware required to support them, as listed below:

1. High-performance computing (i.e., supercomputers),

2. Functional and affordable computing (i.e., desktop computers), and
3. Embedded computing (i.e., integrated CPU and DSP chips).

With time, future workstations and PCs will eventually attain the computational capabilities that today can only be obtained from a supercomputer, and many of the high-performance applications will migrate to the functional and affordable class. Of course, there will always be problems large enough to challenge a supercomputer, because it is easy to expand the problems to match the capabilities available, by increasing the number of variables or the resolution or the number of interacting subsystems to be simultaneously simulated, and so on.

Because of the specialized application for embedded computers as real-time subsystems that are fully integrated into the sensors and the controls of radars, sonars, missiles, torpedoes, UAVs, UUVs, and other high-performance systems, these applications tend to remain distinct from those that find primary support from the other two categories of computers. Often, however, these same systems also contain a general-purpose computer, which is frequently a specially packaged, ruggedized version of a desktop computer. Also, there is widespread compatibility of the lower-level hardware components between all three classes. The multiprocessor supercomputers currently under development generally consist of distributed arrays of cooperating CPU chips that are identical to those that power the desktop computers, e.g., PowerPC or Pentium Pro chips. These same chips may find their way into embedded applications as well. The era of the special-purpose computer or signal processor for military applications has just about vanished, as the commercial developments have consistently outperformed these processors, frequently rendering them obsolete long before the systems are deployed. No doubt, the use of commercial chips in military systems will continue as newer, more powerful computational engines are developed as COTS products for the information-revolution markets.

High-performance Computing

Originally motivated by the demands of nuclear weapons design, supercomputers have found wide application in the physics, astronomy, and engineering communities for the investigation of difficult physical problems and the simulation of large, complex, engineering systems. Through finite difference and finite element approximations, the underlying mathematics is reduced to problems of linear algebra, frequently involving arrays defined on multidimensional grids and requiring computationally intensive matrix operations. Fortunately, these kinds of problems are compatible with parallel operations, and all supercomputers rely on this approach in a variety of ways: vector units, distributed multiprocessors, massively parallel architectures, and so on.

At least three application areas of high-performance computing are of direct interest to the Department of the Navy:

1. *Computational fluid dynamics*—for the design of underwater and surface vehicles and aerial vehicles as well as for the modeling of atmospheric weather and ocean currents;
2. *Modeling of complex engineering structures*—for the simulation-based design of ocean platforms, weapons, and other naval structures and systems; and
3. *Real-time virtual reality support*—for viewing large, situational awareness databases, for training, and perhaps for control of complex systems.

Modeling of complex engineering structures and real-time virtual reality support are discussed in *Volume 9: Modeling and Simulation* and *Volume 3: Information in Warfare,* respectively, of this nine-volume study. Only the first topic, CFD, is discussed in detail here.

CFD represents the use of modern computational techniques to model fluid flows. Partial differential equations embodying approximate physical descriptions of the flow and related transport processes are reduced to differential equations on a discrete space-time grid, which, together with other equations describing fluid properties or processes, and further averaging approximations, are expressed in algorithms for computation. The grids can be fixed with respect to either the flow or boundaries and cover a wide range of scales. Gridding today remains a human-intensive task, although progress is being made in automatic gridding. CFD has developed along with progress in high-performance computing, for which it is one of the major challenges. Physical and engineering insights are essential inputs.

Vehicles of naval interest generally operate immersed in air and/or water, and dynamic and viscous resistance forces resulting from fluid flow absorb all of their effective propulsive energy. Dynamic fluid-flow reaction generates the lift for airplanes and certain kinds of ships and is the principal source of steady and time-varying forces and movements for vehicle or missile maneuver. Fluid flow transports energy in power plants such as gas turbines and nuclear power plants, as well as in many auxiliary systems. And fluid-flow processes dominate the meteorological and oceanographic environment in which naval forces must operate. Physical models of flow, including wind tunnels and ship towing tanks as well as full-scale tests, have long been employed in design and will continue to be used, but are inadequate or uneconomical for many important purposes.

CFD is used extensively in aerodynamic design of modern aircraft, missiles, and gas turbines and is coming into greater use in ship and submarine design. It is also fundamental to models employed to predict environmental effects. To make it computationally feasible for high-Reynolds-number cases of engineering importance, turbulence in the flows is approximated with models, and the limitations of these turbulence models are a principal constraint on practical applications. Another problem is the difficulty of building an appropriate mesh of grid points over which to solve the differential equations that approximate the differential equations of flow. Still another obstacle is the lack of high-quality experimental data for CFD validation. As a result of these and other deficiencies, current CFD models for engineering use are not reliable in cases of extensive turbulent flows, separated flows, and flows with strong vortices. However, detailed numerical simulations of flow near boundaries have been successful in elucidating some mechanisms of turbulent drag reduction.

Except at low Reynolds numbers, detailed calculation of fully turbulent flows is likely to be beyond the capacity of even the most powerful computers projected for 2035. But improvements should move some approaches, such as large-eddy simulation, from a research technique to practical engineering application. An interplay between experiment and computation will continue to be needed because of their complementary strengths, with a trend toward greater reliance on computation.

There is reason to believe, based on physical grounds, that there is a great deal of room to improve the performance and economics of naval vehicles through better prediction and control of fluid-flow phenomena. For example, achievement of laminar, rather than turbulent, flow over the entire surface could, in principle, reduce parasitic drag by as much as an order of magnitude for some classes of vehicles. CFD analysis will permit speedup and reduced costs of design and manufacturing and will improve prediction of performance. The goal of solving the inverse problem, developing an optimized design based on specification of performance parameters, will be brought closer to practicality through computational fluid dynamics and advanced visualization techniques. Beyond this, interaction between CFD and other computational approaches to modeling structural response, control systems, stealth features, and propulsion will allow a closer approximation to a fully integrated modeling approach, with the goal of overall configuration optimization through simulation-based design.

Many improvements are needed to extend the applicability and ease of use of CFD. These include the following:

- Software for parallel computers, including methods for automatic adaptive grid generators, approaches to rapid setup and execution, and results visualization;
- Modeling of flows that deal with arbitrary geometries and that can resolve all important scales;
- Modeling of strong vortex interactions with control surfaces and propulsors for submarines and missiles, e.g., tail and missile tail control surfaces are inside their wakes; aircraft strong vortices are outside the tail;
- Modeling of high-lift transient and separated flows for vertical short takeoff and landing (VSTOL) vehicles and UAVs;
- Modeling of separation postponement and drag mitigation by boundary modification, and;
- Coupled multimedia flow modeling for ships and air-sea interaction, and modeling of combustion turbulent amplification.

The fundamental need, however, is for rapid, affordable, but highly accurate and detailed computations of very complex flows. The crucial technological driver is more adequate turbulence modeling with appropriate physics approximations. This need is well recognized in the CFD technical community and is the subject of active research, but more work is needed. In particular, strong Navy Department guidance and support are essential to ensure that suitable turbulence models will be developed to meet the full range of needs that are of most direct concern to the naval services, such as strongly separated flows, flows with strong vortices (for submarines and missiles), and flows involved in powered-lift vertical landing. This will require a vigorous, focused, and integrated program of theoretical investigation, experimentation, and computation.

Normal development, fostered by industrial and academic concerns, will ensure that CFD will be an increasingly important tool for design and performance prediction of Navy systems. But CFD will realize its full potential to improve the efficiency and performance of naval vehicles only if the Navy and DOD sponsor and guide intensive, high-quality efforts to develop and test turbulence models applicable to the flow characteristics most important for naval applications. This should be a research priority for the Department of the Navy.

Functional and Affordable Computing

With a computer on every desk today, the applications of functional and affordable computing are familiar to us all. Just about any computational task can be undertaken on a workstation or PC, when fast, hard, real-time performance is not an issue. Mathematical calculations, word processing, spreadsheets, database management, graphics, CAD, finite element analysis, thermal or mechanical analysis, use of the Internet, and so on are all commonplace and well supported by the available hardware.

In addition to all of these, and of particular interest to the Navy and Marine Corps, are those applications related to achieving complete situational awareness, that is, data fusion and information mining. As sensors multiply on the battlefield, multiple streams of sensor data will inevitably converge at various collection points, where the information from all sources will be merged or fused into a single composite view of the battle space, which will be far more detailed and complete than can be obtained from any single sensor. The computers to support this activity will come from the functional and affordable class, that is, workstations or PCs. Algorithms and software to best accomplish data fusion are needed and are a Navy Department responsibility, as the commercial world is unlikely to address this uniquely military requirement.

With the proliferation of sophisticated sensors comes the danger of data overload; that is, the volume of the data and the rate at which they are generated are so large that the users cannot keep up and the useful information contained in the data is never recognized, extracted, and utilized. To avoid or to minimize this problem, techniques of information recognition and extraction, i.e., data mining, should be developed. This is a task that the human performs with ease if given the time, but one that often leaves the all-too literal computers at a loss. Today the subject of university research, concepts for algorithms and software that will permit computers, without direct human intervention, to recognize what is important in any given data set need to be found.

Embedded Computing

Embedded computing is one of the most critical applications of computation, for it is directly coupled to the predicted evolution of sensors (see Chapter 4) and will be largely responsible for much of the anticipated performance improvements. For many of the long-range imaging sensors such as radars, sonars, and infrared optical systems, improved performance will come from increased processing of the sensor data with more complex, adaptive algorithms that cannot be implemented fast enough today to meet the hard real-time requirements of such military systems. With the exponential growth of computing capabilities, processing-based sensor performance improvements will follow. Automatic target recognition (ATR) capabilities will improve, digital beamforming of radar, sonar, and optical beams will become commonplace, and the all-digital radar will emerge with significant size and weight reductions and superior performance through extensive use of digital processing.

As computer capabilities continue to increase, while simultaneously decreasing in size and cost, the computational support for many sensors will merge with the sensing element; that is, both components will be implemented on the same semiconductor chip, giving rise to miniature smart sensors or sensors-on-a-chip. In this way, one can expect to see tiny, inexpensive MEMS-based inertial navigation systems in the future with accelerometers and solid-state gyros and the associated signal processing and computational logic integrated on a single chip. When coupled with advanced high-performance and inexpensive integrated imaging sensors (also with built-in computer capabilities for target acquisition, aim point selection, and guidance), these miniature inertial systems can lead to future generations of affordable, accurate weapons with the possibility of sizable reductions in acquisition and logistic costs.

Alternative and Breakthrough Computing Technologies

Molecular Computing

In many circumstances, biological systems act as massively parallel adaptive computers. Since the 1970s, computing models of biomolecular information-processing systems have been explored and numerous advances made in technologies that can contribute to the implementation of molecular computing systems.[1] These include such topics as artificial membranes, recombinant DNA technology, protein engineering, and numerous biosensors for detecting biological substances and processes—e.g., protein, glucose, and fermentation.

Molecular computing differs fundamentally from digital computation in that information is processed as physical patterns rather than bit by bit, and the system learns dynamically as it adapts to its environment. Since the human brain is a form of molecular computer that is small in size, uses very little power, and performs extremely well—particularly on recognition tasks and ill-defined memory queries—it has often been conjectured that, ultimately, molecular rather than digital computation will be the more effective computing paradigm. However, just as our aircraft systems do not directly imitate the physical principles and configurations of birds, bats, and insects, so also it seems unlikely that we will abandon the unique high-speed, error-containing capabilities of digital computing. More likely, molecular computing will provide a way of augmenting digital-processing systems with biological-like capabilities, in terms of pre- and postprocessing pattern recognition, while the central processing tasks continue to be performed digitally. Although the future of molecular computing is cloudy at the moment, given the progress in the biological sciences anticipated for the oncoming decades, this potential breakthrough technology deserves careful attention in the coming years.

Quantum Computing

Of all the technologies discussed so far, quantum computing[2] comes closest to what might be considered a true breakthrough technology. If successful, it would be a completely new approach to computing. Based on a novel and thoroughly quantum mechanical concept of elementary logic, quantum computing envisions a two-state logic structure, known as a qubit, which has the uniquely quantum mechanical virtue of being able to exist in a superposition of both states as long as no measurement is made. Using quantum dynamics to design computations that interfere constructively or destructively, remarkably powerful algorithms become possible. Analyses indicate that using Shor's prime factoring algorithm, quantum computers can solve, in polynomial time, problems that have no polynomial time solution on any classical machine. To date, rudimentary qubit logic has been demonstrated with a single or a few logic operations at a time and generally with a large amount of macroscopic hardware, i.e., not physically small, at this stage. Extending this to computationally useful complexities, how-

---

[1]Conrad, Michael. 1996. "The Lure of Molecular Computing," *IEEE Spectrum*, 23(10):55-60, October.

[2]DiVincenzo, David P. 1995. "Quantum Computing," *Science*, 270(5234):255-261, October 13.

ever, poses a serious problem in the form of loss of coherence. Decoherence of the quantum mechanical wave function, through unavoidable coupling to rest of the world, however weak, threatens to undo the whole scheme. At the moment interest is high in this unique technology, and whether it can, or will, lead to practical quantum computers is not yet known. At the very least, much insight into subtle quantum phenomena will be generated.

FUTURE IMPACT ON NAVAL OPERATIONS

The impact of computation on future naval operations is expected to be enormous. Computers and computation will be everywhere—an inevitable reflection of the information revolution. Combined with advanced distributed sensors, computation will be the primary enabler for achieving and exploiting complete situational awareness. For most sensors, the primary source of improved performance over the next 40 years will come from the application of more and more computational power to the processing and interpretation of the digitized sensor signals. In many cases, as computer hardware continues to become more powerful, smaller in size, and less costly, the sensing elements will become fully integrated with their supporting digital computer hardware to produce smart sensors or sensors-on-a-chip, such as a complete inertial guidance system with multiple MEMS-based accelerometers and gyros on a single silicon chip. Many important sensors, such as radars and sonars, which cannot be shrunk to a sensor-on-a-chip, will become almost completely digital, with analog-to-digital (A/D) converters at each transponder element and all internal signals transmitted and processed only in digital form, e.g., digital beamforming. With increasing amounts of computational power available, more systems will become adaptive, processing in real time the observed signals and altering their system parameters in response to the observations to optimize their overall performance.

Applied to missiles, future computational resources will permit the creation of very smart weapons capable of ATR, precision aim point selection, and extremely accurate terminal guidance performance. With high accuracy, i.e., very small circular error of probability (CEP), the number and size of the weapons needed to counter any given target can be reduced, thereby offering large future savings in acquisition costs and logistics. At the next level, functional and affordable desktop computers will facilitate the netting of resources, the fusion of data from multiple sources, the extraction of the meaningful information contained therein, and its display to provide the degree of situational awareness and real-time control needed to optimize the effectiveness of future naval forces operations. The other side of the coin, so to speak, will be a continuously increasing vulnerability to information warfare technology through both physical effects, such as electromagnetic pulse (EMP), or software attacks.

At the top level, high-performance computing will contribute in numerous ways. The ability to perform vast numbers of elementary computations per second will permit timely, accurate investigation of large, complex physical systems, such as the fluid dynamics that is the subject of CFD or the modeling of the physics of warhead-target interactions. Detailed and accurate simulations of complex components and systems will permit simulation-based design and acquisition. And high-performance computers will provide support for highly realistic, real-time, virtual-reality interfaces for viewing situational awareness databases or simulation-based designs.

DEVELOPMENTS NEEDED: MILITARY VERSUS COMMERCIAL

Although today's commercial computer industry appears self-sustaining, its primary focus is functional and affordable computing. Because the markets for them are more limited and specialized, high-performance supercomputers and embedded signal processors are emphasized less than PCs by commercial developers. Several microchip manufacturers are developing concepts for multiprocessors, and workstations powered by very high performance chips have been developed to bridge the gap between high-performance computing and desktop computing, although the PC seems to be on a path to overtake the workstation within a few years. For embedded computing, several families of DSP chips have been created and could grow in the future at exponential rates, if pursued.

Generally, computer hardware components with ever-increasing capabilities will be supplied by the commercial sector, but the government and the Navy Department will still need to fund basic research, especially the research needed to advance the fundamental underlying technologies. Support for algorithms and software designed to implement applications specifically relevant to Navy and Marine Corps needs will be the responsibility of the Department of the Navy. Both high-performance computing and embedded computing will require special efforts. High-performance computing, because of its limited market potential, will require ongoing government support. The Department of the Navy will continue to rely on high-performance computers to carry out computationally intensive tasks such as CFD, atmospheric weather modeling, and simulation-based design and acquisition of new naval platforms.

In the application of computing to sensors and missiles, the Department of the Navy must take the lead in both writing the software and developing the militarized and special hardware configurations needed for sensor and weapons applications. There is no large commercial market to provide these.

STATUS OF FOREIGN EFFORTS AND TRENDS

Although Asia and, to a lesser extent, Europe compete successfully with the United States in microelectronics components and consumer electronics, the United States dominates the world in computer hardware technology and generally leads substantially in all aspects of computer applications. There is little doubt that the United States will continue to run ahead of the rest of the world in computer hardware development and use for decades to come.

On the other hand, Europe has exhibited significant strength in software, particularly on the conceptual and theoretical side. Such critical issues as formal proofs of the correctness of software and the possibilities for software logic synthesis are the subject of intensive, competent research. Ironically, the language Ada, which is in fact a good language for the disciplined development of large software programs, has been completely ignored by those U.S. software developers and universities not under a DOD mandate. Ada has found profitable use in commercial development in Europe. Excellent air traffic control software, written in Spain in Ada, has found wide use in U.S.-produced air traffic control systems.

India also has become very strong in software and currently serves a good portion of U.S. commercial computer users through nightly debugging of code problems, exploiting the reversal of the day-night periods between the United States and India. Software factories in India are a real competitive possibility.

For the far future, it is less clear that the United States will always be ahead in technology for computation. Given Asia's demonstrated microelectronics development and production capabilities, there is no reason to believe that others cannot catch up, as long as they are willing to provide the substantial investments required. The explosive possibilities of the information revolution will make this option commercially attractive very soon. As for software, anyone can compete. It takes no large investments in capital equipment—just smart, educated people with modest computer resources. Computation technology is likely to be available everywhere in the far future, and success in exploiting it in warfare will go to the diligent, not automatically to the United States.

TIME SCALE FOR DEVELOPMENT AND DEPLOYMENT

The time scale for the growth of computation over the next 40 years is illustrated in Figure 2.2. The simple exponential extrapolations for both high-performance computing and functional and affordable computing can be relied on to represent the order of magnitude of computational capabilities available in any given time frame. This will be true for the near term, i.e., the next 5 to 15 years, and perhaps throughout the whole of the next 40 years. For example, there certainly will be affordable gigaflop computers on the desktop well before 2005. Whether the petaflop level will be reached by 2025 is far less certain, but explicit efforts aimed at petaflop computing have already been initiated, and quite a lot can happen in 30 years.

On the pessimistic side, serious obstacles are anticipated in optical lithography as the wavelengths move further into the ultraviolet and conventional optical concepts begin to fail. This situation could slow the progress toward smaller dimensions in microelectronics fabrication and consequently slow the increase in single-processor clock speeds. On the other hand, growth in computational throughput is accomplished not only through faster hardware, but also through the use of more parallel units in multiprocessor architectures. Any slowing of the growth rate of computational power owing to lithography limitations could be compensated for by parallel hardware.

Optimistically, several multiprocessor supercomputers, planned for operation in the next few years, are already targeting performance levels well above the extrapolated growth curves shown in Figure 2.2. Given the 40 years of consistent historical growth, it seems premature to conclude that this rate will change in the future, for either the better or the worse. For planning purposes today, it is best to assume that the extrapolations validly represent the level of performance that can be expected to be available in the time frame of interest. It should be remembered that these extrapolations do not indicate, and cannot predict, what detailed concepts or technology will permit these levels of performance at any given time. Nor can these extrapolated levels of performance be expected to be achieved continuously. Individual implementations grow rapidly at first to achieve and extend the technology envelope for a while before they reach saturation, as they encounter their limitations. The result is that the actual history of performance exhibits a step-like rather than a continuous growth curve. Since the individual steps are not predictable from past history, the smooth extrapolations must be treated as indicative of general performance capabilities rather than as predictive of actual performance.

RECOMMENDATION

Computer technology will be a major enabler of future naval operations. Computers will enable enhanced situational awareness, realistic modeling and simulation, faster warfighting decisions, more effective weapons, lower-cost platforms, and more efficient and effective use of people. The Department of the Navy should exploit the continual evolution of commercial computer technologies into robust computational systems.

3

Information and Communications

The technologies covered in this chapter include distributed collaboration, human-centered systems, intelligent systems, planning and decision aids, software engineering, communications and internetworking, and offensive and defensive information warfare. These technologies are listed in Figure 3.1 with some rough indication of interdependency.

MILITARY CONTEXT

Many studies of future military needs and capabilities have made the following points, which are reiterated here because they informed the panel's deliberations as it considered information and communications technologies for the Navy and Marine Corps of the future:

- The size of the military will decrease—hence more will have to be accomplished with fewer people, platforms, weapons, and supporting systems.
- The nature of military missions will become increasingly more varied, with many crisis situations being of short duration and high intensity, and occurring simultaneously with other operations.
- The predominant mode of crisis intervention will be via joint task forces, often combined with international forces.

The above points require that distributed collaborating teams share a consistent view of the situations they face, allowing rapid decisions and coordinated action. The military leadership recognizes that information dominance is key to military success and that even the most advanced weapons cannot be leveraged

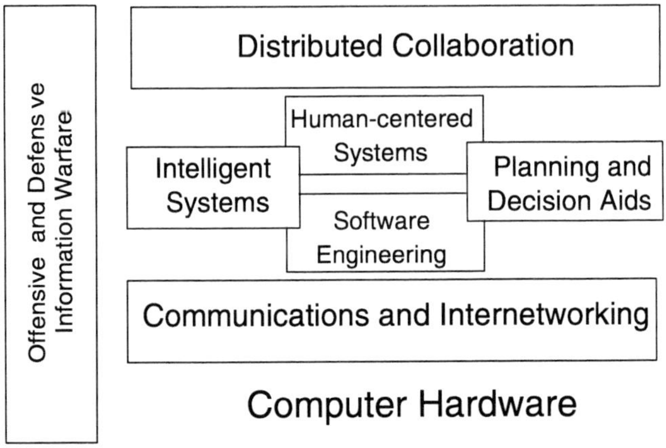

FIGURE 3.1 The key information technologies.

effectively without it. Given the naval forces' unique mission of force projection from the sea, they face unique challenges in the sphere of information dominance. Even these unique challenges, however, must be addressed within the context of interoperability with other Services and the forces of other nations.

DEFINITION OF INFORMATION TECHNOLOGIES

The technologies included under the heading of information are vast and strongly interrelated. Many information technologies feed other technologies and at the same time are dependent on the ones they feed. Because these interrelationships are so complex, it is difficult to organize the relevant technologies into some simple grouping with straightforward relationships to one another. Therefore, the following description should not be considered as a universal truth, but simply as one organizational perspective. Broadly speaking, information technology can be thought of as the underlying capability of computation and connectivity, and their interface with human operators, to apply reasoning power and distributed knowledge to various types of problem solving. Figure 3.2 illustrates the relationships of these elements. The first (inner)-level foundation is the computational hardware and the connectivity. The succeeding levels get at the raw physical capabilities of computation and communication through software. The third layer is that of intelligent systems combined with planning and decision aids and then the abilities of these systems to form effective partnerships with humans. These layers together provide cognitive support for distributed collaboration, which is in the outermost layer. The issues of offensive and defensive information warfare, illustrated in Figure 3.2 as bugs in the system, ultimately pervade the entire set of technologies as they work together to provide a knowledge-rich environment for the warfighter.

DISTRIBUTED COLLABORATION

Description

Distributed collaboration refers to the use of human-computer interfaces, networks, and computer hardware and software technologies to mediate person-to-person interaction and communication. Three key goals of distributed collaboration are (1) the gathering of appropriate problem solvers together across time and space for rapid response in time-critical situations, (2) providing access to and ensuring availability of appropriate information resources across time and space within the context of a task, and (3) enhancing the effectiveness of collaborating problem solvers. In addition, there are two long-range goals for distributed collaboration. The first is to achieve a feeling of "as good as being there" through remote-presence technology. The second is to achieve a level of distributed collaboration efficiency that is "better than being there" through information augmentation. This second goal is important for improving collaboration efficiency even when the collaborators are co-located, particularly when the collaborative task takes place over time. It is expected that both of these goals will be met for normal commercial situations by 2035. The challenge for the Department of the Navy is to ensure that these goals can be met for situations unique to naval forces such as cooperative engagement.

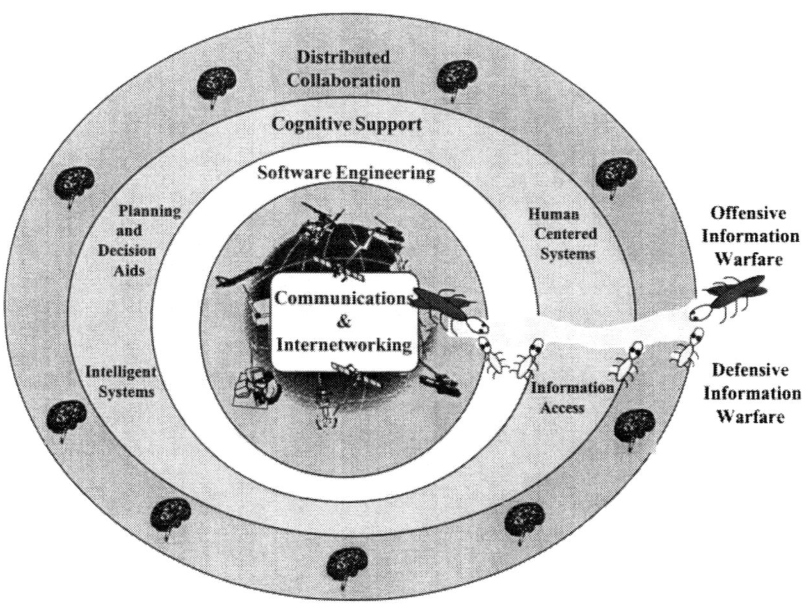

FIGURE 3.2 The layers of elements that make up information technologies.

Relevance

The information revolution has brought individuals the capability to access information through wireless and high-bandwidth wireline technologies with nearly ubiquitous coverage. Because of the large distances that often separate people and resources, the Department of the Navy especially stands to benefit from developments in this revolution through the effective use of people and information regardless of their location.

Technology Status and Trends

Today, naval communications are primarily accomplished via low-bandwidth voice and text. The operational use of distributed collaboration technology in its simplest forms, such as video teleconferencing, is uncommon. Higher-bandwidth communications via satellite will enable video teleconferencing among ships and shore assets. Commercial collaboration tools, such as "groupware" and shared whiteboards, will enable limited sharing of information in distributed groups. The commercial collaboration marketplace is rapidly evolving in four primary areas: groupware, such as Lotus Notes, newsgroups, and chat functions; two-dimensional and three-dimensional multiuser domains (MUDs), such as the Time Warner Palace and products from Worlds Inc.; teleconferencing tools, such as video teleconferencing and Internet teleconferencing; and World Wide Web-based tools, such as Netscape Navigator. Currently, these technology areas appear to be on somewhat divergent paths.

Future Impact of Technology Trends on Naval Operations

Higher-bandwidth interaction will enable widely distributed individuals to meet and understand time-critical information in a shared electronic environment that provides an adequate illusion of shared presence. Future command centers will exploit a range of technologies that are now under development in both commercial and military sectors, including speech recognition and synthesis, high-resolution projection displays (two- and three-dimensional), shared information visualization, resource and mission planning and simulation, and information finding and fusion. These technologies will be integrated in a cohesive system that provides the means for groups of individuals to understand complex situations. Together, shared presence and advanced information understanding will make distributed collaboration much more effective than any form of collaboration today. Most decisionmaking processes that involve multiple individuals will be affected.

Development Needs

Commercial collaboration will continue to target the mass market using low-end capabilities of computers and communications to address collaboration. Other key government-funded initiatives will be required to utilize high-end components and services and provide support for mobility between varying levels of capability. Basic technology advances needed include high-quality, low-cost, three-dimensional rendering; multimedia processing; speech recognition; speech synthesis, avatar (human representation) animation; three-dimensional MUDs; greater network bandwidth; global networks; symbolic compression; and improved network programming languages such as Java. Looking forward a few years, lower-cost, higher-resolution displays will make immersive interfaces more widespread, including personal (head-worn) displays and large-screen projection theaters.

The developmental challenges for the Department of the Navy lie in three areas. The first is rapid prototyping of collaborative support systems for naval-specific situations. An example would be sailors on multiple ships and marines on land collaborating over a crisis-resolution game plan. These prototyped systems would use primarily commercial components complemented with some capabilities coming out of the government laboratories. A second development challenge is the need to support collaboration with lower, rather than higher, bandwidth, which is contrary to commercial trends. The third challenge is the need to support research in those focus areas that, during the prototyping efforts, are found to have weaknesses. The emphasis should be placed on problems considered to be unique to the Department of the Navy. These would most likely include metaphors for collaboration that are different from the commercial trend of room-based metaphors.

Time Scale for Development and Insertion

Military use of distributed collaboration technology is currently under way and is benefiting greatly from rapid commercial growth. The key issues that must be addressed in the use of commercial technologies are difficult and will require significant R&D to overcome. These include information and battle-space visualization, personnel, network and information security, new methods of interaction, and global wireless coverage. Simple prototypes are being field-tested today and can become operational for the Department of the Navy within the next 5 years. Improvements in collaboration technology are dependent on improvements in many other technologies such as human-centered systems, communications and networks, displays, and communications.

Time Scale for System Insertion

Increased effectiveness through distributed collaboration will occur incrementally as the technology matures and advances in the commercial realm are incorporated into naval systems. Window-based video-teleconferencing tools will be incorporated as newer computing resources replace outdated equipment. Virtual presence and "better than being there" capabilities should begin to appear within the next 10 years and continue to improve during the next 20 years. The challenge for the Department of the Navy is that the major benefits from successful insertion of collaboration technology will not be achieved without accompanying changes to naval processes. It has been demonstrated many times over the last several decades that simply automating or improving the current process with information technology provides only marginal improvements. Major gains are achieved only when new processes are developed to take advantage of the new technologies.

Foreign Technology Status and Trends

Significant resources are being allocated in Europe and Japan to develop tools and infrastructure to support distributed collaboration for commercial business. Telecommunications companies in Japan and Europe are conducting efforts in virtual teleconferencing. Large foreign investments are likely to continue and to grow in Web-based technologies that support collaboration.

HUMAN-CENTERED SYSTEMS TECHNOLOGY

Description

Human-centered system (HCS) technology refers to the broad spectrum of hardware, software, and human factors disciplines that enable humans to effectively interact with information systems, computers, sensors, machines, and other humans. This technology draws from a variety of software fields (e.g., visualization, information presentation, human-computer interaction, computer graphics, human tracking, graphical user interfaces, distributed and object-oriented systems, speech understanding), hardware devices (e.g., visual, auditory, tactile, force displays, tracking sensors, graphics and general-purpose computers, networks), and human factors disciplines (e.g., experimental psychology, cognitive science, physiology). HCS technology involves multiple human modes of interaction and sensing. The goal is to enable information-processing technology to be of greatest benefit to humans. Achieving it requires completely natural means of interaction and the design of effective cognitive support.

Relevance

Any hardware or software system in the Department of the Navy that interfaces with humans will rely on HCS technology. The sophistication of modern technology has outpaced the capacity of humans to use it effectively. Information overload is a major problem that will only become more critical. For example, tactical combat information centers (CICs) are designed to provide operators with information from a bewildering range of sensors, including radar, surface radar, weapons radar, sonar, infrared-sensitive devices, and cameras, as well as the naval "infosphere" that includes maps, audio transmissions, and written information. Commanders and operators must rapidly integrate these data and make decisions based on real-time assessments of their situation. The time-critical, high-stress nature of the environment, in addition to the large volumes of information, can result in data overload and decreased efficiency and effectiveness. HCS technology aims to prevent human overload by enhancing human performance, complementing human strengths, and compensating for human weaknesses.

Technology Status and Trends

The Department of the Navy uses human interface technology in command-and-control systems, weapons system interfaces, mission planning aids, simulators, training, logistics, missile guidance, intelligence information access, as well as the traditional administrative functions.

The majority of human system interfaces in use today are based on the windows, icons, mouse, pointer (WIMP) interface or its derivatives, first developed in the late 1970s and early 1980s. The trends in human-centered systems are toward the use of multimodal interaction; visualization in three-dimensional, more immersive, virtual-reality systems; and intelligent assistants.

The Department of the Navy uses variants of the WIMP interface that typically include displays (sometimes color), function keys, and trackballs. Leading-edge efforts are under way to develop and evaluate multimodal interaction techniques and three-dimensional visualization for naval operations such as the Air Force Threat Evaluation and Weapon Assignment (TEWA).

Future Impact of Technology Trends on Department of the Navy Operations

Future system effectiveness will depend on human-centered systems approaches that enable humans to interact using systems in more natural ways with three-dimensional vision, hearing, speech, gestures, and touch. Improved situation assessment will result from full human involvement in the information environment, characterized by immersive virtual-reality techniques. Collaboration will be facilitated via electronic means that enable individuals who are distributed around the world to interact and manipulate information jointly. Widespread information availability will require vast improvements in the interfaces for mobile and wearable computing systems.

Development Needs

The current ongoing revolution in computing and communications will drive dramatic improvements in machine and human interfaces, which will be required to realize the benefits of increased information and communications capabilities. These improvements will be enabled by technologies such as high-resolution miniature image sources for personal (head-mounted) displays and projection systems; wireless human tracking and monitoring technology; continuous, speaker-independent speech recognition; and low-cost, portable rendering hardware. Information presentation will be a key component of future systems and requires extensive research and development to identify modes and methods of visualization and multisensory presentation.

Much of this technology will be developed in the commercial realm because there are large consumer markets driving rapid development. On the other hand, the use of these enabling technologies in applications of interest to the Department of the Navy, such as command-and-control operator workstations, will require extensive adaptation and testing to be effective, much more so than in previous interfaces. This is a result of increased coupling and synergy between the human-computer system and the task at hand. The key to exploiting the advances in technology will be to take a human-centered systems approach to the creation of specific applications, exploiting the strengths of both the human operator and computer. We have only barely scratched the surface in the development of revolutionary new methods of interaction, and careful evaluations of iterative improvements of these systems are almost nonexistent. The Department of the Navy needs to take a rapid iterative approach to refining the development of human-centered systems. This would include human-factors design and rapid-interface prototyping based on immersive multimodal interaction, followed by evaluation and reprototyping in a number of potentially high payoff applications.

The commercial world will drive the development of enabling technologies such as displays and CPU performance. It is unlikely, however, that sufficient effort will be aimed either at the basic science underlying human interaction or at the specific application of new technologies to the Navy Department domain. The Department of the Navy should fund both types of efforts. In particular, the DOD has not focused sufficient resources on the use of immersive interaction technologies outside of the training domain.

Foreign Technology Status and Trends

Both Europe and Japan are investing in next-generation, human-centered system technology including visualization, virtual environments, and hardware. Significant application use of advanced HCS technology in foreign commercial markets such as entertainment and education is expected within the next 5 years.

Time Scale for Development and Insertion

Currently, evaluation of a large number of next-generation human-centered systems in a rich naval infosphere is not possible. Prototypes that span the spectrum of possible approaches should first be evaluated without resorting to operational situations on board ships. Virtual prototypes of human-centered systems provide a means for testing concepts and envisioning future possibilities (such as the Naval Research and Development [NRaD] Command Center of the Future); however, rich, interactive simulations of the naval infosphere that can support evaluation of revolutionary new approaches are not accessible.

Post-WIMP interface technology is beginning to be developed for Department of the Navy applications and should be available in fielded prototypes within 5 years. Much more research is needed to exploit the vast potential of human-centered systems.

Virtual environment (VE) technology is beginning to be developed for naval applications (other than training) and will accelerate within the next 5 years as more immersive displays and more stable tracking systems are developed commercially. Prototype three-dimensional visualization systems, the first step toward the use of VEs, have been tested in naval applications (e.g., Air Force TEWA) and should become more prevalent within 5 years. Evaluations that identify benefits of these technologies are needed.

More advanced interaction techniques such as direct brain interfaces are possible in the 20-year time frame but will require significantly greater understanding of brain function to be of general use.

INTELLIGENT SYSTEMS

Description

In general, intelligent systems are considered to be those systems that do not follow a simple set of procedural instructions and do not provide a single predictable solution to a problem, but instead involve heuristics, sometimes have natural language understanding as an element, are predicated on some set of captured knowledge, and frequently have the ability to learn.

Relevance

Intelligent systems share the ability to be of greater assistance to people in a greater range of complex situations. Intelligent systems will also be used for machine control, but the emphasis here will be for people-in-the-loop systems. Many of the software systems of the future will be considered to be intelligent, and hence everything said for software engineering also applies.

Technology Status and Trends

Intelligent system components show up in a variety of places today. Some examples are pattern matching for automatic target recognition (minimal success here), image understanding for assistance to image analysis, scheduling and resource allocation, sensor fusion, and some expert systems in simulation and maintenance assistance. Some experimental command-and-control systems today use knowledge-based database access.

Future Impact of Technology Trends on Naval Operations

In the future there will be extensive use of better agents in a wide range of areas. Speech and gesture recognition will be used as a common human-system interface along with head-mounted virtual-reality displays. Machine translation will be used to provide real-time translation of military messages from one language to another. Unmanned underwater vehicles will become common, ATR will continually improve, and medical needs on board manned ships will be fully met with telemedicine and telesurgery. Intelligence will become a part of almost all software applications.

One of the limitations of intelligent systems has been the great difficulty of capturing better knowledge in order to have a base of common sense. Another limitation is the difficulty in establishing a hierarchy of knowledge. If knowledge has been captured for one application, it is difficult to take advantage of that knowledge for a new application. These difficulties are currently being addressed, and the expectation is that significant improvements will be forthcoming. Intelligent agents are software packages that are endowed with a certain amount of knowledge that makes it possible for them to carry out tasks autonomously and to respond to changes in their environment. There is an increase in the availability of systems that have embodied in them a knowledge base and can respond to a greater range of situations rather than simply incorporating an embedded sequence of instructions. Intelligent agents may have a role to play in military mission planning and rapid response to changing events. The complexity of the battlefield is such that ordinary software cannot provide much assistance to a commander in the course of rapid replanning.

Future systems will include intelligence systems support enabled by intelligent agents or other forms of intelligent systems technology. All of them will require layers of intelligent systems, each of which will have embedded in it significant knowledge. Intelligent agents will be continually monitoring the battlefield situation and will be able to make predictions and take the appropriate action.

Impact of Trends on Naval Operations

The role of ships in land-strike missions will continue to grow in conjunction with the increased accuracy and potency of smart munitions. Precision-strike capability, increased battlefield awareness, smaller crews on ships and submarines, better collaboration with joint forces, and in general a higher level of preparedness will be enabled by the proliferation of intelligent systems.

Developments Needed

In order for the Department of the Navy to capitalize on the general advances in intelligent systems technology, the Navy must invest in application domain centers of excellence and begin to develop shared ontologies and common knowledge bases. Groups at these centers must insist on vendors using common knowledge representations so that they all can utilize the shared ontologies and ever-expanding sets of knowledge bases.

The development of intelligent systems technology will be driven by a combination of the commercial world and DOD. The Department of the Navy must collaborate with other elements of the DOD to help in developing the necessary standards or at least a useful level of interoperability across ontologies and knowledge representations. The Department of the Navy must then do the application domain work for itself; coupled with work in software architecture, this will allow the Navy Department to conduct intelligent acquisition, guiding vendors in exactly the direction it needs, with each acquisition complementing the others.

Time Scale for Development and Insertion

Ontology work for a selected small domain could begin now, with prototypes being ready in a year or two. In the near term, the use of agent-based programming and the Java programming language to implement agents is progressing rapidly. The Department of the Navy can begin to insert some of this technology now. It should be pointed out that the implementation of Java will give rise to significant security challenges that must be addressed.

In the mid term, over the next 10 years, advances in the area of knowledge capture will reduce the cost and time required for the Navy Department to develop and evolve expert systems. The hardest problem now is that of building hierarchical intelligent systems. Such capabilities as knowledge reuse, knowledge interoperability, and new ways of accomplishing reasoning over large knowledge bases will probably not be ready for naval operational use for another 20 to 30 years.

Vendor systems using Navy-built ontologies and simple knowledge bases could be prototyped 1 to 2 years after the prototyping of the ontologies. These systems could be ready for advanced concept technology demonstration (ACTD)-level use 2 years after that.

PLANNING AND DECISION AIDS

Description

Planning and decision-aids technology is intended to support a wide variety of decision-support applications. These include logistics, resource allocation, inventory planning, mission planning, force structures selection, creation of the master target list for air campaigns, and so on. A significant amount of planning and decisionmaking will be done by distributed collaborating teams. This section does not address general distributed collaboration support technology but concentrates instead on lower-level algorithm support for decisionmaking. The types of technologies that can have a significant impact are generative planning, case-based reasoning, inferential reasoning, neural nets, genetic programming, information fusion, optimization, and scheduling.

In the past the military has focused on increased communication to support planning and decisionmaking. Today, there is an emphasis on the broader aspects of information access to support situation awareness. Even with accurate and timely situation awareness, however, plans and decisions must be made in the context of complex, rapidly changing situations, where frequently the best course of action is not obvious. The volume of details and options can be overwhelming and, even with the best information-accessing capabilities, commanders will be faced with incomplete or ambiguous information.

Relevance

A frequently cited goal of the military is to be inside the enemy's decision loop and to increase the tempo of war to a level at which the enemy cannot function effectively. To accomplish this requires high-quality rapid decisionmaking and planning in the face of many uncertainties. This is true even in many non-warfare crisis-resolution missions, which require rapid reassessment of plans and decisions in the face of changing conditions. The key point is that there is significant value in being able to reduce the time it takes to plan and replan courses of action and to increase the quality of such plans so that they are definitely executable, require fewer resources, cost less, and endanger fewer people.

Technology Status and Trends

Over the last decade, significant research on planning has been carried out by both the operations research (OR) and artificial intelligence (AI) communities. The AI community is pursuing techniques that require the use of captured knowledge and many algorithmic and statistical approaches have been developed by both the OR and AI communities. At this point, however, practical success has occurred only in very focused areas such as resource allocation, task scheduling, some logistics planning, and pattern-matching with neural nets. The future will bring longer-range reflective planning and near-real-time reactive planning much closer together in a seamless way. To achieve faster replanning when an original course of action becomes inappropriate, lessons learned from real-time plan updating for robotic control will be combined with longer-range planning techniques within a single intelligent-agent-based system. These advances would not be possible if it were not for the interdisciplinary efforts that have begun recently in the areas of OR, AI, and real-time control.

The above advances will be complemented by advances in knowledge representation and reasoning, which in turn will improve generative planning and case-based reasoning. The hallmark of AI research has always been the effort to develop systems that can learn. In practice such systems have had little success. However, along with increases in computational power, memory size, and other software advances will come advances in machine learning and automatic system adaptation. These advances will provide significant advantages for systems that support fast mission planning and replanning. By 2010, we should begin to see practical success with human-in-the-loop automated planning in a number of areas that will directly contribute to increasing the tempo of warfare to a level at which the enemy cannot function effectively.

Current Impact of Technology Trends on Naval Operations

The Commander in Chief, Pacific (CINCPAC) has participated in several advanced technology demonstrations testing software to aid in distributed collaboration. These systems contained little analytical help for developing courses of action or decisionmaking, however. The Air Force has experimental systems that aid in some aspects of air campaign planning. Transportation Command (TRANSCOM) during Operation Desert Storm began using some advanced software from DARPA to plan logistics. The Atlantic command is now participating with DARPA in an ACTD that assesses force readiness.

Future Impact of Technology Trends on Naval Operations

A large number of possible future applications could benefit from technology to aid planning and decisionmaking. Some examples include general mission planning, courses-of-action assessment, optimization and selection, logistics, medical care planning, and semiautomatic shipboard disaster response. Courses-of-action assessment includes target selection, force-application selection and coordination, resource allocation, results prediction, and so on. Many of these applications are aspects of dynamic mission planning and are discussed in Chapter 10, Technologies for Enterprise Processes.

Developments Needed

The Department of the Navy should work closely with DARPA and the other Services to ensure that the base technology is continually developed. Integrated feasibility demonstrations and advanced technology demonstrations then need to be sponsored that focus on the emerging planning technologies in ways that best benefit the Navy and Marine Corps. The underlying tools will often be developed commercially, but naval applications will likely be highly specialized and domain-specific. This leads to a need within the Department of the Navy to support the development of experimental planning applications.

Time Scale for Development and Insertion

The DARPA-sponsored research coupled to commercial efforts can be leveraged more effectively to meet naval force needs if the Department of the Navy establishes a testbed to facilitate the continuing identification of metrics and collection of data from the experimental use of decision-support technology. With such a testbed in place, it should be possible to increase the rate at which new operational capabilities are introduced into the fleet.

SOFTWARE ENGINEERING

Description

Software, the set of instructions that governs the execution of digital hardware, includes (1) embedded software for real-time systems with execution response times of milliseconds and with the software and hardware tightly bound together to drive some piece of equipment and (2) nonembedded real-time software, which typically governs the execution of standard COTS hardware such as microprocessors and larger processors. The following sections touch on both categories of software but focus on nonembedded software.

Software can usually be divided into layers. The lowest-level layer consists of compilers and operating systems and is usually referred to as system software. The layer above that is middleware—the layer of services that rides on top of the operating system and provides support for the applications that are in the top layer. Of the three layers, applications, middleware, and the systems, this discussion focuses on the top two. The state of the art today is such that very seldom does any customer ever develop his or her own compilers or operating systems at the lowest layer.[1]

Software engineering encompasses the overall approach, strategy, processes, and tools for developing software. This section focuses on those related strategic issues and technology trends that the Department of the Navy must pay particular attention to if it is to ride the wave of technological developments to the best advantage.

Relevance

Nearly all of the electronics-based products in the Department of the Navy's current inventory have a significant portion of their functionality delivered by software. Even missile systems derive a great deal of their functionality from software. Perhaps even more important is software's great impact on missile evolution, which can be thought of as enhanced functionality over time. Software is dominant in almost every type of system the Department of the Navy uses, from administrative software for payroll, inventory control, and health care to the mission-critical software for command and control, surveillance, intelligence gathering, and processing of imagery. With so much functionality and cost contributed by software, it is vital that the naval forces take the best advantage of current technology trends. There is a great deal of discussion today within the Services about using COTS software, and there is no doubt that COTS technology will play an important role.

Technology Status and Trends Today

A number of exciting software technology trends are under way, including distributed systems (peer-to-peer and mobile agents), agent-based programming, services-oriented layering, loosely coupled systems, reusability, domain analysis, software generation, and high-confidence systems.

The productivity of digital designers has long been considered to be greater than that of software engineers. In both disciplines, the key to increases in productivity is reuse in one form or another. An important aspect of reuse is enabling engineers of one specialty to use components developed by engineers from another specialty. Historically, software developers have been particularly poor at this—typically, each developer has adopted a unique approach.

Lack of software reuse has held true both for large defense contractors and for commercial software development companies. This is an area in which the

[1]At times there may be some debate about what is referred to as a compiler. The software generators at a higher level that are discussed below might be referred to by some as higher-level compilers.

hardware world has been ahead of the software world. Achieving greater reuse of software will be critical for the Department of the Navy. Some positive examples of current software reuse are evident. Today, nobody would write his or her own word processing program or develop his or her own database management system. Spreadsheet software is another wonderful example of a common problem-solving paradigm encapsulated in common reusable software. Today only a handful of commercially available spreadsheets systems prevail. The goal, then, is to extend these examples and develop reusable software modules for command and control, mission planning, and the like. The goal in systems development should be to enable reuse of many types, not just code for major components.

Several different trends are discussed below as if they were separate, but in truth they are tightly interrelated.

Distributed Software

Over the years client-server software has become extremely common. It is in use in many areas of the Department of the Navy and will continue to be so in the future. An emerging trend, however, is the move from client-server to peer-to-peer software where the same piece of software might function as a server in one scenario and as a client in another scenario. The key to distributed software technology is that pieces of software that render certain functionality can collaborate to deliver even higher levels of functionality. These software peers can also be referred to as objects or perhaps even agents. In the past, object-oriented software tended to be objects within the context of tightly integrated systems. Those objects are now trending to behave more like automata and are often referred to as agents. Such agents can be distributed in a network with a peer-to-peer type of arrangement across a distributed network.

Agent-based Programming

Agent-based software is an advance over object-oriented software. An agent is considered to be a self-contained object that understands its environment and can react to and interact with it, as opposed to only responding to messages that are passed to the agent itself. Although some consider that agents will be capable of solving all imaginable problems, agent-based software will be no more of a panacea than object-oriented software was. It takes a step further the notion of having software as separate and somewhat independent entities that can carry out higher-level tasks on behalf of their requester. Many agents will require embedded knowledge and will also be endowed with the ability to learn. A key attribute of agents is mobility; that is, both the state and behavior of the agent can traverse a network and execute on remote nodes. This feature helps address system-load balancing, communication bottlenecks, and constraints on access. However, some approaches to agent-based software limit the functionality available to the mobile agents in order to solve security issues, e.g., to inhibit software viruses that may be disguised as agents. Agent authentication and authorization mechanisms will be needed and are being developed.

Services-oriented Layering

The concept of software architecture in layers has not always mapped directly onto the realized architectures. Today, however, the notion of layers is increasingly more representative of the actual structure of software packages. A module that is considered an application today might be a server tomorrow, and hence other software could use it in the process of running an application. This paradigm exerts a major influence on the manner in which software is written. For the first time, developers are thinking about how the software they are developing might be utilized by others in ways that may be different from what was originally intended. In other words, software development is moving toward the production of software that is inherently more flexible. The DOD is designing this software layering into the next generation of command-and-control systems.

Loosely Coupled Systems

Historically, a good architecture was represented as a system that would be tightly integrated. The more tightly integrated, the higher the performance. The problem is that the more tightly integrated the system, the harder it is to make any changes. It is almost impossible to extract a module from a tightly integrated system and replace it with a newer one. The trend today is toward much looser coupling. Objects, agents, and peer-to-peer protocols are all examples of software partitions or segments that are loosely coupled to those that call them or to the ones that they in turn call. This trend has been enabled by the availability of faster hardware. Looser coupling is achieved through message passing that is typically much slower than other means of module communication. Thus, performance is dependent on improving hardware performance capability, particularly processor and bus speeds.

Reusability

The exciting aspect of all of the above technology trends is that, for the first time, it may be more feasible to reuse existing modules for new applications, rather than to develop new ones. Achieving full reusability requires that attention be paid to software architecture and standards.

Domain Analysis

Another trend that builds on all of those noted above is domain-analysis and domain-oriented software development, which involves planning for reuse at the macro level as opposed to worrying about micro-reuse—the reuse of small software functions such as linklist manipulation. Experimentation with domain analysis and domain-level reuse in the military context started with the System Threat Assessment Report (STAR) program in DARPA.[2] Many commercial companies or defense contractors that do business in a given domain have already recognized the value of this method and are practicing it extensively. Unless software is developed at the domain level, many of the technological advances available in the near future will not be able to be realized by the Department of the Navy. Within the STAR program, it was determined that nearly 80 percent of the command-and-control software in Cheyenne Mountain consisted of more than 20 systems that could be reused. Significant reuse is also feasible for various families of ships where the software is very common and when great savings can be achieved if a domain approach is taken.

Software Generators

Another trend is the development of software generators for common functions that occur frequently within a given domain and are identified as being worth the additional engineering effort to create the ability to automatically generate the software for those particular functions. In the area of scheduling, for example, software has been generated in a fraction of the time needed to generate it by hand, and it has produced schedules that were 10 times faster than those of the software that was manually constructed. Domain analysis is required to identify functions that may warrant automatic software generation.

Generators can also be thought of as very high level domain-specific languages that are produced by analyzing domains, determining exactly what kinds of processes and functions occur frequently, and embedding that knowledge into the software system, thus enabling the application developer to specify the desired functionality at an extremely high level. This is but another example of reusable software. Reuse can be achieved either by inserting a complete module as it stands or by using generators or high-level languages.

High-confidence Software

One last important technology trend is the development of trusted, or high-confidence, software. One problem for the military is the need to test and certify weapons systems control software. Historically, some minor success has been achieved in the area of formal proofs. It is now possible, however, to analyze mission-critical software and determine which aspects must be verified. The

[2]Various reports were prepared by DARPA during the period from 1986 to 1996.

Services should take maximum advantage of the most recent developments in this area to reduce the cost of recertifying weapon systems control software after even small changes are made. It is expected that formal proofs will be adequately advanced in the next 20 years to allow for automatic recertification of reasonably sized critical segments of fire-control software.

Future Impact of Technology Trends on Naval Operations

Software engineering is expected to become more component based with the ability to dynamically configure systems for the task at hand. Future systems will likely be more distributed and agent-based and will have more flexible coupling between modules. Architectures and high-level languages for generating components will likely be based on domain analysis. Progress has been made in the last 20 years, but it will occur more rapidly in the future. How well the Department of the Navy deals with resolving software issues will have a significant impact on its competitiveness and on its ability to leverage limited resources.

Development and Adoption Needed

A number of developments are taking place today that, if successfully deployed, will have the potential of decreasing the cost of software acquisition. Two examples are the maturing of object-based and client-server products, and the evolution to agent-based, Web-based, and peer-to-peer software. New advances will be made in dynamically reconfigurable systems—systems in which modules advertise their capabilities and systems that have task-driven controls and that can find and link together modules with advertised capabilities.

Life-cycle software costs will be greatly affected by an organization's ability to take advantage of the above capabilities and to utilize the same components across an entire domain instead of developing stovepipe systems. The Department of the Navy should begin to develop centers of excellence for key application domains and key software technologies. An application center should perform domain analysis and develop domain-based architectural strategies that are updated as the domain needs change and new technologies become available. Such a center of excellence would obtain expertise in the use of the state-of-the-art tools for domain analysis and the capture of formal representations of requirements, types of modules that would be needed and their potential interplay, and reference architectures that facilitate plug and play of the various kinds of modules, whether they are commercial or not.

The centers would understand the acquisition of knowledge about the domain, which in turn could potentially be used for software generators and higher-level, domain-specific languages. An important aspect of such centers would be the creation of testbeds that would facilitate rapid experimentation. Contractors or vendors for the Navy Department could use the testbeds to determine the best architectures and components for naval systems.

There is a need for highly reliable, trusted, and certifiable systems. This is a major bottleneck in the recertification of weapons control software. Aggressive use of formal proof techniques should be pursued. Europe is far ahead of the United States in this area. Capabilities exist today for being able to certify small systems from the hardware up through the operating system to the application for at least portions of the system. It is not necessary to certify all aspects of a system but perhaps only the critical features. This is an area where the Department of the Navy should aggressively work with DARPA and the other Services to reach critical investment levels. Such critical funding does not exist today.

The commercial world will be the key driver in all areas except certification of weapon systems software. DARPA will be a player, but its investments will have a greater impact on systems survivability and the concept of taskable agents.

Time Scale for Development and Insertion

The technologies discussed above can be incrementally developed and continually prototyped. The key issue for the Department of the Navy is to adopt domain analysis, particularly when it must be done jointly with other Services. The benefits of domain analysis have been demonstrated by the DARPA STAR program. The real benefit of the emerging technologies will not be realized unless the Navy commits to establishing teams that (1) control the acquisition of software across a domain and (2) perform the prerequisite domain-based engineering. After domain analysis begins, beneficial results could be realized for any application domain in 2 to 3 years.

Foreign Technology Status and Trends

As indicated above, the one area where European technology, in particular, is ahead of the United States is that of formal proofs and high-assurance software. The United States could probably gain from a greater collaboration with European research centers in this arena.

COMMUNICATIONS AND INTERNETWORKING

Description

The defining requirement of naval communications is ship-to-shore and extended-range (beyond line of sight) ship-to-ship communications. The extended ranges and extended durations of ship deployments create unique challenges and complexities. These must be met, in general by satellite communications (SatCom) resources, in an environment of constrained budgets and dynamic (usually increasing) requirements. For example, satellite video teleconferencing has grown in popularity, and some would see a requirement on almost every ship for off-board diagnostics and repair assistance. This is consistent with commercial experience, where videoconferencing has grown exponentially in recent years. It has been reported that nearly 80 percent of all switched 56-kbps and switched 64-kbps data calls in 1992 were for videoconferencing.

Other requirements for additional communications capacity result from the increasing tactical availability of overhead and other sensors that are not organic to the battle force. This trend is certain to continue. Supplying a dedicated channel to each communications task is increasingly untenable.

Prior to the launch of the maritime satellite (Marisat) (Gapfiller) 20 years ago, ship-to-shore communications consisted of a few 75-bps semireliable high-frequency (HF) radio circuits and sometimes-available HF voice. At this time, ultrahigh-frequency (UHF) satellites (Gapfiller, fleet satellite communications, and successors) have become the backbone of naval ship-to-shore communications. The fleetwide increase in total communications capacity has been revolutionary. Yet the standard 2400-bps circuits now seem constraining, especially in a time when commercial users commonly employ link rates 10 to 12 times higher still for data and facsimile. More than 200 ships have been equipped with INMARSAT (follow-on to commercial Marisat) terminals, making ship-to-shore telephony readily available. This service has become nearly indispensable to the Department of the Navy.

With the new capabilities afforded by UHF SatCom and INMARSAT, general-purpose communications (voice, data, and fax) use by fleet units will begin to resemble that of modern commercial users. With INMARSAT, for example, the shore termination is to the public switched telephone networks. Use of commercial equipment and standards as well as networks is a natural development. Costs of service and effectiveness to be gained are the driving factors governing usage.

Each of these mainstays, however, suffers critical vulnerabilities in conflict situations. For example, any current UHF SatCom transponder can be seriously degraded by as little as a determined hobbyist anywhere in the Earth-coverage footprint using an inexpensive scanner, a surplus military UHF transceiver, and a homebuilt antenna. HF radio offered at least a measure of jam resistance, despite its various shortcomings, by the multiple operating frequencies and geographically varying propagation conditions.

Use of INMARSAT is burdened by treaty provisions limiting the Intelsat Consortium services to peaceful purposes. Substantial discussion with treaty signatories has not resolved the issue. U.S. Commander in Chief, Pacific (CINCPAC) policy issued in 1993 is that Department of the Navy use of INMARSAT in a conflict situation for more than safety and rescue purposes is permitted if the action being taken is pursuant to a United Nations Security Council resolution. But what about situations where U.S. action is taken unilaterally? The risk was illustrated recently when the terminal of a West Coast-based ship was deactivated by INMARSAT (fortunately on the way home) because of a billing dispute. Both of these examples point to vulnerabilities that need some resolution.

A Vision for Future Naval Ship-to-Shore Communications

A vision for the future of naval ship-to-shore communications must seek to reconcile powerful opposing influences. New capabilities (data, fax, imagery, video) with vast increases in communication channel rates must be added to traditional requirements for content security, jam resistance, and protection from service disruption. Moreover, increased emphasis on nontraditional operations suggests a need for interconnection with the global public switched telephone network complex. At the same time, the outlook is for continued budget pressures, reduced ship manning, and reduced margins for error.

The possibility seems remote that every ship can be equipped for the most demanding requirements (e.g., high-quality imagery and full-motion video), and a natural delineation might be those ships with and without embarked commands. In this environment, a reasonable vision might consider the following principles:

1. *Seamless normal and conflict communications.* Operational effectiveness as well as personnel efficiency would be well served by a smooth continuum of communications resources and practices between normal and conflict operations. This suggests that issues of potential service degradation or interruption be considered beforehand and suitable resolutions determined in advance. This principle would imply that every ship would possess some ship-to-shore capability with at least moderate jam resistance, and ships with embarked commands would possess substantially more. In general, some reduction in total channel capacity would result from the tradeoff between communications rate and vulnerability level.

2. *Seamless underway and pierside communications.* Personnel efficiency can be improved by smooth transitions of both data and voice communications when ships arrive or depart pierside. This implies operating procedures and physical interfaces on board and on the pier for shifting from satellite to wireline interconnect to the public switched telephone network and to data networks. Already under way in some areas, these measures are considered useful for fleetwide implementation.

3. *Parity of general-purpose (voice, data, fax) capability with similar-sized modern commercial units.* Aside from unique communications requirements of a battle force, efficient conduct of operations is leading to a convergence in the use of general-purpose communications. This trend is assumed to continue and implies continued or expanded interconnection to public switched telephone networks and available channel rates in the near mid term (3 to 5 years) of at least 64 kbps for data, facsimile, and moderate-quality video conferencing on all ships.

Both naval and commercial requirements for these applications can be expected to continue growing. For example, trends toward graphically oriented, versus text-oriented, communications might require channel capacities of 1.5 Mbps on every ship during the forecast period ending in 2035.

4. *Maximized/optimized military SatCom capacity.* Current requirements and trends suggest that planning should envision ship-to-shore channel and total capacity increases of 30 to 100 times the current capability, similar to the capacity increases resulting from the transition to satellite communications in the 1970s. This would be reflected in ship-to-shore channel capacities of 64 kbps, increasing to perhaps 224 kbps during the next 6 to 10 years.

Some of these needs will undoubtedly continue to be fulfilled by INMARSAT or other commercial offerings. However, military satellite resources at UHF, superhigh frequency (SHF), and extremely high frequency (EHF) offer vital and unique advantages, including dedicated spectrum, DOD system control, and jam resistance (in some instances). For these reasons, military SatCom is expected to continue its role as the backbone of ship-to-shore communications. It follows that these systems should be optimized for capacity increases and operational improvements.

5. *Virtual access to off-board databases.* Modern commercial practices today employ a wide array of means for remote access to databases and electronic data interchange for parts and supplies management. The Department of the Navy uses some of these innovations, and it is envisioned that considerable potential remains to exploit this developing application area.

Relevance

Nontraditional missions, budget pressures, reduced manning, compressed decision time lines, and reduced margins for error combine to produce extraordinary pressures for communications improvements. Technology advances, whether in the military or commercial sectors, offer valuable potential. Because satellites figure heavily in naval ship-to-shore communications, innovations in that transmission means are particularly relevant. Also, applications developments that improve efficiency or offer new capabilities to remote units are similarly relevant. Examples include the terminal equipment and data compression algorithms employed in video conferencing.

Technology Status and Trends

Technology in communications may be described as falling into four principal areas: transmission, networks, applications development, and termination/application equipment. The areas overlap, as might be expected, and each is experiencing rapid and continuing development. In general, communications technology today is driven primarily by the commercial sector.

Communication transmission has progressed from its wireline analog beginnings such that it is useful to distinguish network trunk transmission from subscriber access loop from intrafacility wiring. In developed countries, trunking is now all digital and conveyed via optical fiber or digital microwave. Satellite trunking is increasingly limited to access to developing countries and remote cities. Subscriber access continues mostly on wireline, but with trends toward a digital format and a wireless medium. Intrafacility wiring means now include coaxial cable and optical fiber as well as copper wire.

Networks are now electronically switched and have progressed from circuit-switched hierarchical configurations for telephony and data to packet-oriented data networks and new signaling control mechanisms for telephone and other stream applications. Communications applications and related termination equipment now form a virtual continuum, expanding from traditional messaging and telephony to data, facsimile, imagery, and live video. With these applications have come continuing fundamental changes in how business (and military operations) are conducted.

Rapid advances in communications technology are currently paced by commercial interests, and in many areas commercial technology is more advanced than that used in the military arena. For example, ordinary cellular phones can access public switched telephone networks and, by extension, such public networks worldwide. Progress in encoding methods for data compression continues, and asymmetrical approaches are being made in many applications, wherein brief queries to databases, for example, elicit voluminous responses of graphic or other data.

Some developing areas considered to have key implications for naval communications include the following:

1. *SatCom PCS.* Several proposed SatCom personal communications systems (PCS) promise global service for handheld telephones. One of them, called INMARSAT-P, is a private-venture spinoff of INMARSAT and may constitute a solution to the current vulnerability of INMARSAT for Navy use. Others are proposed by larger U.S. firms and aim for initial service availability in the late 1990s. Planning has also begun on at least one satellite system for more advanced needs that is capable of worldwide voice, video, and high-speed data links.

2. *ISDN-related equipment.* The long-envisioned integrated services digital network (ISDN) has spurred development of high-quality facsimile equipment and moderate-quality videoconferencing equipment operating at the ISDN basic rate (64 kbps) for medical and other imagery needs. Expanding applications are following. Nearly 80 percent of all switched 56-kbps and switched 65-kbps data calls in 1992 were for video conferencing.

3. *Asymmetric channels.* Much of modern communications is highly asymmetric in character, highlighted by uses of the Internet, where brief requests for

information commonly result in downloads of voluminous graphical data. Some alternatives being examined employ a relatively lower-rate channel for executing the request and a high-rate, one-way channel (cable modem or direct broadcast satellite) for transmitting the response. For example, a recently announced service employs a regular phone line in one direction to elicit large volumes of data for rapid delivery by satellite.

4. *Intranet/multilevel access.* Intranets employ the infrastructure and standards of the Internet and World Wide Web but are cordoned off from the public Internet through software programs known as firewalls. Access is granted to designated corporate interest groups such as customers or employees. Intranets (as well as the Internet) have grown geometrically. Established by many of the largest firms in corporate America, uses range from furnishing order or shipment status data to customers to supply ordering via an electronic catalog to disseminating employee information. Key to continued success is reliable access control, and development in this area is ongoing.

Future Impact of Technology on Naval Operations

In many respects, advanced communications constitutes an enabling capability for nontraditional littoral missions where ships may frequently be called upon to operate independently. Experiences of the past decade highlight risks to the ship, political risks of errors in targeting and other decisions, and the much compressed time line of engagements. Real-time operating orders and mission coordination are critical requirements, especially in stress periods when jamming may be present and commercial networks congested. Communications technology trends should have a favorable impact on Navy objectives of increased efficiency and reduced manning through more effective coordination of replenishment, repair, and personnel matters. In radio hardware technology, future progress in phased array and multifunction antennas could help to reduce superstructure congestion on surface ships.

Development Needed

Following is a list of developments and actions considered by the panel to hold unusual promise:

1. *Commercial SatCom Alternatives.* Several low-Earth-orbiting (LEO) SatCom systems in development promise near-global service for mobile or handheld telephones. These systems are driven by commercial interests, but an early Navy investigation of the alternatives is recommended, with the objective of equipping every ship with a minimum general-purpose (i.e., not necessarily jam resistant) communications capability consisting of global telephone access and 64-kbps data/fax capability.

Since required communications capabilities seem certain to continue increasing, and since satellites will continue to be the principal means for ship-to-shore communications, a new look at how the SatCom capability is provided will likely be required. This would include examining the mid- to long-term potential roles of LEO systems in providing general-purpose ship-to-shore communications now provided by INMARSAT and military geosynchronous satellites. Such an examination should also include whether such candidate systems should employ military frequency allocations.

2. *UHF satellite system improvements.* Military SatCom is expected to continue its role as the principal resource in ship-to-shore communications during much of the present forecast period. In particular, the UHF follow-on satellites offer expansion potential. Current UHF transponders could support at least 10 times the current 2.4-kbps bit rates, and design changes in later flights could readily support 64 kbps or more. Exploiting the asymmetry in some needs and EHF/UHF cross-banding should provide additional improvements. Developments here will naturally be driven only by Department of the Navy initiative.

Reduction of vulnerability of critical links is another area in need of development. Vulnerability of the UHF SatCom uplinks to jamming or interference was recognized at the outset and the risks considered acceptable. In the fleet communications satellite and successor UHF satellites, only the fleet broadcast is cross-banded with an on-board processor to assure a jam-resistant capability. Experience has demonstrated the utility of this configuration. The general awareness of electronic technology worldwide, however, strongly suggests that some mitigation of this shortcoming should now be given some priority. A second SHF-to-UHF cross-banded broadcast channel, but with higher capacity than the present fleet broadcast channel, might be part of a solution in the shore-to-ship direction. In the meantime, signal processing and much reduced channel rates could provide a modest degree of protection in the ship-to-shore direction, where uplink power is limited. Low-rate EHF channels might play a wider role in ship-to-shore communications, especially if employed in an asymmetric channel allocation regime, e.g., an EHF request channel for SHF-UHF broadcast response.

3. *Exploiting asymmetry in channel requirements.* Pressures for even more capacity in ship-to-shore SatCom channels motivate an examination of the limiting factors. Two prominent limitations are available bandwidths in satellite transponders and, in some cases, uplink transmit power in the ship-to-shore direction. Focusing on possible improvements suggests considering asymmetry. Leaving aside the transmission of any on-board sensor system outputs, the normal flow in ship-to-shore communications involves a much larger volume of data bound for the ship than originated by it. Yet ship-to-shore links are commonly full-duplex, whether UHF or SHF, with equal capacity in both directions. Technology will support exploiting these asymmetries to achieve greater efficiency as well as robustness. Although different approaches to asymmetric applications

are under way, largely driven by commercial interests, pursuit of specific naval applications will depend on the Department of the Navy development effort.

4. *Off-board databases.* Several developing technology areas could assist in reduction of the weight and volume of hard copy records carried on board and potentially increase the efficiency of maintaining them. Some exploratory efforts are understood to be in progress, but the total scope could range from personnel and medical records to stores and parts and maintenance manuals. Approaches could range from virtual access to remote databases to electronically stored records on board backed up by master records at, say, the ship's home port. The technologies of databases and intranets for controlling dissemination and access to corporate data are well developed and continuing to advance.

Foreign Technology Status and Trends

Communications technology is very much a global development effort and capability is largely equal among the developed countries. Ironically, the status of commercial communications in some developing countries is more advanced because of the absence of previously installed infrastructure. In general, wireless telephony, which includes soon-to-be ubiquitous SatCom-based telephony as well as cellular technology, is expected to be adopted by the world's military forces as well, for reasons of economy and convenience. This should, in turn, offer more readily available means for joint mission coordination using commercial services.

On the threat side, the generally increased worldwide awareness of communications technology may represent an increased risk of jamming of vulnerable resources.

Time Scale for Development and Insertion

Time periods required for development and adoption for widespread fleet use vary widely among the recommended developments. In the following estimates, normal priorities and budgeting procedures are assumed. Evaluation and selection of a commercial SatCom service to augment or replace INMARSAT are expected to be required 1 year after the candidate systems are in service. After selection, budgeting, procuring, and deploying significant numbers of units throughout the fleet are expected to take about 3 years.

Different time scales are anticipated for UHF SatCom system improvements that are ground-segment oriented and those involving space-segment modifications. Studies and test programs for channel capacity increases and vulnerability reduction should be able to reach useful conclusions in 1 to 2 years. Adoption on, say, two satellites and a significant number of ships might require another 3 years.

Changes in UHF satellite designs for wider-bandwidth transponders or cross-banding options may require more development time. A preliminary estimate would be 3 years for studies and engineering tests, followed by 3 years to incorporate the design changes into a spacecraft. To get the new designs in orbit in two satellites in a routine spacecraft replenishment schedule would likely require an additional 2 to 4 years, with another 2 to 4 years for fleetwide coverage.

Exploiting channel asymmetries and remote use of offboard resources might be most effectively accomplished by adapting commercial sector developments in these areas. A concerted program to identify, adapt, and test candidates might require 1 to 2 years to develop fleet-ready applications, and another 2 to 3 years for deployment.

Changing future requirements and continuing rapid advances in communications technology can be expected to influence future technology development needs, especially toward the end of the forecast period of 2035. For example, trends toward graphically oriented versus text-oriented communications can be expected to require ever-greater overall channel capacities. Satellites will continue to serve as the principal means of communication because of the fundamental geometries of ship-to-shore communications.

However, budgets will surely continue to be constrained, and ships will continue to have limitations on power and topside space. A new look at how SatCom capability is provided will likely be required. It is useful to observe that commercial communications suppliers are shifting to LEO systems to provide voice, fax, and data capabilities (these now have limited channel rates, but higher-rate capability is planned). Numerous issues pertain, but some in the SatCom field believe that the era of large, expensive, geostationary satellites may be ending for such applications.

Navy Department examination of the issues and tests using commercial LEO-based services when they are available could require most of the next decade. If the conclusion favors a LEO system over the current geostationary UHF follow-on system, a dedicated system employing military frequency allocations would likely be desirable. Designing, procuring, and deploying such a system would likely require most of another decade, ending in the 2015 to 2020 time frame.

Summary and Recommendations

Naval communications, like commercial communications, has progressed in recent years from basic voice and messaging to additional functionality, such as facsimile, other than imagery and video conferencing, which requires higher channel rates. This added capability has supported new efficiencies and in some cases has enabled new ways of doing business that have transformed whole business segments. Similar benefits have been experienced by the Department of the Navy, and communications is expected to have a pivotal role in accomplishing new nontraditional missions with reduced manning and within continued budget constraints.

The trend in requirements toward increased channel rates is expected to continue into the foreseeable future. As a result, an increase in total capacity will be required in the next two decades or so comparable to that achieved by the shift from HF radio to SatCom in the past two decades. At the same time, it will be necessary to mitigate vulnerability to service disruption.

The following are recommendations for utilizing and expanding available communications technology:

1. Pursue the objective of equipping every ship with some measure of ship-to-shore jam-resistant communications and a general-purpose (not necessarily jam resistant) channel rate capability of at least 64 kbps in the near- to mid-term and perhaps 224 kbps during the next 6 to 10 years.
2. Investigate and test newly developing SatCom alternatives for obtaining global access to public switched telephone networks to augment DOD SatCom facilities for general-purpose voice, data, and fax needs.
3. Undertake a multiphase improvement program for UHF SatCom systems to increase channel capacities and reduce vulnerability.
4. Examine what combination of SatCom capability (LEO versus geosynchronous and military versus commercial) is most appropriate for general-purpose ship-to-shore communications.
5. Undertake a development and test program to exploit the asymmetry in certain applications to achieve greater communications efficiencies.
6. Expand programs for virtual access to remote databases and automated electronic data interchanges.

OFFENSIVE AND DEFENSIVE INFORMATION WARFARE

Background

This section discusses emerging technologies and issues related to information warfare (IW) and electronic warfare (EW). The key results regarding offensive and defensive IW and EW of importance to the future naval forces are as follows:

- The proliferation of more sophisticated digital microelectronics into Navy and Marine Corps information and communications systems will make these systems more capable but will also make them more vulnerable to external disruption efforts.
- Actions to make information systems more interoperable by making better use of commercial hardware and software initiated by the promulgation of a DOD-wide, common, joint technical architecture for command, control, and com-

munications (C^3) systems will introduce many new vulnerabilities and present new attractive targets for enemy disruption.

- Forcing all military command-and-control systems to have a high degree of architectural commonality and incorporating commercial software and hardware into trusted military command-and-control systems will introduce additional vulnerabilities that can be exploited by the enemy.
- Emerging transmitter technologies capable of generating extremely high peak power ultrawide-band radio-frequency impulse signals pose a significant threat of disruption of microelectronic digital systems.

Background

Over the past few years, primarily since the Gulf War, many articles and books have been published regarding the nature of IW and its possible implications for the reliability and integrity of both civilian and military information systems.

Following the issuance of DOD guidance on information warfare by the Office of the Secretary of Defense (OSD), the military departments and the civilian telecommunications industry have tasked their various advisory or management oversight activities to carry out in-depth studies about the future implications of IW on their operations. For example, in 1994, the Defense Science Board (DSB) examined the manner in which IW is utilized as well as possible defensive IW actions and has made a number of recommendations[3] to the OSD regarding future management, organizational, and technology issues associated with IW.

Since the issuance of the OSD IW policy, the Department of the Navy asked the Naval Studies Board (NSB) of the National Research Council (NRC) and in a separate action the Navy Research Advisory Council (NRAC) to examine the nature of IW and recommend approaches that would optimally exploit IW for naval force purposes. The NSB IW study[4] was of quite broad scope, whereas the NRAC study[5] focused primarily on defensive IW concerns. Both studies made a number of important observations and recommendations regarding strengthening the Navy's ability to carry out both defensive and offensive IW activities. In large part, the recommendations from both studies are still valid and constitute an appropriate background for consideration by the Panel on Technology.

---

[3]Defense Science Board. 1996. *Report of the Defense Science Board Task Force on Information Warfare—Defense (IW-D)*, Office of the Under Secretary of Defense for Acquisition and Technology, Washington, D.C., November.

[4]Naval Research Advisory Committee. 1986. *Information Warfare*, National Academy Press, Washington, D.C.

[5]Naval Research Advisory Committee. 1986. *Information Warfare—Defense*, Office of the Assistant Secretary of the Navy for Research, Development, and Acquisition, Arlington, Va., September.

In part stimulated by the results of the studies noted above, President Clinton recently chartered a new Commission on Critical Infrastructure Protection[6] that will attempt to bring military and civilian knowledge about IW issues together to ensure that the critical national infrastructure—telecommunications, transportation, power, medical services, and so forth—receives appropriate attention with respect to protection from both physical and IW threats. Because of the Navy and Marine Corps' operational dependence on the nation's infrastructure, the Department of the Navy should participate in this activity and assist in the development of appropriate protection measures.

Clearly the viability of information systems in use today, in both the civilian and military activities, is of considerable concern in light of IW threats—particularly as they are exploited by other nations that might be potential adversaries of the United States.

Categories of Computer and Information System Disruption

To establish the context for introducing the emerging disruption technology associated with very high peak power UWB RF impulse technologies, the scope of the variety of major computer and information system disruptions is described below.

Software Attacks

System disruptions of information networks and their supporting computers may be caused by use of external computer terminals—such as those used by hackers. The media have given particular attention to these kinds of disruptions because of the perceived and often colorful role of the hacker in carrying them out.

System disruptions may occur as a result of improper software code, either inserted in the system as it was developed or inserted into the system after it became operational, perhaps by a hacker. The improper code may be the result of software errors of commission and omission or the result of tampering. Software is particularly vulnerable to tampering during the development process because many people may have access to the low-level code during development. Security and integrity of the software development process were serious issues in the Strategic Defense Initiative.

Some system software threats associated with IW are categorized below:

- *Computer virus.* Probably the best known of these threats, although not usually considered in that context, computer viruses are those programs that stealthily infect other programs and then replicate and spread within a computer

[6]Commission on Critical Infrastructure Protection. 1996. Executive Order 13010, July 15.

system or network. Typically small, they are difficult to detect; some of the more recent versions have active antidetection protection measures. Modern computer viruses may be encrypted, compressed, or polymorphic to reduce the probability of detection.

- *Covert channel.* A communications channel that allows information to be transferred in a way that the owners of the system did not intend. Variations include cover storage channels and cover timing channels. Use of a covert channel could be a way to insert a virus into a computer system.
- *Data manipulation.* With the increasing technological capability for manipulation of data come opportunities to use those capabilities for nefarious purposes. The composition and content of pictures as well as databases are vulnerable to advanced techniques of manipulation.
- *Flaw.* A flaw is defined as an error of commission, omission, or oversight in a system that allows protection mechanisms to be by-passed. The insertion of a flaw into a system could be enabled with other IW weapons, such as the use of coherent RF weapons, over networks from remote sites or could be designed in by an agent-in-place.
- *Logic bomb.* A logic bomb is a piece of code buried within a larger computer system that executes when a specific system state is realized. An example could be a program that checks for the presence of a piece of data within a file, say, an employee's name within a payroll list, and when the specified logic state is reached—the existence of the employee's name is false—the bomb explodes; for example, the software program may command that the entire system memory be erased.
- *Logic torpedo.* Weapons like viruses are essentially uncontrolled. A weapon that can be aimed at one or more specific systems and then released through cyberspace to hunt down its target, known as a logic torpedo, would be very useful.
- *Time bomb.* Similar to the logic bomb, this type of software program waits for a specific time to be realized and then executes.
- *Timing weapon.* Also thought of as insidious clocks, these weapons affect the timing of internal clocks to throw off system synchronization.
- *Trap door.* A trap door is a hidden software or hardware mechanism that permits system protection mechanisms to be circumvented. It is activated in some nonapparent manner—such as a special random key sequence at a terminal. Anyone who has ever programmed knows that despite best efforts, no program ever works correctly the first time (or even the first 10 times). Trap doors are a common safety step used to make sure there is always a way into the program to fix bugs, no matter what the problem may be. The utility of a trap door to provide access for IW purposes is self-evident.
- *Trojan horse.* As the image invoked by its name implies, a Trojan horse is a computer program with an apparently or actually useful function that contains additional functions that surreptitiously carry out tasks that the user of the

program would not necessarily execute willingly. For example, a spreadsheet program could contain additional logic to make surreptitious copies of all of the data files on a system. The user of the spreadsheet program would not be aware of those activities occurring while working with the legitimate functions of the program.

• *Worm.* Similar to viruses, worms are self-replicating but not parasitic; that is, they do not attach to other programs. As demonstrated dramatically by the Internet worm of 1988, they can deny legitimate users access to systems by overwhelming those systems with their progeny. Worms illustrate attacks against availability, whereas other weapons may attack integrity of data or compromise confidentiality.

Disruption Caused by External RF Signals

Radio-frequency energy at the right amplitude and frequency can affect the performance of microelectronics hardware used in computer and information systems. Depending on the nature of the RF signal, a wide range of deleterious effects can occur such as burning out electronic components, disruption of internal logic that may require rebooting or reinitializing the system, interference with the reception of desired signals (jamming), and the introduction of spurious and confusing information (spoofing).

RF signals that can disrupt information and communication systems can be classified as follows: extraneous radio signals from other RF systems, EW jamming and spoofing signals, nuclear-produced EMP signals, high-power microwave energy (HPM), and as a subset of HPM, the very high peak power UWB RF impulse signals. UWB RF impulse signals have both an offensive (disruption capability) and defensive aspect (mitigation)—both aspects need to be considered.

Extraneous Interfering Radio Signals

Computers and information systems can be disrupted by radio signals coming from other systems. Although the disruption may not be purposeful, these interfering radio signals can nevertheless be extremely deleterious and may not be mitigated easily. Some of these interfering signals may be generated by local oscillators in adjacent systems or by static electricity. Lightning-generated impulses have been found to interfere with the guidance computers on space launches; static electricity generated by carpets has caused the failure of computers, and precipitation static generated by aircraft has interfered with the functioning of aircraft computers when not properly dispersed by appropriate wicks.

For example, the Global Positioning System (GPS) operates with virtually a zero signal-to-noise ratio. Almost any other signal that falls within the bandwidth of the GPS receiver can disrupt the ability of the GPS receiver to provide position, track, and velocity information. In one instance, the 13th harmonic from a local oscillator in an aircraft very high frequency omnidirectional range (VOR) navigation receiver was of sufficient strength to totally disable the functioning of the GPS system. Because of concerns about interference, the British have expressed alarm over the U.S. Federal Aviation Administration's plans to discontinue the support of the existing VOR navigation systems in favor of using augmented GPS in the Wide Area Augmentation System (WAAS). This is scheduled to happen in 2000, long before more capable satellite-borne high-power GPS transmitters will be available.

The vulnerability of computers and information systems to disruption by high-power interfering signals being radiated for other purposes is a vulnerability to modern information systems that must not be overlooked. To address this issue, the government has put in place a special organization, the Electromagnetic Compatibility Analysis Center (ECAC), to ensure minimal disruption of government systems resulting from improper use of allocated frequencies.

Electronic Warfare Signals

Electronic warfare has been an effective warfighting technology for some time. For the purposes of this report, EW activities related to the intercept and ion exploitation of enemy signals and spoofing are not discussed here as they do not by themselves lead to disrupted computers and information systems but involve the collection of electronic signal intelligence.

EW technologies underwent rapid evolution during World War II when new warfighting radio communications and radar capabilities used by both the Axis and Allied forces gave rise, in turn, to systems to disrupt these new communications and remote-sensing capabilities. Since World War II, EW capabilities have continued to evolve—prompted, in part, by perceived new radar and communication system threats, and as a result of the experience gained during the Korean, Vietnam, and Cold wars.

EW signals can be used to spoof, disrupt, or jam RF systems that are integrated into information systems. Disrupting signals can find their way into the RF front ends of the information network receiving systems or into the RF front end of sensor systems. The EW approach takes advantage of the fact that extraneous RF signals can disrupt functioning of the system by raising the noise level of the system being attacked, by introducing extraneous signals that overwhelm the processing capability of the system, or by using EW signals to spoof the ability of the system to carry out its intended functions.

EW offensive actions offer the tactical warfighter the operational capability to defeat the effectiveness of enemy electronics-based communications and radar sensor systems. The ability to operate U.S. information systems in the presence of jamming, especially by sophisticated signal wave forms, will require more attention by the developers of information systems.

Nuclear Weapon Electromagnetic Pulses

The detonation of nuclear weapons causes a number of different types of energetic radiation to be emitted. The high-energy x-rays, neutrons, and other emissions can disrupt the functioning of many solid-state electronic components. This type of disruption is labeled as transient radiation effects on electronics. Another type of disruptive emission is the very high energy EMPs produced by the detonation of nuclear weapons.

EMP concerns have prompted extensive Department of Energy (DOE) studies of how nuclear weapons generate such emissions and their possible effects on military systems. Following serious disruptions of power and other electronics systems during the U.S. nuclear test programs, the DOD subsequently initiated efforts to find ways to prevent EMP signals from entering C^3 facilities and burning out electronic equipment. Based on many years of EMP interference measurements, a special set of handbooks was prepared and made available to system designers that describes in detail how to harden systems and facilities against the disruptions that can be caused by EMP signals. As a result of this engineering know-how regarding EMP protection, many important U.S. C^3 facilities are appropriately hardened against EMP, and some satellite systems also incorporate EMP-hardening attention (e.g., MILSTAR).

High-power Microwave Disruptions

The identification of the existence of nuclear EMP signals during U.S. nuclear test programs raised the feasibility of developing nonnuclear HPM sources to destroy electronic systems. This concept encouraged a wide range of DOD-supported R&D activities in HPM transmitter developments, with research carried out in both government and industrial laboratories. These efforts have evolved, expanded, and matured into what has become known today as the High-power Microwave effort.

Until quite recently, virtually all of the HPM developments have focused on obtaining very large average power transmitters capable of disrupting or destroying electronic components. Laboratory-based experimental HPM facilities are very large—consisting of rooms full of equipment—and are not suitable for use in tactical warfighting scenarios. HPM sources may, however, have practical applications at fixed antiballistic missile (ABM) facilities as adjuncts to conventional missile-based ABM approaches.

In addition, interest in potential HPM weapons was stimulated by intelligence reports about Soviet advances in HPM-generation technologies—many still consider the Russians well ahead of the United States in HPM research. Although operational HPM weapons have entered the development process, to date none has been successfully fielded. As with EMP disruption, a set of engineering handbooks has been prepared that deals with the various techniques of protecting electronic systems from HPM disruption signals.

Ultrawide-band Radio-frequency Impulse Disruptions

Ultrawide-band radio-frequency impulse signals consist of very-short-duration, very-high-peak power RF pulse signals. The duration of the impulses is on the order of one or two RF cycles (a nanosecond or two), and they may have peak signal powers of a few hundred kilowatts to hundreds of gigawatts, depending on the antennas used to radiate them. Most UWB RF impulse transmitters can generate repetitive impulses with pulse rate frequencies up to 100 kHz.

Some consider UWB RF impulses to be part of the HPM family. Proponents of HPM technologies have found that some electronics can be affected if the HPM signals are modulated with specific frequencies that match those in the electronics to be targeted. The panel has chosen to consider UWB RF impulse signals as a separate category primarily because they disrupt rather than attempt to destroy or burn out electronic circuitry. UWB RF impulse transmitters generate extremely high peak power signals, whereas most HPM transmitters attempt to generate high-average-power signals. Also because of the small-size, lightweight nature of the UWB RF impulse transmitters, they are more readily deployable than are HPM transmitters.

UWB RF signals can disrupt most of the computers and information systems in use today. UWB RF-induced disruptions are typically transient, but they can cause a system to require rebooting before it can again become functional. Variations in the pulse rate frequency can significantly affect the nature of the disruptions observed.

UWB RF impulse signals tend to be disruptive regardless of the existing internal clock rates, but their disruptive potential may be enhanced by matching the pulse rate frequency to the internal clock speed of the targeted equipment. The wide bandwidth of the UWB RF impulses—as much as several gigahertz in some cases—enhances their ability to disrupt a variety of systems, although detailed understanding of how UWB RF pulses couple to electronic systems is not currently available.

One type of fundamental UWB RF impulse technology is the bulk avalanche semiconductor switch (BASS). BASS operates on the principle that a solid-state chip, when illuminated by a laser, will short-circuit in a few picoseconds. When some 10,000 to 12,000 volts are momentarily placed across a GaAs semiconductor device, and the switch is placed in a simple RF circuit, the resulting avalanche of current produces approximately 1 megawatt of RF energy that is essentially one RF cycle in duration. A single BASS module, complete with timing circuits, is about 2 × 2 × 6 inches in size and weighs less than a pound. All it requires to function as an RF source is an external trigger generator and a battery.

The pulse repetition frequency that can be achieved with BASS devices is on the order of 10 kHz. As a result, the average energy in the repetitive pulse train is small, and it is the peak power that affects digital microelectronics functioning. RF impulses with center frequencies from 50 to 2,000 MHz have been generated using the BASS technology. Through the use of precision timing, several BASS switches can be combined to form a much higher peak power system. When configured in the form of an antenna array, peak power levels of 1 GW have been obtained, and because of the gain of the array, a 100-GW ERP can be obtained. The individual BASS units making up the array can be timed—equivalent to phasing in a more conventional phased array antenna—so that the beam can be steered over a 60° angular cone.

Because of the high bandwidth of UWB RF impulses, they have noise-like characteristics when detected by most narrow-band receivers. Using UWB RF signals as a jammer source introduces the possibility that enemy systems could be jammed, whereas friendly systems could be designed to accommodate the periodic impulses. A possible application of this concept is the ability to jam GPS and deny it to an adversary but still allow friendly forces to continue to use it without degradation—a scheme that might be called intelligent jamming.

Because UWB RF impulse transmitter systems are relatively small and lightweight and require only modest levels of prime power, they appear to be suited for tactical applications. They also appear to be the only RF disruption technology compact enough to be deployed on satellites.

Information Warfare Related to Space

The possibility of conducting offensive IW actions against space systems has been considered, including interfering with the functioning of ground terminals and interfering with the functioning of satellites by using ground-based HPM systems. In most instances, such electromagnetic-based antisatellite (ASAT) efforts have been based on HPM technologies that are capable of burning out or destroying electronics in the satellite systems but that require huge facilities and copious amounts of power. Moreover, there are treaties and space laws that preclude such destructive actions. The recent emergence of new transmitter technologies that permit the generation of UWB RF impulses opens the door to new disruption capabilities—and using such transmitters in space seems particularly practical.

Do UWB RF Impulse Systems Pose a Threat to U.S. Information Systems Now or in the Future?

Countries other than the United States have extensive background in the development of RF weaponry. For example, the former Soviet Union pioneered the development of UWB RF impulse technology, and this technology is now being offered to other countries by both Russia and the United States. It is expected that in future conflicts the United States will encounter enemies using UWB RF weapons to disrupt U.S. information systems, and this will require that U.S. information systems be appropriately protected. Some protection approaches are available from prior EMP hardening efforts, but additional methods are needed to counter extremely high peak powers and high-frequency components of UWB RF impulses.

Other Related Technologies

High-average-power transmitter systems from the HPM program may still have practical utility, especially when such microwave signals are modulated in ways to have a maximum effect on digital electronics. High-average-power HPM transmitter systems that are best suited for large ground-based facilities may find use in ballistic missile defense applications.

The existence of very-high-power HPM RF transmitter sources will allow for other kinds of directed-energy weapon (DEW) disruptions, e.g., linear accelerators that can be used to generate a focused beam of secondary x-rays—where RF shielding and filtering protection approaches are totally ineffective.

Development Drivers

The U.S. military does not have to develop all of the defensive IW capabilities independently of the defensive IW technologies available from the commercial sector. In most cases, the technologies that will support military defensive IW needs in the hacker and software categories are being developed more rapidly in the civilian information system environments than they are in the military environments. Defensive and offensive IW activities associated with emerging UWB RF impulse technologies will be more suitable for military development than commercial. A major increase in efforts is warranted to develop defenses for information and communication systems that offer protection against this new capability. Offensive developments of this technology not only should be considered for disrupting enemy information infrastructures but also should be used to evaluate the vulnerabilities of U.S. information infrastructures.

The panel believes that the Navy Department must make every effort to work on both the defensive and offensive IW development programs in close coordination and cooperation with the other military departments, with joint chiefs of staff, and with operational commanders.

The Department of the Navy should establish close working relationships with the recently formed presidential commission for the protection of critical infrastructure and undertake, where appropriate, cooperative efforts to protect infrastructure critical to naval operations, even though such infrastructure may be in the commercial sector.

Time Scale for Development and Insertion

There are some defensive and offensive IW prototype concepts involving the

UWB RF impulse technologies, but to date little has happened to bring them to the operational naval forces. As such hardware and software offensive and defensive IW concepts are brought into existence, the Department of the Navy should undertake an expanded set of IW-related advanced technology demonstration exercises in which these new concepts, ideas, and capabilities are brought together and experimentally evaluated and verified.

Development of both offensive and defensive IW capabilities is going on now at a modest pace. Developments should be accelerated and expanded and should be expected to continue indefinitely into the future for some time to come. Today, the United States probably has better offensive IW capabilities than it does defensive ones. The proliferation of more complex information system architectures with attendant large-scale integrated circuit electronics has caused the defensive posture associated with Navy and Marine Corps information systems to currently be far short of the robust levels needed for achieving information dominance and winning future information wars.

Concluding Observations

The panel presents the following observations:

1. Emerging new UWB RF impulse transmitter systems are small and lightweight, require relatively low levels of prime power (average), and are suitable for tactical or terrorist applications.
2. Because of their very high peak power signal, UWB RF impulses have unique capabilities to disrupt nearly all digital microelectronics hardware used in information systems.
3. UWB RF impulse signals have the potential to pose a very serious threat to the cyber portion of the critical infrastructure.
4. UWB RF impulse transmitter technology is being made available outside the United States and is becoming available to terrorists and potential enemies.
5. UWB RF impulse transmitters have such low cost and complexity that many nations who might want to take down or disrupt both military and commercial information and communication systems can readily acquire the technologies to do so.
6. Concerns exist about the paucity of defensive capabilities to mitigate these emerging UWB RF impulse weapon threats.
7. To date, there are no confirmed examples of any use of these emerging UWB RF impulse technologies to disable U.S. military or commercial information system hardware.
8. It is probably only a question of time before a UWB RF impulse disruption is experienced.

RECOMMENDATIONS

Information and networking are expected to play a dominant role in future military engagements. Rapid access to appropriate knowledge at all levels will optimize warfighting and crisis response capabilities. Within this context a number of key information-processing technologies are discussed here. Recommendations regarding these technologies for Department of the Navy consideration are as follows:

- The Department of the Navy should exploit evolving commercial technologies in knowledge extraction, data management, and data presentation, together with unique military technologies in data fusion and automatic target recognition to deal with the increased complexity and tempo of warfare.
- The Department of the Navy should engage in early exploitation of the rapid growth in commercial communications capabilities, including satellites and fiber-optic communications, to acquire the necessary increased bandwidth and diverse routing for future networking needs. The Department of the Navy should also prepare for graceful degradation of these systems at times of warfare.
- The Navy Department information systems should be protected against increased software and electromagnetic information warfare attacks and other vulnerabilities. The Department of the Navy should develop offensive information and electronic warfare technologies to find, identify, and attack adversary systems and to strengthen naval systems.

Specific recommendations regarding IW are as follows:

- The segment of the Navy Department that is involved in and concerned about the survivability of the its information and communications systems should become familiar with possible disruption threats associated with UWB RF impulse signals.
- Ways to use these new emerging UWB RF impulse signals as offensive IW weapons should be integrated into operational systems by the EW technical community, which has an established track record of working with the military forces.
- The Department of the Navy should recognize both the disruption and protection aspects of the UWB RF impulse threat. Successfully solving the protection problems requires understanding how UWB RF impulse signals disrupt systems.
- To ensure that the United States maximizes its ability to conduct both defensive and offensive warfare, the current restrictions separating the two activities should be eliminated so that both defensive and offensive IW aspects can be considered in an integrated fashion.
- All hardware and companion software associated with information sys-

tems should have their vulnerabilities to UWB RF impulse disruption signals evaluated and mitigated where practical. Particular attention should be given encryption systems and emerging microelectromechanical systems technologies.

- The Department of the Navy should encourage its own intelligence activities as well as National-level[7] intelligence activities to make it a high priority to track the proliferation of UWB RF impulse technologies.
- The Department of the Navy should make every effort to enter into cooperative activities in support of the recently formed presidential commission for protection of critical national infrastructure components as their protection will ensure that naval force uses will continue to be robust in times of military conflicts.

---

[7]The term "National" refers to those systems, resources, and assets controlled by the U.S. government, but not limited to the Department of Defense (DOD).

4

Sensors

INTRODUCTION

Although many factors contribute to the success of any military operation, it has long been recognized that information is one of the most important—information in many different forms and acquired on many different time scales.

Information

During conflict situations several different kinds of information come into play. At the highest top-down level is situation awareness. Warfighters must understand everything that they can about the nature of the opposing forces—their current positions, their movements, their composition, their infrastructure, their capabilities, their communications, their weapons, their threats, their plans. The more the better—in real time on a scale that ranges from minutes to hours—and not what they were doing a day ago, but what they are doing right now. Obviously, for maximum cooperative effectiveness the United States ought to have the same complete picture of its own forces, distinguishing accurately between friendly and adversarial forces to minimize or eliminate friendly-fire incidents. To complete the scenario, an accurate, real-time picture of the environment is needed—e.g., the details of the terrain, the current and anticipated weather or sea conditions, the presence or absence of mines, and so on. Generally, in real conflicts, only a few of these factors are actually known to the degree desired—knowledge of some may be stale and out of date, and other factors may only be guessed at and some completely unknown. The better the job U.S. forces do in achieving valid situation awareness, the larger the potential competitive advantage they can enjoy.

On a shorter time scale, from the bottom-up point of view, effective utilization of weapons requires detailed and timely—fractions of seconds to minutes—information about both the targets and the weapons themselves. Targets must be recognized as such, their positions localized instantaneously, their motion measured to high precision, their most vulnerable aim points identified, and so on. Similarly detailed continuous information about the weapons is needed to aim or guide them to a successful interception of target—e.g., weapon position, inertial parameters (such as orientation, velocity, and acceleration), and environmental parameters (wind and tidal currents). On a much longer time scale, outside actual combat situations, information is needed in several forms to provide for such necessities as equipment maintenance and support and overall logistics. For example; it is necessary to know what has failed or is about to fail, or what the weather will be tomorrow.

Sensors

To acquire desired information, measurements of all kinds of physical parameters must be made. The devices that permit these measurements are known broadly as sensors. The term "sensors" encompasses an enormous range of technologies and devices. Some can be as simple and old fashioned as the direct measurement of a local temperature by means of a thermocouple, and others can be as modern as the detection of a biological agent by a miniaturized mass spectrometer, or as complex as a synthetic aperture radar (SAR) all-weather imaging system. In all cases, whatever the sensor, an interaction between the sensor and its local physical environment results in the generation of some kind of signal, generally a form of electrical response of the sensor's physics, chemistry, and biology to the physics, chemistry, and biology of the outside world. The interpretation of these sensor signals through signal processing, data fusion, and the like leads ultimately to the extraction of the desired information.

Sensor Categorization

Operational Modes

Sensors, whatever information they are attempting to collect, can be broadly classified into two categories of operation, passive and active, which are defined as follows:

- Passive sensors simply measure and report on, via their response signals, whatever they detect in their local environment. In a sense they just listen. A thermometer and video camera are good examples of passive sensors. Even though a passive sensor responds only to local physical phenomena, the information may well be coming from a distance—perhaps in the form of photons of light that travel from the objects or scene of interest and impinge on the sensor's detectors to produce the necessary local interactions. From a military point of view, passive sensors have the great advantage of producing valuable information without emitting any signals of their own that might give away their position and expose them to possible retaliation.

- Active sensors, on the other hand, typically stimulate the environment by generating and emitting known signals, which propagate out to the objects or targets of interest, interact with them, and reflect or scatter energy back to the sensor, which then responds as in the passive mode. Because the self-generated signals have known properties, it is often possible to use signal processing to extract very weak emitted signals returning from the objects of interest from the inevitable competing noise and clutter-generated signals in the sensor. Although operating in an active mode reveals the location of the emitting source, it is sometimes the only practical way to collect the desired information. The most familiar example is radar. If an object of interest is itself not emitting electromagnetic radiation, the only practical way to provide the desired geometric information about the object's existence, location, motion, size, shape, identification, and the like is to illuminate it deliberately. Although frequently the transmit and receive functions are combined in the same hardware, they do not have to be physically co-located. In the context of radar, this variation or active sensing is known as bistatic; in missile guidance, such a configuration is referred to as semiactive, in that the missile itself operates in a passive mode, whereas the target is actively illuminated by a separate source.

Sensors that operate on the basis of wavelike, propagating physical phenomena such as electromagnetic waves, e.g., radar and laser detection and ranging (LIDAR), or acoustic waves, e.g., seismic sensors and sonar, can be operated in either passive or active mode to collect different kinds of information. Sensors based on nonpropagating phenomena such as those that sense chemical compounds or accelerations (initial sensors) operate only passively.

Physical Phenomena

The range of sensor types of interest to naval forces is enormous. Box 4.1 lists the basic physical phenomena that underlie the types of relevant sensors. As can be appreciated from this list, the subject of sensors is very complex, involving a large number of apparently quite disparate technical disciplines.

Generic Sensor Model

Each of the physical phenomena listed in Box 4.1 is discussed below in more detail in the context of the current state of the art and projected possible future

BOX 4.1 Physical Phenomena Underlying Sensors

- Electric only
  Direction
  Field strength
- Magnetic only
  Direction
  Field strength
- Electromagnetic: active and passive
  High frequency
  Microwave (L, S, C, X, Ka, Ku)
  Millimeter wave
  Optical (IR, visible, UV)
  X-ray, Gamma-ray
- Acoustic
  Air—sound
  Water—sonar
  Solids—seismic
- Mechanical
  Inertial—acceleration, linear and rotational
  Gravitational—direction and weight
  Flow of fluids
  Position
  Stress and strain
- Chemical
- Biological
- Nuclear
  Alpha, beta, neutron
- Environmental
  Atmospheric parameters—temperature, wind, humidity, visibility
  Ocean parameters—temperature, currents, salinity
- Time

performance capabilities of the sensor technology associated with each. Although useful for better appreciation of the individual sensors and sensor technologies, this approach can be quite confusing when it comes to projecting the future because of the great variety in sensor types and related details of technology that mask their underlying common features. This section discusses sensing in general and the common concepts and issues that characterize all of sensor technology. Five critical technologies—semiconductor, superconductor, digital, computer, and algorithm technologies—appear to be common to all sensor types, and careful delineation of these greatly simplifies the difficulties of projecting future sensor capabilities. Insofar as it is possible, for each identified common critical technology, quantitative performance projections based on today's historical performance growth patterns are discussed below as clues to future potential. A model of a generic sensor is shown in Figure 4.1.

Sensor Interface with the External World

The first thing that must be considered is the interface between the sensor and the physical phenomena to be sensed. For some classes of sensors, e.g., chemical or biological sensors, physical samples of atoms or molecules or chunks of material must be collected and inserted into the sensor's detection mechanism. The design of this kind of interface permits a good deal of flexibility and no doubt will vary considerably over time.

Sensors such as LIDAR designed for propagating phenomena collect samples

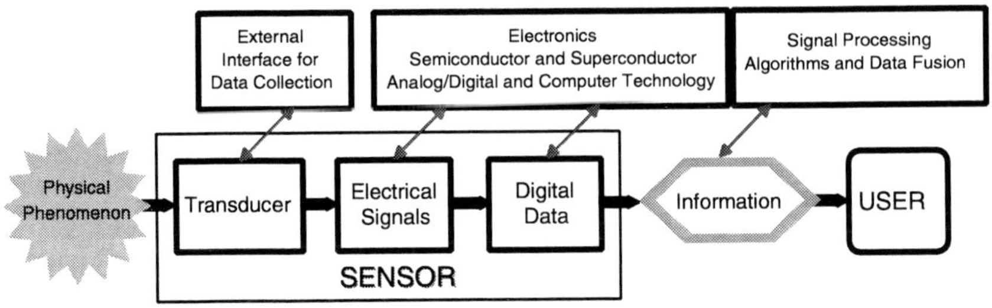

FIGURE 4.1 Generic sensor model.

that are less material in character and more wavelike; e.g., the detected objects are photons or phonons. However, the interfaces for such sensors are highly constrained by the free-space wave-propagating physics of the phenomenon being sensed: Whatever the details of the implementation technology, the sensor interface to the outside world must provide an appropriate wave equation impedance match. The details of the sensor implementations can be altered by the designer, but the outside world's physics is whatever it is and is not under the control of the designer or the sensor. When sensor performance capabilities are projected into the near and far future, the interface constraint remains invariant—e.g., although it will be possible, with time, to compress more and more digital and computing capabilities into ever-decreasing volumes, the dimensions and characteristics of the radar aperture needed for a given task will remain basically the same. Beamwidth requires so many free-space wavelengths across the aperture, and grating lobe suppression requires a certain minimal spacing of elements, again in terms of the free-space wavelength. Certain characteristics of the interface are, and will always be, independent of the technology used to implement the sensor hardware.

For sensors—e.g., inertial, gravitational, and time—as the sensors are totally and inextricably immersed in the phenomena to be sensed, the concepts of external world interface and collection are simply moot.

Detection

Once properly interfaced to the external world, the sensor must selectively detect the manifestations of the phenomena of interest and produce signals that can be used to quantify and convey the desired information. Although in the past many simple sensors used purely mechanical means of indicating the detected signal, as, for example, an automobile thermostat or a thermometer based on the motion of a bimetallic strip or membrane, most sophisticated sensors of interest to the Navy and Marine Corps indicate the results of detection as an electrical signal—a modulated voltage or current. Even though the physical phenomenon being sensed may not be directly electrical in nature, but rather chemical or biological or acoustic, the detection is usually accomplished by using the phenomenon of interest to generate or move free electrons, thereby transducing the physical manifestations into electrical signals. For example, microwave signals are detected by causing the associated time-varying electromagnetic fields to induce currents in electrons already available in conducting elements of the sensor. Optical signals generally use the large energy inherent in each photon ($E = h\nu$) to kick loose bound electrons to create free electrons that are then collected and/or moved to produce measurable charges or currents. Other sensors use mechanical components that move under the influence of the physical phenomenon to be sensed but that are also part of an electrical capacitor or other circuit element so that the resulting motion alters the circuit parameters, thereby modulating an electrical signal in an interpretable way. Chemical and biological sensors use yet other techniques to produce electrical detection outputs.

The technology for processing signals that are electrical in form is well understood, and the future capabilities of many diverse sensors can be projected in terms of electrical signal processing technology. Electrical signals are now uniformly dealt with via electronics. Modern electronics is uniformly based on semiconductor technology, and projected progress in semiconductors is often directly translatable into valid projections of improved sensors and sensing. In addition, superconductor technology offers many attractive ways of dealing with electrical signals that operate on the basis of quantum effects that are quite different from those encountered in semiconductors. Although still very much in development, because of the great promise and the recent progress demonstrated, progress in superconductor technology must also be carefully assessed for application to future advanced sensors.

Although electrical signals from independent sensor elements are often combined directly today in analog form, increasingly they are converted immediately to digital format. It is virtually certain that this will be the only approach considered in the future because of the many proven advantages of digital technology. Thus, projected progress in digital technology will be directly translatable into projected improvements in sensors and sensing.

Information Extraction

The stream of digital data emerging from each individual sensor element, e.g., each pixel of an IR focal plane array or each receive element of a phased array radar, must be assembled, stored, processed to extract the desired information, and communicated to a user—sometimes a human and sometimes another mechanical/electronic device—that can use this information for guidance and control or some other purpose. Progress in all aspects of computer technology, and particularly algorithm technology, will translate directly into improved, more capable sensors.

RELEVANCE: WHAT DO SENSORS DO FOR THE NAVAL FORCES?

That sensor technology, in its myriad embodiments, is critical to the success of all naval force operational tasks or missions is obvious (Box 4.2). Whenever information is required, sensors are utilized to make the physical measurements from which the desired information is extracted. Radar, optics, and sonar sensors, through the active or passive exploitation of the physics of wave propagation, give information about distant objects that is useful for general surveillance and situation awareness as well as for more specific purposes, such as real-time target location and weapon guidance. Other sensors, such as position-sensing devices or inertial sensors, produce useful real-time local measurements that can be used to control all kinds of platforms, including whole ships, steerable radar or communication antennas, and gun mounts on ships, or even individual missiles in flight, depending on just where the sensors are located. Yet other sensors produce measurements for which the long-term variations in the measured parameters provide the useful information. For example, temperature or atmospheric pressure sensors can supply inputs for short- and long-term weather prediction, whereas acoustic sensors mounted on rotating machinery can provide evidence of bearing wear or imminent gear failure, thus triggering needed repair and maintenance procedures.

In short, naval forces are heavily dependent on the use of sensors today, and the future seems to promise even broader use of sensors as the technology continues to evolve toward more capable performance and the demand for more and better information escalates. Future sensors, as compared with existing implementations, promise to cost less, have higher sensitivity and precision, be available in much smaller, lower-power packages, and perhaps be capable of measurements currently unimagined (i.e., be completely new types of sensors).

TECHNOLOGY STATUS AND TRENDS

Despite the dangers in attempting to project tomorrow's technology entirely in terms of what we see today, doing so can still impart valuable lessons. Preced-

BOX 4.2 Sensor-dependent Operational Tasks and Missions

Situation awareness
General foe/friend information
Surveillance
Threat detection, recognition, and localization—general or specific
Weapons targeting—offensive or defensive
Logistics and maintenance

BOX 4.3 Technology Trends Common to all Classes of Sensors

- Solid-state technology
  - Miniaturization
  - Lower power
  - Integrated circuit manufacturing
  - Integral packaging
  - Microelectromechanical systems
- Atomic-level manipulation—nanotechnology
  - Designer materials
  - Quantum wires and dots
- Digital implementations
  - Analog-to-digital conversion/digital-to-analog conversion
  - Direct digital synthesis
  - Computers and signal processors
- Distributed systems and implementations: fault tolerance
  - Networking
  - Data fusion
- Distributed systems (continued)
  - Data compression
  - Societies of microsensors—sensors plus computers
- Multidimensional signatures (clutter rejection, detection, automatic target recognition)
  - Multispectral
  - Hyperspectral
  - Data fusion
- Multifunctional sensor systems; common transmit and receive apertures for:
  - Radar
  - Communications
  - Electronic warfare/electronic support measures

ing the panel's discussion of the five technologies critical to all modern sensors is a brief review below of the technology trends that are evident in today's developments and that are shared, in some way, by all classes of sensors (Box 4.3). Toward the end of this chapter, the panel speculates on the impact of sensor technologies on tomorrow's naval forces, touching on other possible future directions, not so evident today but desirable from the user's point of view and not obviously incompatible with some law of nature.

Solid-state Technology

The most obvious overall trend of significance in technology today is solid-state technology's dominant role in both analog and digital electronics. Today's digital circuits are solid-state—the semiconductor transistor, in one form or another, is the workhorse of the industry and the foundation for all practical digital IC implementations. Its one evident competitor, lurking in the background but always gaining ground, is another solid-state technology—superconductors, e.g., Josephson junctions, superconductor quantum interference devices (SQUIDs), and RSFQ logic.

Even in the analog world, solid-state technology has come to dominate the audio and video amplifiers in entertainment electronics. In the last decade or so, even the generation of microwave power for radar and communication applications has come to be realized increasingly often in solid-state form. Although many legacy microwave systems still generate RF by means of tubes, all new radars are automatically assumed to be some form of solid-state phased array, and soon the only place tube technology will be found is in microwave ovens. Today, magnetrons are still cheaper than the equivalent power transistor, but that may not last as solid-state electronics continue simultaneously to improve in performance and fall in price.

Given this widespread trend, it seems likely that all future advanced sensors will process their detected electrical signals with some form of solid-state circuitry. It can be expected, then, that advanced sensors will share in the continually improving aspects of solid-state technology, that is, increasing miniaturization, higher speeds, decreasing power per function, increasing device density and complexity via IC manufacturing techniques, integral packaging concepts, and decreasing unit costs. A spin-off application of semiconductor manufacturing to three-dimensional micromachining of silicon (Si), enabling the development of MEMS, has already produced a range of novel sensors and actuators with significant performance and cost advantages over the conventional forms.

Atomic-level Manipulation

One of the most exciting recent technology developments is the growing ability to manipulate matter at the atomic level. Largely because of the efforts applied to the fabrication of solid-state devices and integrated circuits, through mastery of thin-film deposition techniques and the physics and chemistry of surface phenomenon, it is now possible to control material and structural fabrication at the level of the individual atoms. These skills have already been used to create artificial materials with unique characteristics, e.g., alternating-layer semiconductor structures for advanced microwave devices, integrated multilayer Bragg reflectors for photonic applications, magnetic structures with as many as 50 alternating layers of iron and chromium for giant magnetoresistance sensors, biologically inspired self-assembling layered materials of polymers and ceramics with unusual properties, and even artificial high-temperature superconductors with monomolecular layers of alkaline earth atoms (calcium, strontium, or barium) alternating with copper dioxide (CuO_2) layers.

These techniques allow the tailoring of materials and devices at the nanostructure level,[1] i.e., accurate growth and placement of clusters of a few or a few

[1]Robinson, E.Y., H. Helvajian, and S.W. Janson. 1996. "Small and Smaller: The World of MNT," *Aerospace America*, 33(9):26-32, September.

tens of atoms down to the positioning of single atoms. They will provide completely new sensor materials and the quantum wires, dots, and single-electron transistor devices that are likely to be exploited to continue the long-term growth trends in solid-state electronics into the future for decades to come. It seems clear that sensors and sensor systems of all kinds will benefit from these capabilities, getting continuously smaller and cheaper and more capable with time. The implementation of microscale or perhaps even nanoscale self-contained entities with integrated sensors, computers, and actuators will certainly become possible over the next several decades, and such devices will probably represent a mature, widespread technology by 2035.

Digital Implementations

Another very obvious characteristic of modern electronic technology is the inexorable march toward all-digital implementations. The significant advantages of digital versus analog implementation in terms of flexibility of processing, error containment, and robustness against drift have long been recognized, but cost, speed, and other obstacles have hindered the conversion in many areas. With the performance and costs of digital computers and signal processors improving exponentially with time, i.e., by factors of 5 to 10 every few years, the obstacles are vanishing and the digital implementation of all sensors and sensor systems in the near future appears inevitable.

Analog-to-digital conversion (ADC) technology and its converse (DAC) are currently making such rapid strides in sample rate, number of bits, size, and power, that systems that involve the direct digitization of microwave radar and communication signals at gigahertz rates, with adequate dynamic range and low enough size, power, and costs to be considered practical, are already under development in many places, with field deployments expected in less than 5 years. These kinds of ADC/DAC capabilities, combined with accelerating computational capabilities, will permit the implementation of advanced adaptive processing algorithms, e.g., digital beamforming and space-time adaptive processing (STAP), and effective ATR algorithms, as well as the exploitation of multisensor data fusion techniques. Given the ability to digitize and digitally process almost any radar waveform, it is clear that the time is ripe for the application of digital techniques to other aspects of the system, such as the generation of arbitrarily complex radar transmission waveforms with performance features that go well beyond the simple continuous wave (CW), linear frequency modulation (FM), and occasional phase-coded waveforms that have dominated radar technology to date.

Distributed Systems

As individual sensor and computing elements grow ever smaller in size, power, and cost and simultaneously more powerful, the temptation to combine a large number of them into a super complex of distributed, intercommunicating elements becomes irresistible. The trend today is toward distributed phased-array antennas for radar and communications, multiprocessor supercomputer architectures, distributed power conditioning, the Internet, and so on. And there are enormous advantages to be gained—more information-bearing signals can be collected, more overall computer throughput can be achieved (but always by a factor less than number of elementary processors combined), and much higher overall reliability with graceful degradation characteristics can be obtained. Since it is always possible to make things complicated faster than it is possible to improve the reliability of the individual elements, fault-tolerant redundancy techniques must be explicitly addressed for graceful fail-safe degradation. Equally obvious is the need for efficient, high-bandwidth interelement communications, probably fiber optic and wireless, in which data compression techniques, both lossless and lossy, will come into play.

Moreover, as the Internet has shown, information collected from many, perhaps widely dispersed sources, with the proper communications technology and protocols in place, can be profitably correlated to allow a fuller understanding of a subject or situation. Data-fusion and data-compression algorithms will play a key role in the application of these concepts to sensing and to the achievement of the desired battlefield situational awareness, ATR, damage assessment, and related capabilities. An obvious danger lies in the potential flood of data that a capable multisensor distributed data collection system can create. It is possible to generate overwhelming amounts of data that can cause a total shutdown of the human users, whose performance is notoriously nonlinear and prone to catastrophic collapse. (Degradation is definitely not graceful for overloaded humans.) Concepts and algorithms to permit the recognition, extraction, and viewing of only the minimally required information from large distributed databases must be developed—an area of significant future need for research and development.

Eventually powerful miniature sensors will be combined in a single package with on-board, integral computational capabilities to form mini-systems-on-a-chip. An intriguing prospect for the future is the notion of interacting armies of small, capable, autonomous entities—microrobots that fly or crawl or swim—that combine miniature sensor packages with integrated computers, actuators, power sources, and wireless communication capabilities. These assemblies might be capable of functioning as ant societies do, with each participant acting locally on the basis of mostly local information, and the whole assembly functioning effectively to reach some global goal. This kind of implementation of sensors suggests the possibilities of higher overall performance in surveillance, for example, through adaptive, autonomous spatial repositioning of the individual sensors. The development of single, small, flying sensors of this sort is already under way. Successful artificial societies of this type will require the development of a deep understanding of what the appropriate rules of behavior should be and their implementation in software—another topic for future R&D efforts.

Multidimensional Sources of Information

Yet another striking trend in modern sensor system technology is the use of multiple simultaneous sources of information—e.g., spatially dispersed multiple sensors of the same type or perhaps single-sensor systems operating on multiple spectral bands such as several IR bands, several RF bands, one IR and one RF band, and so on. Dual-band IR focal plane arrays with precise pixel alignment between the band images have been produced, and several RF/IR advanced missile seekers (i.e., multispectral seekers) and optical sensors for satellite platforms that collect pixel-aligned data in hundreds of small (e.g., 10 Å) bandwidths across a large range of the optical spectrum—known as hyperspectral systems—are under development at the present time.

The higher the dimension of the information that can be collected from a pixel or an object, the better the chance of correctly detecting and recognizing it. Clearly, effective data fusion and ATR techniques are needed, and these are already areas of active research. On the other hand, significant increases in computational memory and throughput are required and offer additional challenges on the path to achieving high performance and affordability.

Multifunctional Configurations

The final technology trend of significance to the future growth of advanced sensors and sensor systems is the broad and growing interest in the implementation of multifunctional configurations—that is, sensors capable of performing several different functions via shared hardware. It has long been common to combine the search-and-track function on a single radar by using time-sharing of different waveforms and beam scan patterns, thereby gaining certain implementation advantages in size and perhaps cost over the alternative of building a separate radar for each function. For a number of reasons associated with the available real estate and the growing interest in and need for electromagnetic signature control, there is considerable interest in the Navy in providing multifunction capability in single locations.

It is always easy, in principle at least, to share computer resources between multiple functions. What is often more difficult is to share the interface with the external world and the microwave elements between functions, because many of these functions put quite conflicting requirements on these components and, as was mentioned in the introduction to this chapter, the properties of this interface are strongly constrained by the physics of the propagating phenomena, whether electromagnetic or acoustic. For example, radars can utilize time sequencing of the transmit and receive functions to achieve the isolation needed to prevent the high-power transmitted energy from leaking into the sensitive receivers and saturating or destroying them, but microwave communication systems typically require continuous simultaneous transmit and receive and cannot use the time-sequencing approach. Electronic support measures (ESMs) systems, which might also utilize the same or adjacent portions of the microwave spectrum as do the radars and communications, have similar requirements for continuous, sensitive, passive reception and again cannot easily exploit time sequencing.

In addition, if radars of quite different frequencies (say, L- and X- or K_u band) attempt to share the same aperture (such as phased-array elements), extreme linearity and difficult-to-implement high-Q microwave filters in the amplifying electronics and wideband radiating elements are required to prevent deleterious cross-coupling of signals between bands, even if time sequencing can be used to implement transmission/reception isolation. Even more difficult is the problem of combining such widely separated frequencies as conventional RF microwave bands with millimeter waves or either of those with electro-optical systems. The physical implementations suitable for each of these spectral regions differ so much that it is next to impossible to usefully share transmit and receive resources. The amplifiers and filters suitable for one region simply do not work for the others, and often, rather than being able to share resources (particularly at the critical, constrained interface with the outside world), the different implementations conflict. There is always room for ingenuity here, but many of the obstacles are fundamental.

CRITICAL COMMON TECHNOLOGIES

The discussion above of the generic sensor model identifies five key technologies as common to all modern sensors and sensor systems and as absolutely critical to their performance potential. Understanding the current state of the art of these individual technical areas and the growth patterns that can be extrapolated will allow reasonably confident prediction of the kinds of performance achievable in the future for the different classes of sensors and the kinds of new naval force applications that might be enabled.

Below, each of the critical technologies (Box 4.4) underlying advanced sensors in general is discussed briefly, and historical growth curves are presented where possible. Each technology is extremely important, in its own right, for

BOX 4.4 Critical Common Technologies

Semiconductor
Superconductor
Digital
Computer
Algorithm

many applications beyond sensors, and each is discussed also in Chapter 2 of this report.

Semiconductor Technology

Conventional Semiconductors

Semiconductors constitute an enormous topic[2] that can be surveyed here only briefly, however critical it is to the future development of sensor technology. Since the invention of the transistor some 40 or more years ago, 3 single-component and at least 10 binary semiconductor materials, as well as a number of tertiary materials (e.g., mercury cadmium telluride [HgCdTe] and aluminum gallium arsenide [AlGaAs]), have been exploited for the implementation of a variety of electronic devices and applications. These single and binary semiconductors, ordered by the magnitude of their bandgap, are listed in Table 4.1. Others not listed here may someday be exploited, including mercury telluride (HgTe), manganese selenide (MnSe), gallium antimonide (GaSb), indium nitride (InN), scandium nitride (ScN), aluminum nitride (AlN), zinc selenide (ZnSe), and boron nitride (BN). Also indicated in Table 4.1 is the nature of the bandgap, that is, whether it is direct (D) or indirect (I) gap material, as this determines to a large extent just what kinds of applications the semiconductor may be suited for.

First to be exploited, germanium was soon replaced by silicon, clearly the most widely used semiconductor material so far and likely to remain so for the foreseeable future. The reasons for silicon's dominant position are numerous: It is readily available in large quantity; it is mechanically strong; it is a good thermal conductor; it can be easily grown into large-diameter, ultrapure, defect-free crystals; it forms stable insulating oxides of excellent quality; and it is nontoxic and easily fabricated, via a wide range of patterning, etching, implanting, and diffusing techniques, into devices with literally millions of circuits per chip and hundreds to thousands of chips per wafer with high yields.

These virtues, and an enormous multidecade investment in time and resources, have led to the explosive proliferation of digital and microelectronics fabrication technologies that characterize and enable the rapid growth in computer and information technology. Linewidths continue to decrease exponentially with time, with optical lithography still performing effectively at submicron dimensions that were thought to be beyond its capability only a few years ago, and with finer, although less convenient, x-ray and electron-beam techniques waiting in the wings to continue the fabrication down into the regime of quantum dots and wires and ultimately to single electron logic structures—about as far as can be imagined today. With finer dimensional capabilities in hand, and as defect

---

[2]Singh, Jasprit. 1994. *Semiconductor Devices: An Introduction,* McGraw-Hill, New York.

TABLE 4.1 Common Semiconductor Materials

Material	Bandgap (eV)	Bandgap Type
InSb	0.230	Direct (D)
InAs	0.354	D
Ge	0.664	Indirect (I)
Si	1.124	I
InP	1.344	D
GaAs	1.424	D
CdTe	1.475	D
AlAs	2.153	I
GaP	2.272	I
ZnTe	2.394	D
SiC	2.416	I
GaN	3.503	D
C	5.5	I

densities continue to be reduced, the number of circuits that can be placed on a single chip with reasonable yield grows exponentially with time, causing the cost per operation to spiral downward while performance, in terms of clock speeds and throughput, continues its exponential upward growth—a pattern of factor-of-10 improvements every 4 or 5 years, which has been consistent for at least a decade and a half and shows no signs of slowing as yet. Figure 4.2 illustrates the exponential growth in the total number of transistors on a single chip from 1970 to the present and also extends the average observed growth pattern to the end of this study's time frame, 2035. Since there are no obvious fundamental physical laws that limit the number of transistors achievable per chip, and given the almost 30 years of observed consistent exponential growth, extrapolation of the observed pattern into the future appears to be reasonable. There is little doubt that this prediction will be quite accurate for the near future, say, the next 5 years, but it is obviously far less certain for the distant future. The obstacles to further growth are generally practical rather than fundamental; e.g., because of material absorption, ultraviolet imaging systems are difficult to implement via conventional optical concepts, until the predicted fabrication linewidth dimensions reach interatomic dimensions near 2035, as is discussed below. With sufficient motivation these obstacles may be overcome and the extrapolated pattern may continue into the future much further than may be evident today.

For all its virtues, silicon is not a perfect semiconductor material for all applications. Its bandgap is rather small, limiting its performance at elevated temperatures. Its bandgap is indirect, which inhibits its use as a laser or light-emitting diode (LED) source and makes its ability to absorb, and hence detect, optical photons weak near the band edge energy; this does not mean that silicon

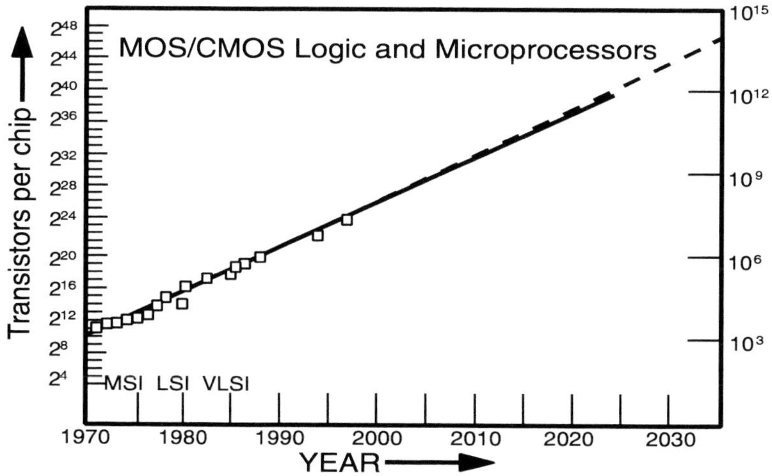

FIGURE 4.2 Historical trend in silicon technology (following Moore's Law). SOURCE: Data points from Yu, Albert, 1996, "The Future of Microprocessors," *IEEE Micro,* 16(6):46-53, Figure 3, December.

cannot be used as an optical detector but rather that it requires a larger thickness than direct bandgap materials for the same detection effectiveness. Finally, the charge carrier mobility and saturation velocities are rather low for silicon compared with some of the other semiconductors, such as gallium arsenide (GaAs), indium phosphide (InP), silicon carbide (SiC), and carbon (C; diamond), seriously limiting the speeds at which silicon devices can operate.

A number of approaches have evolved to address these limitations, some of which are mature or close to maturity and others less so but which show great promise for the future. Other semiconductor systems offer different inherent physical parameter characteristics that can be exploited, as is discussed below, and alternate device configurations, difficult to implement in silicon, are possible in other technologies with a variety of specific advantages.

After silicon, GaAs is the next most mature semiconductor device technology. GaAs transistors enjoy distinct advantages with respect to speed and power over silicon transistors, because of the shape of GaAs's electron velocity versus electric-field curve—an inherent property of any semiconductor material determined by the specifics of its crystal structure and the resulting band structure. The peak electron velocity in GaAs is several times greater than that in silicon and is reached at a much lower value of electric field.

GaAs field effect transistors (FETs) are capable of operation at clock rates as high as 5+ GHz in digital circuits and as analog amplifiers with significant power and gain in the microwave range up to 30+ GHz. High-performance analog-to-digital converters capable of 6 to 8 bits at sample rates of 2 to 3×10^9 per second have recently been implemented in GaAs heterojunction bipolar transistor technology. And for the past decade or so, solid-state microwave developments have been dominated by monolithic microwave integrated circuit (MMIC) technology—based almost exclusively on GaAs.

In addition to its electron mobility-related advantages, GaAs offers a direct bandgap, making it suitable for optical applications as a detector or as a light source, such as an LED or a laser, thereby easing the interface between its microwave and high-speed digital capabilities and the fiber-optical communication links that will be utilized for the transfer of data from some sensors to their associated computational resources. The direct bandgap also helps in that it produces short minority-carrier lifetimes such that undesired electron-hole pairs recombine so rapidly that parasitic effects may be safely ignored in the design of devices leading to simpler structures.

Of course, GaAs, for all its performance advantages, is a much more difficult material system to work with than silicon. Its hole mobility is low, and there is no natural insulating oxide, so that it is not practical to implement the CMOS structures that offer such low-power performance advantages in silicon. Its dielectric constant is almost twice as large as that of silicon, giving it a higher capacitance for the same-area device. Finally, its low thermal conductivity requires very thin substrate thickness for device thermal control, which, when combined with its extreme brittleness, leads to significant fabrication yield losses through handling. Further yield problems are inherent in the fact that as a more complex system than silicon, it is much more difficult to prepare large-diameter, defect-free GaAs substrates.

Bandgap and Heterojunction Engineering

Two of the most significant concepts to emerge from the efforts expended over the past several decades on the development and maturation of GaAs semiconductor device technology are bandgap engineering and heterojunction engineering, both reflections of our increasing abilities to manipulate materials and structures at the fundamental atomic level. The GaAs high-electron mobility transistor incorporates both of these concepts. The bandgap of complex multi-component materials, such as AlGaAs, varies continuously as the proportions of components are changed. In this way it is possible to tune or engineer the material's bandgap to optimize transistor performance or, if the device is an LED or a laser diode, to control the specific wavelength emitted. Through the use of molecular beam epitaxy, metal-organic chemical vapor-phase deposition, and combinations thereof, very thin layers of high-quality, defect-free, properly tuned AlGaAs can be fabricated with electron mobility as much as three orders of magnitude larger than can be achieved in silicon and, when grown on ordinary GaAs, produce beneficial heterojunctions at the interface because of the different bandgaps of the two materials. The resulting heterojunction devices show significantly higher switching speeds as digital circuits and improved power and efficiency performance at high microwave and millimeter-wave frequencies than the more straightforward metal-semiconductor field-effect transistor (MESFET) implementations.

These concepts apply broadly to other material systems, and, in fact, their application to silicon through the bandgap and heterojunction engineering of silicon germanide[3] (SiGe) has produced devices that perform as well as or better than GaAs devices but retain much of the simplicity and ease of manufacturing aspects of pure silicon technology. Indicative of the continuously accelerating growth of technology today, SiGe has moved from a laboratory curiosity less than a decade ago to a commercial reality today. SiGe transistors have already demonstrated cutoff frequencies in excess of 100 GHz, which is better than the best silicon transistors by more than a factor of two and in the same ball park as GaAs and InP technologies. SiGe is an exciting and promising technology and could someday displace GaAs completely.

Next of interest in the III-V family for electronics and MMIC applications is InP. Although behind GaAs in development, InP exhibits even higher electron mobility characteristics than GaAs and also has a direct bandgap that is somewhat smaller than that of GaAs, but just right to permit bandgap engineering of LEDs and laser diodes in the two wavelength bands of most interest to long-distance fiber-optic communication applications (i.e., 1.3 μm for minimum dispersion and 1.55 μm for minimum loss). High-electron-mobility transistor (HEMT) or pseudomorphic HEMT (PHEMT) devices for millimeter-wave applications have shown higher power and efficiency than similar GaAs implementations and are very promising. Unfortunately, InP has many of the same negative features as GaAs technology and currently lags it in overall maturity.

Wide-bandgap Semiconductors

Even more interesting, and still further behind in maturity, are the so-called wide-bandgap semiconductors (WBSs). Although this term is often not used precisely, it generally includes the IV and IV-IV materials and the III-V nitride compounds.[4] The most promising of these are SiC, gallium nitride (GaN), and carbon (diamond), all with bandgaps above 2.4 eV. Because these materials have

---

[3]Cressler, J.D. 1995. "Re-Engineering Silicon—Si-Ge Heterojunction Bipolar Technology," *IEEE Spectrum,* 32(3):49-55, March.

[4]Davis, R. 1991. "III-V Nitrides for Electronic and Optoelectronic Applications," *Proceedings of the IEEE,* 79(5):702, May.

such large bandgaps and much higher thermal conductivity and dielectric breakdown strengths than silicon, GaAs, or InP, they offer promising high-temperature and high-power performance. With low-dielectric constants and high-electron mobility, the prospects for millimeter-wave capabilities are excellent. And as these materials are much harder and stronger than conventional semiconductors, they offer the potential for higher processing yields and lower manufacturing costs. In addition, SiC, GaN, and diamond all offer the potential for blue, green, and ultraviolet light emission and have been proposed for application as LEDs and radiation detectors (visible and ultraviolet) as well as for power and microwave devices (bipolar, MESFET, and IMPATT), thermistor sensors, and high-speed switching devices.

Although SiC was one of the first semiconductors recognized (electroluminescence was reported in 1907), and the electronic properties of diamond were first investigated in the 1930s, most of the progress in WBS technology has come in the last decade, and formidable fabrication obstacles remain. Detailed predictions of SiC and C (diamond)[5] microwave devices have been made by means of basic measured material parameters and device models that correctly predict observed GaAs device performance. Figure 4.3 illustrates predicted performance for SiC and diamond MESFETs compared with measured performances of equivalent GaAs and silicon devices. The GaAs results represent the state of the art for GaAs MESFETs. The predicted performance of SiC and diamond MESFETs is significantly better than that for GaAs, suggesting that at 100 GHz, about 300 mW and 1 W of RF power is possible from SiC and diamond devices, respectively. At lower frequencies, in the more traditional microwave bands, significant power performance is anticipated, i.e., from CW power amplifiers with tens of kilowatts at L-band and below, to 1 kW at S-band, several hundred watts at X-band, and several tens of watts at 35 GHz (Ka band).

The obstacles that remain are related to achieving high-quality, uniform films with controllable properties. The technology for producing single-crystal films of diamond with understanding and control of nucleation, growth, the methods of impurity introduction and activation, and the formation of ohmic contacts with good adhesion is still at an embryonic stage and must be mastered before diamond electronics can become a reality. SiC and GaN materials and fabrication technologies are in better shape and, although GaN is still in a basic research phase, a host of practical devices have already been implemented in SiC. There is little question that, via WBS, high-power, high-temperature, robust semiconductor electronics with interesting potential for short-wavelength optical applications will become available in the next several decades. The growth of micro-

[5]Trew, R.J., J.-B. Yan, and P.M. Mock. 1995. "The Potential of Diamond and SiC Electronic Devices for Microwave and Millimeter Wave Power Applications," *Proceedings of the IEEE*, 79(5):598-602, May.

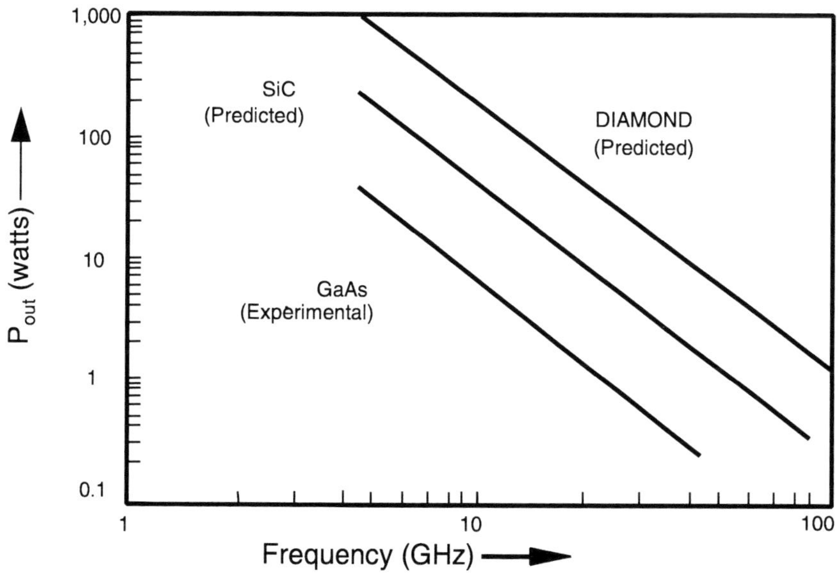

FIGURE 4.3 RF power performance vs. frequency for diamond, SiC, and GaAs MESFETs. SOURCE: Adapted from Trew, Robert J., Jing-Bang Yan, and Philip M. Mock, 1995, "The Potential of Diamond and SiC Electronic Devices for Microwave and Millimeter Wave Power Applications," *Proceedings of the IEEE,* 79(5):598-620, Figure 15.

wave power generation over the past decade and a half is illustrated in Figure 4.4, showing the increase in X-band amplifier power performance.

Extrapolation of these curves into the future and adding the levels of power capability predicted for SiC and diamond suggest that SiC may represent the state of the art by about 2005 and that diamond needs to be fully mature by 2010 to 2015 to be the state of the art at that time. Further progress may simply involve multiple MESFET amplifier chains on a single substrate. With SiC and diamond, it should be possible to maintain the exponential trend out to 2020 and beyond. Of course, the real art at that time may come from technologies not envisioned today—the details are rarely predictable, but it is highly probable that the envelope will persist.

Higher Levels of Integration

Digital Circuits

Semiconductor transistors, particularly those designed for high frequencies, are by nature quite small—the active dimensions of single transistors, whatever their design, are limited by fundamental physical properties of the materials, e.g.,

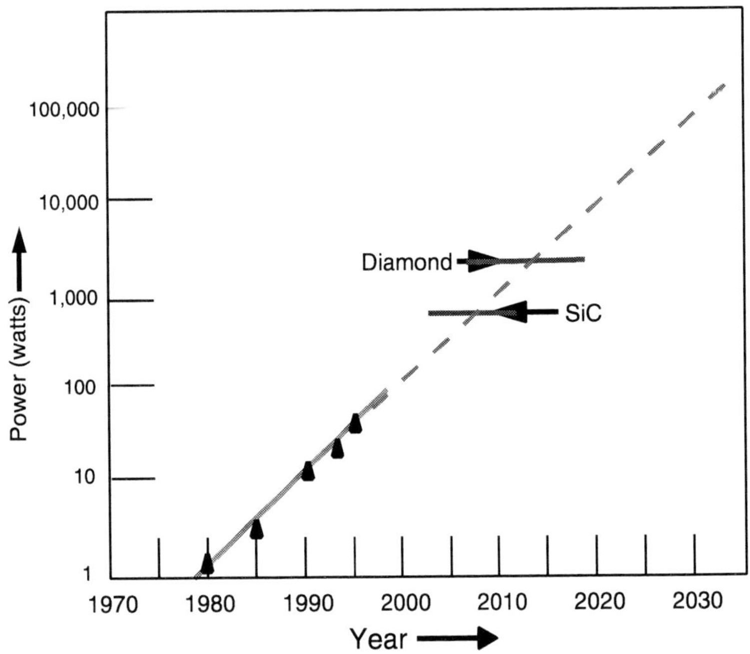

FIGURE 4.4 X-band power amplifier output. SOURCE: Adapted from the U.S. Air Force Scientific Advisory Board, 1995, *New World Vistas, Air and Space Power for the 21st Century, Sensors Volume*, U.S. Air Force, Washington, D.C., Figure 6-6, p. 102 and data shown in Figure 4.3 of this report.

electron and hole mobility, to dimensions typically measured in microns to tens of microns. Combining this feature with the fact that the fabrication techniques employed can frequently be applied uniformly over wafers of many inches in diameter leads to the possibility of fabricating large numbers of transistors on the same small substrate at the same time. In the digital world, this rather obvious extension from single devices to integrated circuits (ICs) took some time to evolve. The first key was to understand how to interconnect the active devices with useful passive components, equally small, that could be fabricated by the same photolithographic, deposition, diffusion, and etching techniques that produced the transistors. The second was to control the fabrication imperfections so that economical production yields of fully functional integrated circuits could be obtained. The final challenge lay in the ability to design and simulate accurately, through software and CAD tools, complex circuits with hundreds to thousands to millions of devices. This improved design and simulation capability was very much a bootstrap operation, as each generation of computer chip enabled the more powerful software tools needed to implement the next generation.

For silicon digital ICs, these endeavors have been quite successful, as is well illustrated by the silicon-technology growth curve (see Figure 4.2) presented earlier. For other semiconductor systems, particularly those with multiple components, although the fabrications techniques are similar to those used for silicon, the systems are generally more difficult to control, and so the maturity of these technologies is significantly behind that of digital silicon. For example, the closest contender, digital GaAs, which has many performance advantages over silicon in speed, in particular, is characterized today by only tens of thousands of gates per IC capability, whereas state-of-the-art silicon technology can produce several million transistors per chip circuit with economical yields. No doubt, some time in the future, when silicon technology's capabilities are saturated—even with quantum devices and single-electron transistors this will happen—other semiconductor systems, such as GaAs, SiGe, and SiC, will catch up and continue the overall digital technology growth envelope illustrated above.

Analog Circuits

Although the same arguments for a high level of integration implementation apply to microwave and optical analog circuits, the obstacles have proven to be different in detail and much more difficult. Although microwave transistors, optical detectors and emitters, and various passive components, including solid-state strip line and waveguide transmission line structures with low loss and good impedance control, can be made by the same microelectronics manufacturing techniques as used for digital electronics, these high-frequency applications cannot approach the level of integration that characterizes digital devices. Not only do these high-frequency applications demand more precise control over dimensions, impedance, and losses, but also the passive components required are physically much larger than those used by the digital implementations. For although integrated, both microwaves and light must propagate finite distances before useful operations can be performed on them by passive structures, e.g., RF inductors and Mach-Zender modulators. Combined with the more difficult multicomponent semiconductor systems, such as GaAs, InP, AlGaAs/GaAs, and InGaAsP/InP, which characterize these applications, the result is that microwave and photonic chips are inevitably limited to device-per-chip densities that are a small fraction of what can be implemented in modern digital silicon technology, i.e., tens to hundreds of components per IC rather than the thousands to millions that characterize today's digital chips.

In spite of this unpleasant obstacle, the microwave and optical ICs are as capable and far smaller and lighter than their conventional equivalents that use discrete components and free-space propagation and, because of the monolithic processing used, offer many cost and reliability advantages. This combination of advantages has already opened up new application areas, the enabling of phased arrays by MMIC technology being a good example. Without doubt, the trend toward smaller size and more capability per analog chip will continue into the future.

Microwave Components

During the past several decades, great progress has been made in applying HF semiconductor technology to the generation of MMIC for a wide variety of useful radar and communication applications—active antenna phased arrays, in particular. MMICs are integrated circuits containing multiple active devices as well as integral passive components such as diodes, resistors, capacitors, inductors, and low-loss controlled-impedance transmission lines, and they perform useful microwave functions such as low-noise amplification (LNA), power amplification (PA), phase shifting, and attenuation. It would be desirable, from a manufacturing cost and perhaps a performance point of view, to be able to fabricate multiple functions on a single MMIC chip—such as a complete transmit/receive module with oscillators, filters, mixers, PA, LNA, circulators, and the like. For many reasons, this is not practical today, and so current practice combines limited-function MMIC chips in a hybrid package, very much like a digital multichip module (MCM), which implies costly, difficult to automate, and often unreliable discrete interconnections from chip to substrate to chip.

The application of MMIC technology to phased-array radar and communications is limited by the cost of individual transmitter, receiver, or transmitter/receiver (T/R) modules. Current efforts are devoted largely to reducing phased-array element costs, i.e., the costs of the T/R MMIC plus the support structure, cooling, radiating elements, and such. To make phased-array radar and communications applications affordable, these costs must be reduced by one or two orders of magnitude. Currently, the total antenna cost divided by the number of phased-array elements ranges in the thousands of dollars. The best approach to affordability would seem to lie in mastering the implementation of single-chip MMIC modules so that low-cost, automated microelectronic manufacturing practices can be applied. This should occur as the technology continues to evolve.

For microwave applications, achieving the highest level of complexity and power-generating capability in the smallest volume possible is not always desirable or necessary. Once a phased-array element can be made small enough to fit the element separation constraints and still give the required performance, it makes little sense to try to reduce its size further if there are no accompanying significant benefits in power, weight, or cost. On the other hand, it may make sense to add additional functional capabilities to enable multifunctional performance (i.e., different functions from the same physical aperture), but this suggests only modest increases in complexity, that is, by the number of separate functionalities needed.

From another point of view, as is discussed below, digital techniques will inevitably move as close to the external interface (antenna) as possible, suggesting the possibility of combined microwave-digital chips of great complexity but with the complexity largely confined to the digital portions. Such hybrids have already been implemented, i.e., MMIC chips with integral GaAs digital control logic on board, although the fabrication of analog and digital transistors differs in details, e.g., impurity profiles, that are often incompatible. Appropriate compromises and techniques will be found, and progress will continue in this direction.

Optical Components

Detectors—Focal Plane Arrays

Semiconductors form natural optical detectors, because incident photons whose energy exceeds the bandgap ($h\nu > E_{BG}$) readily kick electrons from the valence band up into the conduction band, giving rise to measurable electrical responses. The wavelength corresponding to the bandgap is known as the cutoff wavelength (λ_{cutoff}), as all radiation with wavelengths less than the bandgap wavelength will have enough energy to generate a response and wavelengths that are longer produce no response at all. To reduce the effects of thermal noise inevitably present in any semiconductor electronic circuit, the temperature of the detector (T_D) must be kept low enough so that KT_D is well below the photon energy $h\nu_{cutoff}$. In practice, this reduces the necessity to satisfy the approximate relationship, $T_D \cdot \lambda_{cutoff} \sim 550$ K μm. Thus detectors for the 3- to 5-μm IR spectral region must be cooled to about 110 K, which, in practice, implies liquid nitrogen at 77 K. For longer IR wavelengths, say out to 20 μm, 30 K is required, which demands cryo-engine cooler technology. For semiconductor optical detectors, thermal control can be quite a limiting and expensive inconvenience. Fortunately, for the near-IR (< 1.8 μm), visible, and UV regions, optimal room temperature operation is feasible.

Unfortunately, even with appropriate cooling, there is no single, ideal, detector material that gives optimal performance for all IR, optical, and UV spectral bands of interest to sensing. And so, over the years, many different semiconductor detector systems—e.g., various forms of silicon and germanium with different doping for different spectral regions, lead sulfide (PbS), lead selenide (PbSe), indium antimonide (InSb), platinum silicide (PtSi), various compositions of HgCdTe for different spectral regions, and so on—have been employed. Each system requires its own fabrication techniques and today exhibits varying levels of maturity, depending on the specific system. It is no surprise to find that the visible and near-IR regions are currently the most mature as they are typically implemented in silicon.

Although single optical detectors find use in some nonimaging situations, the application of most interest is imaging. The earliest imaging systems employed single detectors in scanning configurations, but with the development of silicon microelectronic IC technology, visible focal plane arrays (FPAs) with multiple silicon detectors in one- and two-dimensional configurations were rapidly developed during the early 1980s. By means of traditional optics, an image is projected on the focal plane where the detector elements respond in parallel, accumulating charges proportional to the image irradiance levels at each detector position. The signals are then read out very rapidly, serially or simultaneously, into some form of readout electronics, for display or storage. The detector elements are then reset and the cycle repeated, typically at video frame rates of 30 Hz or higher. Conveniently, the readout circuitry, often a CCD configuration, could also be implemented in silicon and integrated directly on the FPA itself to form an all-solid-state monolithic imaging detector. These, of course, form the basis for the high-quality and inexpensive video camcorders so widely available today. Sensors that operate in the visible region with millions of picture elements (pixels), e.g., 2,000 × 2,000, are in production now, and formats with several tens of millions of pixels, e.g., 5,000 × 5,000, will soon be possible.

Applying these same FPA concepts to the semiconductor materials needed for optimal performance in the longer IR bands, such as InSb, PtSi, or HgCdTe, is possible but is far less straightforward. Generally, in these technologies, implementing the readout circuitry on the same substrate with the detector elements in a monolithic form is not practical, and so the detector array must be interfaced to an external silicon very large scale integrated (VLSI) circuit chip. In addition, as is the case for digital circuits, it is much more difficult to economically produce large-area structures in these more exotic systems. And so, medium-wavelength IR (MWIR), i.e., 3 to 5 μm, and long-wavelength IR (LWIR), i.e., 8 to 14 μm, FPAs lag the visible somewhat, with MWIR FPAs of 1,024 × 1,024 elements in InSb and HgCdTe and LWIR FPAs of 256 × 256 elements in HgCdTe and extrinsic silicon representative of the current state of the art.

Optoelectronics

Semiconductor optical capabilities are by no means limited to the detection of light but also can be readily configured to generate light at the bandgap energy—that is, at λ_{cutoff}, as LEDs or as laser diodes. With the advanced fabrication techniques, such as molecular beam epitaxy (MBE), metallo-organic chemical vapor deposition (MOCVD), ion beam implantation, and micromachining, miniature high-quality optical waveguides, mirrors, and lenses can be manufactured. Several decades ago these possibilities gave rise to the concept of integrated optics, that is, optical systems that could be implemented in a miniature, monolithic, solid-state form. Although significantly smaller than the conventional free-space optical alternatives, integrated optics implementations are never as small as digital ICs, as the name suggests, because of the necessity of using finite amounts of optical propagation distance to achieve useful amounts of modulation, mixing, diffraction, and so on. The dimensions of integrated optical elements are commonly measured in millimeters or centimeters, rather than in microns, as is the case for digital ICs.

As very thin layer, multiple quantum well (MQW) two-dimensional electron gas and tunable bandgap engineering concepts evolved for the optimization of single transistor performance, these concepts were quickly applied to the generation of efficient LED and laser diode structures for many different spectral regions. The simple discrete laser diodes of the 1960s, with optical feedback provided by polished or cleaved reflective facets at the ends of the active p-n junction region, gave way to sophisticated MQW multilayered structures with gradient index and multilayer distributed feedback (DFB) confinement built in. To meet the needs of long-distance fiber-optic digital communications—the most successful application of optics to electronics so far—CW room-temperature semiconductor lasers, i.e., InGaAsP/InP MQW devices, capable of direct current modulation with bandwidths of tens of gigahertz and with outputs approaching 1 W or so, have been developed for both the 1.3-μm (minimum dispersion) and the 1.55-μm (minimum attenuation) spectral regions of silica fibers.[6] Adding strained layer concepts promises to further improve the power, efficiency, and lasing thresholds of MQW lasers in the future. Because of the growing fiber-optics communication business, commercial sources for 1.3-μm and the 1.55-μm lasers abound.

In addition to long-distance fiber-optic digital-communications applications, there is increasing interest in the possibility of using optics to transfer analog microwave signals within phased-array radars to supply time-delay steering, as well as to distribute the high-data-rate, large volumes of digital data associated with modern digital receiver radar concepts. For these applications, the earlier-developed and simpler AlGaAs/GaAs MQW lasers, which operate in the 0.8-μm spectral region, are ideal, as the increased fiber attenuation and dispersion in this spectral band are of no consequence for the minuscule distances (a few to a few tens of meters) typical of a radar, and the commonality with the GaAs MMIC technology suggests the possibility of fully integrated, monolithic optical-microwave implementations. Interest in these applications should grow in the near future, although the technology of AlGaAs/GaAs MQW lasers has languished up to now in view of its inapplicability to long-haul communications, and there are no commercial sources for these lasers currently available.

The possibilities of combining optical and electronic functions on the same chip bring us finally to the broader subjects of optoelectronics[7] and the optoelectronic integrated circuit (OEIC). The formidable manufacturing technology that has evolved to support digital and microwave microelectronics has enabled optical, electro-optic, and electronic components to be implemented together in monolithic semiconductor form, combining the advantages of each normally disparate element in a single device—the OEIC. Implementing drive electronics on the

---

[6]Lee, T.P. 1995. "Recent Advances in Long-Wavelength Semiconductor Lasers for Optical Fiber Communications," *Proceedings of the IEEE,* 79(3):702.

[7]Higgins, T.V. 1995. "Optoelectronics: the Next Technological Revolution," *Laser Focus World,* 31(11):93, November.

same chip with a laser diode makes an attractive cheaper, faster, smaller fiber-optic transmitter for communication applications and was one of the first OEICs developed. Equally important are the fiber-optic receivers, and it is natural to combine detectors with transimpedance low-noise amplifiers as an OEIC. For long-distance communications, this implies InP digital/analog electronics, whereas for photonic radar applications, GaAs technology is suitable. OEIC chips in GaAs, combining high-bandwidth optical detectors with both matched digital electronics and MMIC amplifiers suitable for extracting digital and microwave information transmitted simultaneously on a single optical carrier, have already been demonstrated in the laboratory. Interest in this technology for the implementation of photonic radars with exceptional properties is running high at the present moment, but formidable practical obstacles still remain, and no photonic-based radar has yet been fielded.

Although the success of the application of optoelectronics to microwave systems is uncertain, there is an area of application that is just building momentum but that seems destined to succeed in the end—free-space photonics for broadband interconnects, e.g., chip to chip or board to board, internal to computers. Progress in computer technology will be of great interest to future sensor systems. Analysis suggests that, for distances of more than about a millimeter and data rates of more than 100 Mbps, less power is required on the chip for photonic interconnects than for electronics. Recent progress with very-low-threshold lasers suggests that the break-even distance may soon be only fractions of a millimeter.

Finally, it has been suggested that optoelectronics,[8] in the form of various device technologies, such as AT&T's self-electro-optic effect device (SEED), offers the potential for logic implementations that operate at the speed of light and that may someday replace the existing VLSI circuit technology on which the digital information revolution is now based. Although the ability of such optical devices to perform logic has been demonstrated, only very simple functions were implemented, i.e., a few gates. Intriguing as this seems, there are good reasons to believe that conventional digital logic cannot be matched by optical logic, the speed of light notwithstanding. The primary problem is that it takes a good deal of effort to generate, modulate, and detect light—which is always done by moving electrons first, as light does not interact strongly with anything else—and this investment must therefore return significant benefits to be worthwhile. For the communication of wideband information over macroscopic distances, whether these be millimeters or miles, the investment is advantageous and optical communications is the preferred path of the future. For implementing logic, however, particularly in view of the shrinking dimensions of digital circuit elements and with the possibility of single-electron transistors on the horizon, optical logic

[8]Udd, E. 1996. "Fiber Optic Smart Structures," *Proceedings of the IEEE,* 84(6):884.

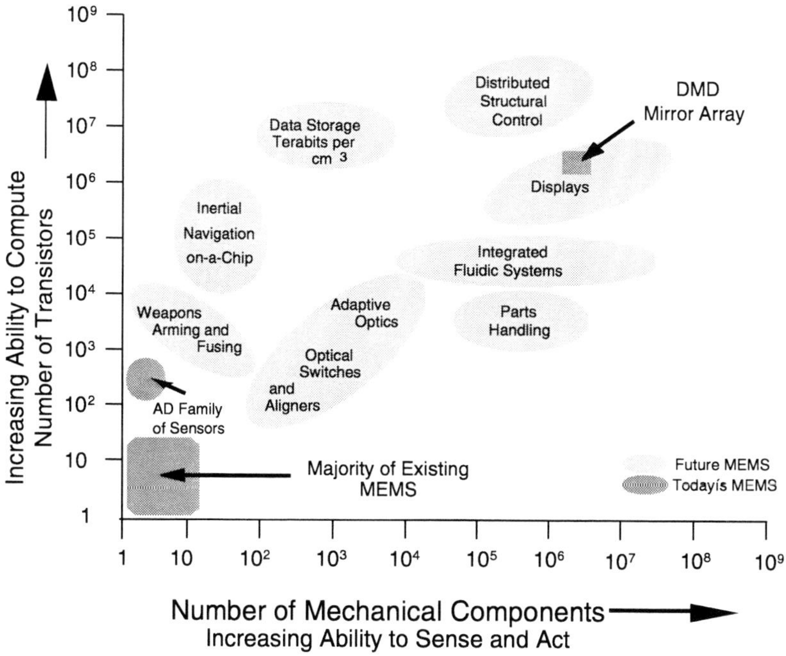

FIGURE 4.5 MEMS technology roadmap. SOURCE: Adapted from Gabriel, Kaigham J., 1996, "MEMS Technology Trend and Roadmap," in the briefing "Microelectromechanical Systems (MEMS)," to the Panel on Technology presented by DARPA, September 13.

seems to require the use of many more electrons for each logic operation than is, or will be, required for the digital alternative.

Microelectromechanical Systems

The final semiconductor topic of interest to the future of sensor technology is MEMS. MEMS technology is an imaginative, but logical, exploitation of microelectronics. Through the use of traditional silicon fabrication techniques, microelectronic circuits and miniature, movable mechanical components with dimensions measured in microns are combined on a single substrate to perform a wide range of sensing and actuation tasks (Figure 4.5). Drawing on the fabrication techniques and materials of microelectronics as a basis, MEMS processes are used to construct both mechanical and electrical components. Mechanical components in MEMS, such as transistors in microelectronic devices, can be fabricated with features that have micron or submicron dimensions, thus enabling the inclusion of millions of mechanical components on a single chip. From the mechanical point of view, silicon is ideal for this application because it is strong and easily fabricated into ridges, valleys, free-standing bridges and cantilevers, cavities, nozzles, membranes, and other structures.

MEMS technology combines the advantages of miniaturization, multiple components, and large-scale integration. Because of the flexibility inherent in the integration of electrical and mechanical components, the potential applications of MEMS technology appear to be limitless. Examples of possible applications include miniature inertial sensing and guidance devices, miniature sensors of all types that can be widely dispersed or gathered into large arrays, and miniature actuators, including steerable mirrors for directed-energy weapons and smart skin structures for aerodynamic flow control. Other specific military uses include inertial guidance for munitions, integrated fluidic systems for biological and chemical analytical instruments and for hydraulic and pneumatic control, miniature DNA detection systems, integrated micro-optomechanical components for displays, IR detector arrays, fiber-optic switches, and vibration sensors for condition-based maintenance. Some commercial applications of MEMS technology are already available:

- *Inertial sensors,* for example, an accelerometer for an automobile air bag sensor—from Analog Devices;
- *IR imaging,* for example, an uncooled IR camera based on a MEMS microbolometer FPA—from Raytheon-Amber; and
- *Projection displays,* for example, a MEMS array where each pixel is controlled by a micromirror—from Texas Instruments.

Although MEMS technology is being actively developed for commercial applications, the defense community cannot rely on the commercial sector to address all of its development needs because MEMS devices are highly application specific.

Superconductor Technology

High-temperature Superconductors

Superconductor technology has shown tremendous potential for application to both ultralow-loss, high-Q microwave devices[9] and to very-high-speed, very-low-power digital circuits—advances that could be incorporated into advanced sensors in the near future through the maturation of high-temperature superconductor (HTS) technology. HTS systems, such as YBaCuO,[10] have superconduc-

[9]Hergenrother, J.M., J.G. Lu, and M. Tinkham. 1995. "The Single-Electron Transistor as an Ultrasensitive Microwave Detector," *IEEE Transactions on Applied Superconductivity,* 5(2):2604.

[10]Tonkin, B.A., and Y.G. Proykova. 1993. "Microwave Properties of Bulk and Thick Film YBaCuO," *IEEE Transactions on Applied Superconductivity*, 3(1):1723-1726.

tor critical temperatures well above that of liquid nitrogen (77 K) and do not require the liquid helium (at 4 K) of the earlier classical systems. Because the superconductor phenomenon is a macroscopic manifestation of quantum mechanical behavior, its properties are often strikingly different from what classical physical intuition suggests. These significant differences in behavior permit the implementation of completely new devices—such as SQUIDs, which have found application in accurately sensing weak magnetic fields—as well as the offering of alternate implementations of familiar devices, e.g., microwave components and digital logic circuits,[11,12] with completely different characteristics, often permitting very-low-power, very-high-speed performance that cannot be obtained in any other way. However, the obstacles to the realization of this potential have proven formidable and with a few exceptions, medical imaging being a noteworthy example, HTS systems have yet to find widespread, practical, commercial application. Contrary to the popular image, the low temperatures required, although sometimes stressing and awkward with the earlier systems, are much less problematic with HTS technology and the newer generations of cryocoolers than are the basic materials and large-scale fabrication issues that remain.

High-performance Microwave Devices

The best known characteristic of superconductivity—the direct current (dc) resistivity of the superconductor abruptly vanishing as the temperature is lowered below the critical temperature—suggests that very-high-Q, i.e., low-loss, superconductor implementations of microwave delay lines, resonators, and filters may be feasible. This is indeed true, although it not nearly as straightforward as might be thought. Microwave losses do not vanish below the critical temperature but do diminish rapidly with temperature and vary with frequency, with the losses diminishing toward the dc limit as the frequency is lowered. The combination of low surface resistance and a frequency-independent penetration depth allows the implementation of compact microwave designs with low insertion loss, large bandwidth, and low dispersion. In the past several years, practical thin-film HTS implementations of dielectric resonators with Qs exceeding 3 million and narrow, tunable high-Q microwave filters, operating at 77 K and capable of handling tens to hundreds of watts of power, have become available. Small-scale integration of receiver front ends and other microwave assemblies, with a few tens of superconductor and semiconductor components on a single wafer, also have been demonstrated but are not yet widely available. Such components, particularly the high-

---

[11]Browne, J. 1993. "Superconducting Circuits Make Practical Strides," *Microwaves & RF*, 32(7):123-126, July.

[12]Van Duzer, T., and C.W. Turner. 1981. *Principles of Superconducting Circuits*, Elsevier, New York.

Q tunable filters, are certain to play a role in systems that utilize digital receiver techniques with conversion at microwave frequencies.

High-performance Digital Circuits

The full realization of the quantum nature of superconductivity in the late 1950s and the discovery of the Josephson effect in the early 1960s led quickly to the exploitation of the quantized nature of the Josephson junction (JJ) response in the form of digital logic. What JJ technology offered for digital implementations was primarily intrinsic switching speed of only a few picoseconds and low signal voltages determined by the superconductor bandgap, i.e., about 3 mV for traditional low-temperature superconductors, which imply power dissipation per junction of less than 1 microwatt.

However, early attempts to develop superconductor computers at IBM were frustrated by material problems associated with temperature cycling failures and by the relatively low speed, i.e., a few gigahertz, latching concepts utilized. In 1985, Russian workers developed an alternative form of JJ logic, known as rapid single-flux quantum (RSFQ) logic,[13] which is much faster, promising 100- to 300-GHz performance capabilities. Rather than switching between a superconducting and a nonsuperconducting state, which the earlier logic used, RSFQ switches superconducting currents between alternate paths, never passing out of the superconducting regime, thereby avoiding the finite transient times associated with the superconducting-to-nonsuperconducting transition. The general concept is familiar from the earlier days of emitter control logic (ECL) semiconductor logic, which utilized the same trick to achieve high-speed performance in exchange for the higher powers associated with never entering the current-off state.

Today, digital circuits and analog-to-digital converters of the RSFQ logic family are under investigation in several laboratories. RSFQ demonstration projects have been built with classical low-temperature materials (superconducting) and, with progress in HTS technology, certainly could be developed into practical special-purpose processors in the near future. A wide variety of ADC of both the Flash and Sigma-Delta[14] architectures have been proposed and implemented to some extent. The first superconductor Sigma-Delta ADC was demonstrated last year, in a niobium-based (low-temperature) technology and obtained a respectable 78-dB spurious-free dynamic range bit over a signal bandwidth of

[13]Likharev, K.K., and V.K. Semenov. 1991. "RSFQ Logic/Memory Family," *IEEE Transactions on Applied Superconductivity,* 1(1):3.

[14]Miller, D.L., J.X. Przybysz, D.L. Meier, J. Kang, and A.H. Worsham. 1995. "Characterization of a Superconductive Sigma-Delta Analog to Digital Converter," *IEEE Transactions on Applied Superconductivity*, 5(2):2453-2456.

at least 5 MHz, through a single-loop modulator clocking at 45 GHz. Performance far better than this is projected for the future.

At present, the most difficult barriers to widespread digital application are the need to fully master the material properties of the HTS systems, which are complex, multicomponent ceramic materials, and the need to adapt the fabrication processes to larger scales of integration than have currently proven practical—a few hundred to a few thousand on a single chip—even in the better-behaved low-temperature niobium systems. Most device demonstrations to date have utilized the low-temperature systems, which remain awkward because of their requirement for more complex cryogenics and are thus less attractive than HTS.

Digital Device Technology

The growth of digital device technology is the single most important factor controlling the foreseeable future of sensor technology, whether it be to the year 2000 or to 2035. The advantages of a digital, rather than an analog, representation for signals or information are many and well known, including immunity from drift, containment of errors introduced by analog-to-digital (A/D) quantization and subsequent digital processing, simpler fabrication requirements (i.e., transistor linearity is not a serious issue), low-power physical implementations, and complete flexibility in defining and changing the processing algorithms, whether they are linear or nonlinear. The march toward a completely digital world has been inexorable over the past several decades. Although most evident today in such devices as the popular cellular telephones, high-definition television (HDTV), and personal computers, the digital revolution is by no means limited to these obvious applications, as is illustrated by the fact that our watches have long been digital, most medical thermometers are now digital, and automobiles, which all run far better and far more reliably than they did 20 years ago, contain multiple embedded digital computers. There is no doubt that all sensors will be digital in the near future—immediately converting the outside world's physics-induced electrical sensor responses to bits for further processing, storage, communication, and display.

Microelectronics

The fundamental characteristics of the growth of digital electronics are most robustly described in terms of integrated circuit fabrication and performance parameters, independent of the semiconductor material systems, the device designs, or the circuit architectures employed or expected to be employed at different time periods to achieve each level of performance. As discussed above, envelopes representing the best of expected technology performance at different times are more reliable as predictors than are attempts to describe what any individual technology option will achieve. This is particularly true for technolo-

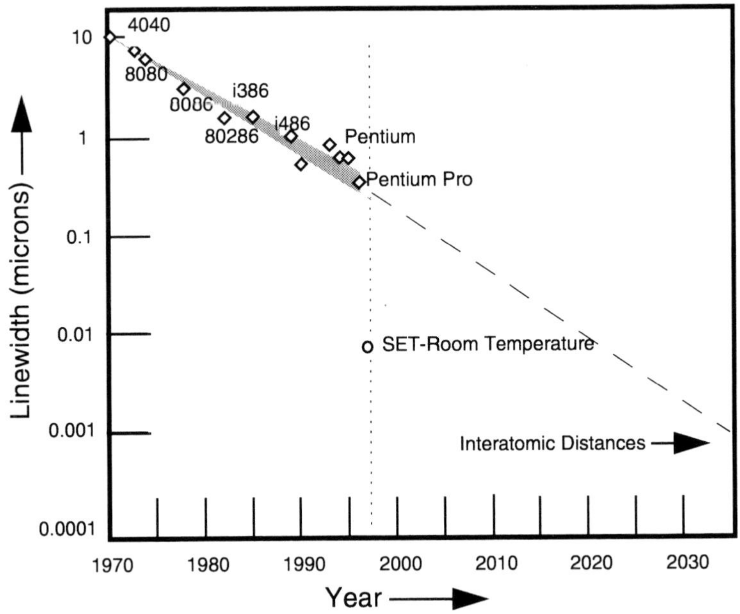

FIGURE 4.6 Microelectronic fabrication linewidth definition. SOURCE: Adapted from Yu, Albert, 1996, "The Future of Microprocessors," *IEEE Micro*, 16(6):46-53, Figure 3, December.

gies that have been advancing exponentially for many decades, as is the case for digital technology. In the absence of contradictory evidence, the panel's best projection for the future is continued exponential advance, recognizing that the advance will likely be stepwise rather than continuous. There is no reason to project lower or higher rates of growth for the future, at least until some fundamental technological barrier can be identified.

Digital technology is best characterized by the minimum achievable fabrication linewidth, the maximum area of defect-free chip that can be economically produced, and the clock speed achievable. The historical growth of minimum manufacturable mask fabrication linewidth projected to 2035 is illustrated in Figure 4.6. In many ways, progress in achievable linewidth explains much of the growth in devices per chip and clock speed—reducing the linewidth definition permits smaller devices to be fabricated, thereby increasing the number of devices we can implement on the same size chip while also encouraging faster switching. Additional benefits accrue from lower voltages and lower power per transistor as the dimensions shrink.

To date, the reduction in minimum linewidth definition shown in Figure 4.6 has been accomplished largely through continuous improvements in optical lithography—an achievement that was totally unexpected 10 years ago and that

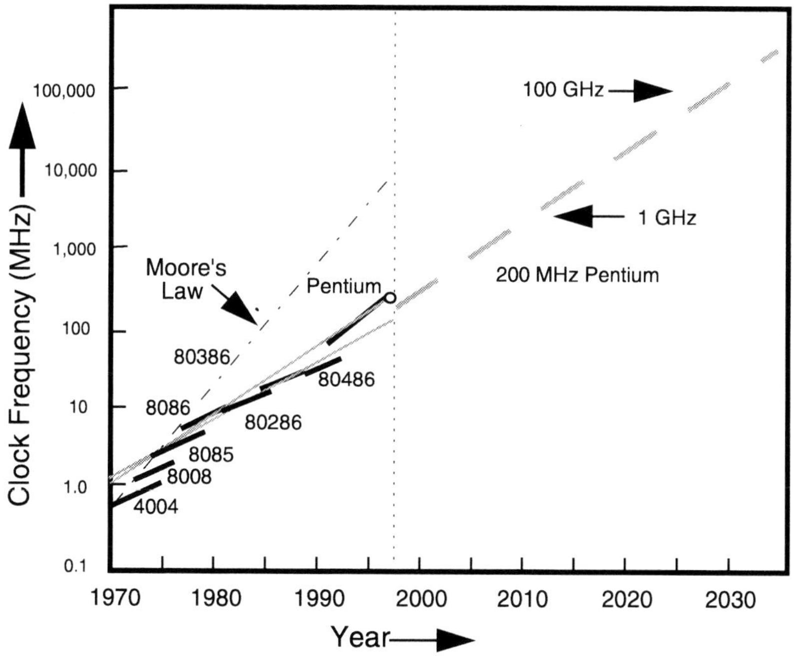

FIGURE 4.7 Clock frequencies of affordable microprocessors. SOURCE: Adapted from Hospodor, A.D., and A.S. Hoagland, 1993, "The Changing Nature of Disk Controllers," *Proceedings of the IEEE*, 81(4):586-594, Figure 10.

illustrates the difficulty of successfully projecting the growth of any specific technology. Although more expensive but finer-resolution x-ray and e-beam technologies have long waited in the wings, optical lithography has continued to maintain its dominance through shifts to shorter wavelengths and the introduction of subtle phase manipulation concepts.

Although data exist for the individual trends, the history of microprocessor clock speed, shown in Figure 4.7, provides an excellent summary of the overall growth of digital capabilities. A simultaneous measure of intrinsic switching speed as well as of the ability to manufacture useful complex circuits, microprocessor clock speed has been increasing at the rate of a factor of two every 2 years since the early 1980s. There is no evidence to suggest that this trend will not continue.

By 2005, the clock speeds of affordable (desktop) microprocessors are predicted to exceed 1 GHz—which is not surprising given that PCs with 100- to 200-MHz clocks are already common. No single approach can accommodate all this growth, yet the specific technology that may achieve the best performance at any particular time is not predictable with confidence. Today's silicon circuitry will not move significantly into the gigahertz range. New technology will appear in the form of new, more capable device concepts, such as quantum dots and wires, and/or new material systems. GaAs is already capable of 10-GHz rates, and SiGe promises similar performance, in an easier-to-fabricate material system. InP devices could possibly operate at 100 GHz, which is anticipated near 2030. Well before this, superconducting RSFQ logic, with its 100- to 300-GHz potential, should become available in HTS technology. By 2035, the end of the time horizon of this study, currently envisioned RSFQ technology may approach fundamental limits and the technology of choice may have shifted to superconductor quantum dots or a family of devices yet to be invented.

Anticipating technology limits is generally a very difficult task, particularly if the deficiencies are of a practical rather than a theoretical nature. Linewidth, in contrast to other significant technology parameters, may encounter some fundamental obstacles to continued exponential improvement. The most obvious is that, although some day it might be possible to manipulate single atoms on the surface of a wafer, it seems to make no sense at all to discuss dimensions smaller than an atom or molecule. Checking the linewidth dimensions expected by 2035, one finds from Figure 4.6 an estimated fabrication capability of 1/5 to 1 nm, which corresponds to only a few interatomic distances for a silicon surface. Because it is not expected that linewidth fabrication technology will reach this absolute limit by 2035, some as yet unknown sophisticated techniques will have to be developed to permit manufacturing at these levels.

New Physics—Quantum Devices and Nanoelectronics

New types of elementary devices[15] based on different physics may become available before atomic dimensional limits are encountered. The exploitation of quantum resonances in one dimension has already found its way into practice through the HEMT or modulation-doped field-effect transistors (MODFETs) and single and multiple quantum well (SQW and MQW) semiconductor lasers.

Precise control of thin-layer deposition to dimensions as small as 5 or 10 nm has been common in semiconductor manufacturing for years. The resulting one-dimensional resonances result in quantum confinement in the vertical dimension, leading to two-dimensional electron gases with a number of useful attributes, such as energy level splitting with restricted energy transfer between levels and a constant density of states.

With the ability to fabricate horizontal structures lithographically with dimensions of 50 to 100 nm, it becomes possible to provide quantum confinement in any or all three dimensions, leading to a variety of quantum plane, wire, and dot configurations. Because of the quantum effects, the number of electrons in

[15] Grabert, H., and M.H. Devoret. 1992. *Single Charge Tunneling,* Plenum Press, New York and London.

the well of a quantum dot is quantized to an integer number of electrons, and even if the number is as large as tens to hundreds, single-electron changes can be observed.[16] Combining quantum dots with very thin insulating layers, single-electron tunneling transistors have been implemented, and the possibility of single-electron logic is on the horizon. Although ingenious designs are being proposed, the less-than-unity voltage gain that characterizes the devices tested to date remains a fundamental unsolved engineering problem.

Research in this promising discipline has been active since its formulation in the mid-1980s. Even quantum molecules—i.e., several closely spaced quantum dots interacting via quantum mechanical tunneling—have been explored. All these single-electron semiconductor concepts apply equally well to superconductor implementations with a single Cooper pair replacing the single electron. The work to date has provided a sound theoretical and experimental basis for the underlying physics. The challenge remaining is to come up with practical, working, manufacturable devices in time to keep up with the exponential growth of digital electronics projected for the time frame beyond 2020. It seems likely that the transition from microelectronics to nanoelectronics and its future high-speed devices will depend on single-charge tunneling effects and the single-electron transistor. The Department of the Navy should pay close attention to this critical technology.

Analog-to-Digital Conversion

Soon, for practically all sensor applications, the detected signals will be digitized as close to the physical world interface as possible. The benefits of a digital signal representation over analog are familiar—fewer, easily miniaturized components for lower size and cost, immunity to component drift, control of computation-induced errors, and great algorithm flexibility. The translation of these analog signals into digital form with adequate signal bandwidth and dynamic range is of critical importance.

Conventional high-performance ADCs are strictly limited by the timing jitter associated with the comparators. Figure 4.8 summarizes the performance of available and projected ADC devices. No conventional ADC performs significantly above the limit indicated for a jitter level on the order of 1 ps root-mean-square.

Analysis indicates that the so-called jitter limit can be succinctly expressed by a single constant value of the product of dynamic range times sampling frequency, that is $(2^{(B+1)} - 1) \cdot f_{sample} = 8 \times 10^{11}$ = 1 ps jitter limit, representing formally the familiar practical tradeoff between these two parameters.

[16]Singh, Jasprit. 1994. *Semiconductor Devices: An Introduction,* McGraw-Hill, New York.

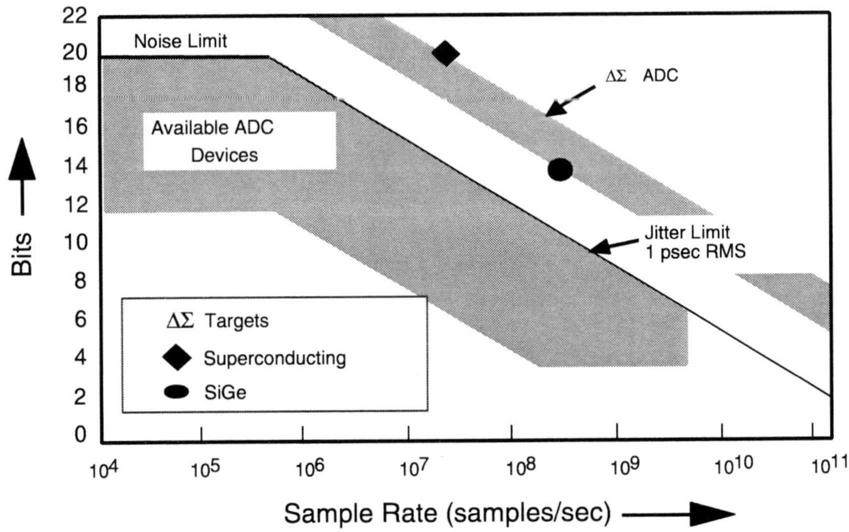

FIGURE 4.8 State-of-the-art analog-to-digital converters. Current development of Delta-Sigma (ΔΣ) devices will add up to 6 bits by the year 2000.

In recent years, although the region of available ADCs has been expanding systematically into the higher sampling rates through developments in GaAs and InP technology, progress across the jitter boundary has moved at a glacial pace—by some estimates, it has been taking about 6 years to gain a single bit. A different concept has been sought—one that avoids the jitter issue entirely, if possible.

Such an approach, the Delta-Sigma (ΔΣ), was already available but as yet unused for RF sample rates. This technology was developed first for audio applications in the 1970s and 1980s and is inherent in the digital audio that underlies CDs and the like today. As a result of this long history, the concepts are well understood today and have proven robust as predicted, permitting high-dynamic-range implementations with inexpensive, modest-to-low-quality components. Today, for the first time, these proven concepts are being extended from megahertz to gigahertz rates and promise to make digital RF systems practical.

In this approach, the signal is greatly oversampled at a rate that permits a valid 1-bit representation of the difference between successive samples, and the resulting digital representation is passed back through a simple digital-to-analog (D/A) conversion and subtracted from the incoming analog signal to produce a form of differentiation. The resulting filter loop structure is used to move the quantization noise away from the spectral regime that contains the signal information in such a way that when the signal and related quantization noise are passed through a band pass filter, the reduced quantization noise levels are equivalent to having digitized the signal directly in the conventional manner with multiple-bit accuracy. The output digital data stream is then reduced to the rates appropriate to the information bandwidth of the signal. Depending on the order of the loop structure, the gain in effective number of bits can be 1.5, 2.5, and 3.5 bits, and so on, per octave of oversampling for 1, 2, 3, and so on, pole filters. In strong contrast, the conventional approaches to multibit A/D, which do not try to shape the noise, achieve only a one-half bit per octave increase in effective bits with oversampling.

The loop averages out jitter effects in exchange for new problems—e.g., the higher the order of the loop, the more effective the bits per octave, but the more difficult it is to keep stable. The ΔΣ approach will take us rapidly beyond the jitter barrier of existing ADC for the larger information bandwidths and dynamic ranges, characteristic of high-performance sensors of interest in this study. Successful development of gigahertz ΔΣ technology will produce a distinct jump in ADC performance in the near future. In view of the projected increases in clock speeds, this technology can be expected to provide several bits, i.e., 3.5 per octave for a three-pole loop, every 2 years or so, as additional octaves of oversampling become available through the continuing growth of digital clock speed capabilities.

In Figure 4.8, the points labeled ΔΣ (Delta-Sigma) represent the performance targeted for several noncomparator ADC architectures currently under development for RF sampling rates. Included are one to be implementated in SiGe and one each to be implemented in high-temperature and low-temperature superconductor technology. Both superconductor efforts are addressing the same performance—20 equivalent bits at 20 MHz.

Packaging and Interconnections

Packaging and interconnections, often underestimated or even ignored, are of enormous importance. It does little good to develop powerful digital chips if they cannot be effectively and reliably packaged and interconnected. Some digital circuits already are limited in performance more by the packaging and signal interconnection characteristics than by the basic clock speeds of the individual chips. Performance and packaging are interdependent, and in recent years these interdependencies have grown so strong that it is becoming increasingly difficult to separate the devices from their packaging. Future generations of sensors will utilize very-high-speed, large-area, extremely complex logic and memory chips that must be interconnected and packaged within severe constraints on volume, weight, power-handling capability, and system operating speeds. The digital electronics will be completely integrated into a single unit, and its design and physical implementation, including packaging, will have to be approached from the point of view of a single entity.

Although monolithic or full wafer-scale integration has been a dream for many years, current practice targets the more attainable goal of hybrid wafer-scale integration. In the MCM approach, bare chips—that is, IC chips without packaging of any kind—are attached to a wafer or wafer-like substrate that contains the passive circuitry to supply the chip-to-chip interconnections, very similar to a small backplane or printed wiring board. Starting with two-dimensional configurations, MCM technology now includes consideration of three-dimensional stacked structures that, although providing the desired small sizes, bring up serious challenges in heat removal, z-axis signal interconnectivity, and test and repair. Since such complex modules are both expensive and very difficult to access internally once fabricated, the sensor architecture should be designed with fault-tolerant elements providing resource redundancies and autonomous reconfiguration capabilities.

Many MCM options have been developed, utilizing a wide variety of substrate materials, chip-to-substrate attachment approaches, electrical-signal interconnect technologies—both in-plane, from chip-to-chip, and vertically, from substrate-to-substrate—and thermal control techniques. As yet, no single MCM approach clearly dominates, that is, is optimum for all applications. It is too early to tell if future packaging will evolve as a refinement of the MCM hybrid concepts or whether fabrication practices will eventually permit monolithic, wafer-scale implementations, or whether something else entirely unexpected will appear.

As clock speeds increase, moving signals from place to place without significant loss or distortion becomes increasingly difficult. Currently, within computers or other digital processors, high-speed signals are routed electrically via controlled impedance lines with impedance-matched connections. Only between widely separated computers are high-bandwidth modern fiber-optic communication links routinely employed. Optical interconnects exhibit much less signal loss with propagation distance than does an electrical interconnect and, because of the high frequency of the optical carrier, easily support very-large-bandwidth signals. However, there are serious efficiency penalties associated with modulating the electrical signal information onto and off the optical carrier that are not present in the direct electrical approaches.

In the future, as clock speeds and information bandwidths continue to grow, free-space and fiber-guided optical interconnects will certainly become the preferred way to transfer the ever-growing number of high-speed signals from board to board or even chip to chip.

Whatever the details, it is extremely important that packaging and interconnection considerations so critical to the success of all future high-performance sensors be addressed and be factored into chip design from the beginning, as part of the overall architecture. These technical factors have to be an integral and explicit part of the Navy's future R&D plans.

Computer Technology

Little need be said here about the future of computer technology and its impact on future sensor capabilities. The exponential growth of computer capabilities can be expected to continue throughout the many decades of interest to this study, fueled by advances in device technology, architectures, and software. As the systems grow in the direction of ever larger distributed networks of interconnected multiprocessors, the implications for reliability must be carefully taken into account and both fail-safe and fault-tolerant mechanisms included so that a single failed component cannot pull down the whole network or completely invalidate the outputs.

A history of the growth in throughput capabilities for both supercomputers and affordable (desktop) computers is described in more detail in Chapter 2 of this report.

Perhaps the most interesting possibilities lie not so much with the large, multicomponent computers certain to be needed for powerful sensors systems, such as SAR radars and sonars, digital adaptive beamforming phased arrays, multisensor surveillance networks exploiting data fusion, and so on, but rather with the potential for combining, or in the extreme, integrating, monolithically on the same substrate, sensors with an ADC, a digital processor, and a communication output port to make a complete sensor system on a chip.

Inevitably, multiple copies of these smart sensors will be configured into intercommunicating networks that will act together to perform as a single composite metasensor, sharing, combining, and fusing the individual sensor's data into a total sensor data view that is more powerful than just the sum of the parts. Adding mobility to the individual sensors, useful ant-like societies of intercommunicating microrobot sensors, functioning as large distributed, spatially adaptive, extraordinarily capable surveillance systems, no doubt will follow.

Algorithm Technology

To exploit the enormous opportunities that the exponential growth of digital computing capabilities will continue to provide in the future requires algorithms. For some sensor applications, many years of effort have already been applied and effective algorithms are known to exist. Some of these algorithms find application today, but others have not yet been introduced into practice because the computer or digital signal-processing support available for operational deployment is inadequate to permit real-time implementation. That is, the real-time configurations permitted by today's technology are too large, require too much power, or are simply too expensive for the intended application. The future will certainly change this picture dramatically.

Adaptive Processing

As available computational resources grow beyond the simple requirements of the sensor, whatever that may be—radar, sonar, and electro-optical—it is natural to apply the additional computational power to improving sensor performance by continuously optimizing its processing in response to the observed data. Several easily implemented analog adaptive techniques have long been employed in radar and sonar—constant false alarm (CFAR) detection, with an adjustable detection threshold to prevent the processing of false alarms from overwhelming the real-time capability of the system, and sidelobe cancellation utilizing an adjunct omnidirectional antenna, to distinguish between main beam and sidelobe returns.

With enough digital-computing resources, much more progress can be envisioned. Interfering signals, which are localized in space or frequency or both, can be estimated from the measurements and explicitly subtracted or canceled to improve the detection of the desired target signals. For array antenna systems, performance in the presence of jammers can be greatly enhanced by estimating the direction of the interfering sources and choosing the weights to be applied to the signals from each element such that a null is placed in the antenna pattern in the direction of the interfering signal. For computational reasons, to date these ideas have been applied only to small antennas, i.e., a few to a few tens of elements, or to very-low-frequency radars with several hundred elements—the Navy's operational, relocatable over-the-horizon radar (ROTHR) uses digital beamforming, for example. As it will soon be practical to digitize the return signals on each receive antenna element close to, or at the front end of, a radar, even at X-band and above, digital beamforming of large arrays with hundreds to thousands of elements can be expected in the next 5 to 10 years for a wide range of applications. These possibilities will be developed further under digital radars.

For clutter suppression, similar concepts apply, with the exception that although jammers are typically spectrally complex but well localized spatially, i.e., point sources, clutter is distributed in both space and spectrum. In the particular case of airborne radar, the ground-clutter returns have Doppler offsets that are determined by the aircraft velocity and by the angle between the line of sight to the patch of ground and the aircraft's velocity vector. The solution is thus to introduce spatial nulls into the received beam patterns separately for each Doppler bin. This approach, known as space-time adaptive processing (STAP), is obviously computationally demanding and has yet to be applied to any operational system but is certainly a candidate for the near-term Joint Strike Fighter (JSF) generation of airborne radars.

One of the most intriguing aspects of digital adaptive beamforming is the potential for enabling the use of less-than-perfect radar antennas or optical configurations that can be digitally corrected to provide near-optimal system performance. In this way, digital processing can be used to alleviate difficult manufacturing issues (associated with high-precision optical components) or to permit antennas to be optimized for other purposes, such as reduced signature.

Automatic Target Recognition

As the volume of sensor-provided data increases and the operational time of advanced weapon systems decreases, automatic target recognition (ATR) becomes mandatory—no human can provide the necessary decisions fast enough. Although certainly challenging, ATR has acquired a reputation for impossibility that does not reflect the facts. It is the case that, given two- and three-dimensional image data visually displayed, and enough time, a trained human is unbeatable in many recognition tasks. Consider, for example, the task of finding a familiar face in the crowd without knowing in advance who it is going to be: A human can do this if the face is familiar to the observer; a computer cannot. But this does not mean that computer-based ATR is impossible. And it does not mean that ATR will not outperform the human in some circumstances.

Several effective ATR applications have already been demonstrated—search-and-destroy armor munition (SADARM) recognizes its specific target automatically through millimeter-wave passive imagery, brilliant antitank (BAT) munition uses a combination of acoustic and IR imagery, and the tactical LIDAR seeker uses LIDAR three-dimensional range-to-pixel imagery to automatically detect and recognize a variety of military targets, such as tanks, trucks, and bridges. In each case a simple form of ATR, tuned to the characteristics of the scenario and simplified to permit real-time implementation, was employed.

Increasing the sophistication and effectiveness of the ATR algorithms faces several obstacles. These challenges for the future include the following:

- First, there are conceptual issues in understanding just what features of the data and what kind of a reference database the human brain uses to achieve superior performance in certain tasks. In a way, this represents the ultimate challenge, and its solution, however incomplete, will no doubt extend well beyond the 40-year window of this study. On the other hand, a computer operates differently from the human brain, performing exhaustive operations far faster and with fewer errors. Just as for the game of chess, completely understanding and imitating the algorithms used by the brain is not a prerequisite for creating a computer program that, with rather brute force techniques, can beat the best human players. In all probability, this will also be true for ATR, and effective algorithms, quite different from those humans use, will be developed.
- Second, many times the ability to perform ATR with a high probability of success and a low probability of false alarm is limited by the information available, rather than by the algorithms employed. If the measurements from an actual target do not differ in any significant way from those of an uninteresting part of the background or from other objects in the scene, then it will not be possible to

distinguish them and ATR will fail. The obvious approach to resolving this kind of difficulty is to collect more information so that significant differences can be distinguished. This represents one of the common technology trends discussed above. The approach, which promises to significantly improve ATR performance under these circumstances, is to collect multidimensional signatures and combine the information. That is, instead of relying on a single sensor operating on a single spectral band, provide multiple sensors, co-located or separated, that can collect data on many different spectral bands and provide some form of data fusion to properly register the multiple sources of information and to resolve any apparent contradictions. This is simple to describe, but difficult to implement. Not only are multispectral sensors required—and there may be serious physical conflicts at the interface to the outside world if the spectral bands are sufficiently far apart (such as RF and microwave, or RF and IR)—but also advanced signal extraction and data fusion algorithms need to be developed, and the overall computational complexity will inevitably become enormous.

Fast Algorithms

Whatever the algorithm, the fewer the operations required to implement it, the fewer are the computer resources required and the more practical the algorithm becomes. Many effective algorithms—singular value decomposition (SVD) is a good example—are languishing today because the computer resources have not yet met their requirements. The development of a fast algorithm can change everything. The best known example of a fast algorithm is the fast Fourier transform (FFT) that, by exploiting the many symmetries of the discrete Fourier transform, reduced computational requirements without affecting the accuracy of the results. Of course, the growth of computer capabilities will eventually enable brute force implementation of any algorithm, but a fast form will always offer advantages by freeing up computer resources for other tasks. The creation of fast algorithms should remain a focus of research during the period from 2000 to 2035 and beyond.

New Algorithms

From time to time completely new approaches to signal or information processing arise and should be watched and carefully assessed for potential application to sensors. Some such advances represent the development of new mathematics, such as wavelets and chaos, which have given rise to interesting concepts in multirate and nonlinear signal analysis with potential for superior sensor performance. Other advances can arise from attempting to understand the superior performance of some biological systems, such as the sonars of bats or dolphins or the navigational abilities of birds or sea turtles. Enhanced sonar performance already has been demonstrated using biologically motivated (e.g., bat) signal processing.

Information and Data Extraction

If sensor technology progresses as envisioned, with multiple, multispectral sensor systems and increasing rates of data sampling, the amount of data collected could easily overwhelm the user. In many ways, too much data is equivalent to no data at all. In either case, no useful information is forthcoming. Although ATR goes a way toward resolving this dilemma through the processing of raw data into compressed information—e.g., target type Q, detected at position x-y, with parameters {a, b, c, ...}, and so on—not all sensor-produced information is a simple list of targets. It may be a picture of what the sensor is seeing, yet what is important differs from one user to another. The challenge for the future is to develop information-identification and information-extraction algorithms that can identify and encode the interesting information in the data, whatever that might be, so that redundant or uninteresting portions of the data may be deleted or so that the full data set may be characterized such that only those subsets that contain a particular form of information need to be examined in full detail. Better means of identifying information contained in large sensor-generated databases are required, as are better search strategies to efficiently exploit this information.

Data Compression

The final algorithm challenge lies in the necessity to communicate between sensors and from sensors to the users. The reality is that communication bandwidths, particularly those in an active battle zone, will always be severely limited and certainly not capable of moving all of the raw data generated by all the sensors everywhere. The data must be compressed for efficient communication of the significant information without losing important portions or introducing artifacts or false alarms. A major portion of the solution to this communication problem lies in the development of information- and data-extraction algorithms as discussed above.

More straightforward are the more mechanical aspects of encoding the data to be transmitted once the critical information has been extracted in minimal form, although there remain some unresolved issues. If the data is to be transmitted in completely uncorrupted form, it can only be compressed by small factors of two or three by lossless encoding, and this gain may be seriously compromised by the need to apply heavy transmission redundancy encoding if the channel is unreliable, i.e., noisy or jammed. For this situation, the real gain is going to be in information- and data-extraction algorithms.

On the other hand, if imagery to be examined by the user is to be sent, lossy compression—sometimes by factors of 40 or 50 to 1—may be acceptable. Experiments with medical imagery at these levels of lossy compression have shown little or no difference from the original images in the diagnostics resulting from these images. For two-dimensional images, efficient and effective compression schemes, which exploit the local continuity of the images, have been developed using various transform techniques, including Fourier transforms and wavelets. For higher-order images, that is, higher-dimensional data sets, such algorithms need to be formulated and proven. This is another important and promising area of research for the future.

Technology Vulnerabilities

For many of the technologies just discussed, improved performance carries with it the potential for enhanced or additional vulnerabilities—to deliberate countermeasures or perhaps simply to the environment. Table 4.2 summarizes the most obvious ones. The sources of most of these weaknesses lie with the twofold trends toward low-voltage electronic implementations and complexity.

From the device point of view, as fabrication linewidths decrease, operating voltages and signal levels also decrease in order to avoid excess electric-field conditions, making the electronics more sensitive to noise and pickup effects also known as electromagnetic interference (EMI). This suggests that future, high-performance smart sensors may be quite vulnerable to weapons that employ EMP generation, unless careful consideration is given from the start to providing inherent shielding as part of the packaging. Similar issues arise in semiconductor systems that exploit superconductor bandgap phenomena, such as SQUIDs and RSFQ logic devices, and as a result use voltages comparable to the bandgap, i.e., millivolts.

From the system side, as individual digital devices grow smaller and cheaper and simultaneously more computationally capable, the systems grow more complex. With increasing clock speeds, storing and distributing the data become

TABLE 4.2 Technology Vulnerabilities

Technology	Vulnerabilities
Low-voltage devices	Electromagnetic interference susceptibilities
Superconductors	Cryo dependencies
	Very-low-voltage circuits, EMI
Computers	Intrinsic reliability, fault tolerance
Software	Intrinsic reliability
	Latent faults, correctness
	Viruses, time bombs
	Malignant applets or beans
Nanotechnology	Tasking, networking, fail-safe operation

BOX 4.5 Classes of Individual Sensors

- Electromagnetic
  Radar
  Passive radiio frequency
  Electro-optics
  Laser, LIDAR
- Acoustic
  Sonar
  Seismic and vibration'
- Inertial
- Chemical and biological
  Time-of-flight micromass spectrometer
  Microelectromechanical systems
- Other
  General local
  Fiber-optic

increasingly challenging. Multiprocessor configurations introduce new issues of coordination and reliability—the systems become increasingly opaque, i.e., harder to understand and to confirm or guarantee correctness, while simultaneously becoming easier to upset by malicious actions. The cooperating societies of sensors envisioned for the future involve even higher levels of complexity and additional communications vulnerabilities, and also raise difficult questions of sensor autonomy, fail-safe behavior, and the potential for insanity—e.g., could a sensor fail, or be induced to fail, to the point of turning a weapon against its owners?

These generic technology vulnerabilities must be addressed continuously during the development of advanced sensor systems and included explicitly in the assessments of the proposed sensor's utility.

INDIVIDUAL SENSORS

Having discussed the state of the art and the observed growth patterns of the five technologies identified as generic to all sensors, the panel now describes the advanced sensors known or thought to be critical to future naval applications that these technologies enable. Each of the individual sensor classes summarized in Box 4.5 is discussed briefly below, with the implications of the general technology trends and projected growth factored into the context of each sensor class.

Electromagnetic Sensors

Radar

Radar, with its all-weather, long-range capabilities for detection and tracking, is the primary electromagnetic sensor in the Navy's tool box and promises to be useful throughout the time horizon of this study. Most of the common technology trends identified earlier are exhibited by radar technology and provide excellent guidance to the potential future capabilities and applications of this key sensor class.

Phased Arrays

For many decades, radars have been evolving toward distributed phased-array configurations. Given the proven flexible, rapid beamsteering capabilities of electronic scan, with few exceptions new radar implementations—whether surface based or airborne—are envisioned to be phased arrays. Complementing this transition to phased arrays has been the steady replacement of classical microwave tubes and waveguide-based analog components by solid-state microwave equivalents—oscillators, low-noise and low-power amplifiers, phase shifters, complete T/R modules, and the like—in MMIC technology.

Monolithic Microwave Integrated Circuit Technology

For the past decade and a half, as the technology matured, MMIC-based T/R modules have increased steadily in performance while declining in cost, albeit more slowly than desired. With GaAs technology, the output power per module has grown exponentially in the common radar bands, while simultaneous exploitation of decreasing fabrication linewidths combined with new material systems, such as InP and soon SiGe, has extended the operating frequencies through the millimeter-wave bands to 94 GHz and higher. Continuing exponential increases in the performance of individual devices are expected as wide-bandgap material technologies (e.g., SiC, GaN, and diamond) mature.

Millimeter-Wave Frequency

Operating at millimeter-wave frequencies, however, brings with it more serious obstacles to phased-array implementations than simple power generation. At 94 GHz, a millimeter-wave frequency that coincides with a spectral region of minimal atmospheric absorption, the wavelength is only 3 mm, and lower-frequency discrete structural-assembly techniques for half-wavelength separation become difficult, if not impossible, to implement in any straightforward manner. New concepts for continuously distributed phased-array implementations become necessary—calling for higher levels of integration for which the distinction between the array elements and the packaging ultimately should vanish. In spite of the inherent implementation difficulties, precisely because of the small wavelengths, millimeter-wave multibeam antennas offer the potential for generating high-resolution two-dimensional imagery from conveniently small physical antenna. The use of passive millimeter-wave imaging systems offers interesting, perhaps even break-through, possibilities for foliage- and camouflage-penetration.

Module Costs

The overall affordability of a phased-array radar is particularly sensitive to the cost of an individual T/R module. To suppress the large spurious sidelobes known as grating lobes, phased-array elements are usually distributed uniformly across the face of the antenna with an element-to-element separation of not more than a half wavelength. On the other hand, good transverse resolution often requires aperture dimensions that are hundreds of wavelengths across. The result of these two conditions is arrays with hundreds to thousands of array elements. Unless the cost of each element is relatively small, phased arrays may simply not be affordable for many applications. For example, today the cost per element (i.e., total antenna structure cost divided by the number of T/R modules) of an X-band T/R module similar to what is used in the Army's ground-based radar is between $1,000 and $2,000, and so the total cost of such a phased-array antenna (~ 20,000 elements) will be millions of dollars. Current efforts target reduction in production costs of an order of magnitude through a combination of increased integration and the manufacturing learning curve. A $300 T/R module has long been the goal, but the current high module costs inhibit the purchase of large, phased-array radars, which in turn reduces the number of T/R modules to be manufactured, thereby limiting the manufacturing learning experience—a chicken and egg situation. There seems little doubt, however, that this situation will eventually be resolved in favor of solid-state phased-array technology and costs will gradually fall into an acceptable, affordable range. Addressing these cost issues over the next several decades as the technologies continue to evolve exponentially should be a Department of the Navy priority.

Multifunctional Systems

Another important trend in radar technology today is the drive toward multifunctional system combinations—operating from the same platform, perhaps even through the same physical aperture. On shipboard or airborne platforms, this could be a combination of different radars for surveillance, tracking, or missile illumination, with RF communication links at different RF frequencies and various wideband EW systems, as is envisioned in the new advanced multifunction RF system (AMRFS) program recently begun by the Office of Naval Research (ONR).

The benefits of such combinations are clear in terms of reduced real estate, opportunities for electromagnetic signature control, and perhaps lower costs, if components can be shared. The obstacles to achieving these advantages lie primarily in the nature of the interface of the sensor to the outside physical world.

Although the internal electronic and digital details of the sensors can be altered fairly arbitrarily, little can be done to change the interface of a radar sensor to the physical world. To project beams, the conditions at the outward surface of the antenna must match the boundary conditions for Maxwell's equations corresponding to the desired mode of propagation, hence the need for half-wavelength array element separations and for certain aperture dimensions. If different spectral ranges are involved, conflicts often arise because of the spectral dispersion of the properties of the materials used to implement the interface aperture—for example, materials that are completely reflective at one frequency might be partially tranmissive at another. Or, if several RF signals of differing spectral bands are passed simultaneously through an amplifier, large unwanted harmonics may be generated through inherent nonlinearity of the amplifier. In other configurations, if different types of RF systems are combined in close physical proximity because of the propagation characteristics of electromagnetic waves, it becomes extremely difficult to ensure that power radiated from one system's aperture or transmit elements does not couple adversely into another's receive elements, and so on.

Conflicts at the interface are fundamental and very difficult to resolve. These issues deserve continuous R&D investments to enable and bring to fruition the very desirable multifunction aperture concepts.

Synthetic Aperture Radar

Synthetic aperture radar, of growing interest and utility because of its high-resolution three-dimensional imaging capabilities, emerged about three decades ago as a creative solution to an interface aperture constraint. Simple diffraction considerations indicate that to obtain, for example, 1-ft radar resolution at several kilometers range from an L-band (1-ft wavelength) radar, the physical antenna must have about the same dimension as the range—i.e., kilometers. Since this was impractical, the solution devised was to exploit the linear nature of Maxwell's equations and create a synthetic phased-array aperture through a series of time samples taken with a small antenna, which was moved sequentially to the physical spatial locations where real phased-array elements would have been located in the physical antenna. The collected samples could then be processed together to produce the desired high-resolution, three-dimensional image. Originally, coherent optical processing was used on the ground, well after the airborne data collection flight took place. In the 1960s, digital processing could not deal with such large amounts of data in reasonable lengths of time, but coherent optics was perfectly matched to the problem and could do it. Gradually, as digital capabilities grew at the factor-of-two-every-2-years exponential rate it still exhibits today, three decades later—and coherent optical processing did not—digital processing took over until today it can easily process most SAR applications in real time, in modest amounts of computer equipment.

As a microwave sensor, SAR combines all-weather and cloud-penetrating capabilities with optical resolution imaging capabilities. It has already found use in airborne and satellite reconnaissance and even in missile guidance. As digital computational capabilities continue to grow throughout the next four decades, SAR concepts will find more and more practical applications.

Digital Radar

Finally, after years of anticipation, a revolution is at hand in radar technology through the exploitation of a combination of modern digital techniques and fiber-optic communications. Not only will the next-generation radars be solid-state phased arrays, but they will also be almost entirely digital, confining the analog microwave portions to the extreme front end interface of the antenna with the outside world. Received signals will be digitized at the element after minimal analog processing—e.g., with an antialiasing filter, a low-noise MMIC amplifier, and perhaps a single stage of up or down conversion—and transmitted in digital form over wideband fiber-optic links to convenient remote locations off the aperture for processing, e.g., digital beamforming, in-phase (I) and quadrative (Q) generation, pulse compression, clutter suppression, target extraction, multi-hypothesis tracking, and so on. Similarly, for transmit, digitally created wave-forms will be generated off aperture and distributed via fiber optics to individual antenna elements where D/A conversion and MMIC power amplification will take place. With all signals in digital form, the phase shifting required by both transmit and receive functions can be implemented digitally by simply delaying the signals to or from individual antenna elements by different amounts. Coarse delays can be obtained by slipping clock cycles and fine delays by digital interpolation. This approach eliminates the need for analog phase shifters in the T/R modules and supplies, without effort, true time-delay digital beamsteering.

There are many obvious advantages in reduced analog drift and enhanced processing flexibility, e.g., digital beamforming and nonlinear filtering, to be gained by digitizing the radar, but the two most significant come through an enormous reduction in the size of the analog receiver hardware needed—a reduction of two orders of magnitude was projected for a proposed digital version of the Navy next-generation fleet surveillance radar—and the architectural flexibility that results from the remoting possibilities offered by digital fiber-optic signal distribution, e.g., possibilities for putting the rest of the radar away from the antenna, anywhere on the ship or airplane. Of course, there are potential penalties, too. A high-speed, high-dynamic-range A/D and D/A converter at every element could be expensive. Creating data samples at gigahertz rates implies that at least some of the subsequent digital processing must be carried out at similar rates—which is admittedly difficult today and possibly power hungry, but is predicted to become easier by a factor of two every 2 years or so. Finally, the flood of data samples converging on the central signal and data processors conjures up visions of multiple teraflop capabilities.

What enables this potential revolution is the recent progress in ADC technology—in particular the great leap forward promised by the application of the proven Delta-Sigma concepts from the audio spectral range to megahertz signals at gigahertz sampling rates—combined with the inexorable growth of digital capabilities and the coincident maturity of wideband fiber-optic communication technology. Delta-Sigma technology, with its 1-bit oversampled signal representation, suggests future implementation and architectural possibilities. Exploiting single-bit arithmetic, such functions as digital beamforming and time delay can be performed before filtering and decimation down to more conventional multibit information bandwidth data rate representations. Single-bit arithmetic is very efficient, requiring only simple shifts and 1-bit adds, and, by delaying the filtering and decimation, the associated hardware requirements are reduced from one filter-decimation unit per element to one per beam—generally a significant savings in hardware without any anticipated performance penalty.

Given its success in audio applications and the lack of obvious obstacles to its extension from megahertz to gigahertz rates, Delta-Sigma seems certain to succeed in radar as well. The possibilities suggested by the single-bit arithmetic approach outlined above deserve close attention in the near future. In the longer range, greatly enhanced performance can be expected through the application of increasingly capable ADCs, exponentially growing amounts of computer power, and the exploitation of sophisticated algorithms for real-time adaptive digital beamforming, which are not yet implementable with today's computers.

Photonics

In recent years, photonics has become a popular topic in the radar community. The word "photonics" is fairly imprecise, often used simply to indicate the use of optical techniques—i.e., lasers, optical communication links (free space or fiber optic), mirrors and lenses, and various hybrid components that consolidate optical, electro-optic, and electronic elements into monolithic structures known as optoelectronic integrated circuits (OEICs).

In the radar context, "photonics" usually refers to wideband analog optical implementations of microwave signal distribution and true time-delay steering. Modulating the microwave signals onto a microwave carrier offers a number of advantages, primarily in the reduction of the physical size of the components and an increase in the flexibility of relocating the signal and data processing off the antenna. Placing the microwave signals on an optical carrier changes the physics of propagation completely so that the awkward, often large and expensive wave-guide structures traditionally used to distribute the microwave signals to and from the phased-array elements are replaced by small, lightweight, inexpensive fiber-optic links. And once the signals are on the fiber, increasing the length of the link to relocate signal generation and processing components that do not have to be on the antenna introduces very little extra penalty, because the loss of typical communications-grade optical fibers is measured in decibels per kilometer. Finally, because the microwave modulation is such a minuscule fraction of the optical carrier frequency, the propagation of all the microwave frequencies takes place without noticeable dispersion and so true-time-delay concepts can be implemented in a relatively straightforward fashion, e.g., by optically switching the signals into different fiber lengths to achieve controllable time delays.

Although many photonic radar configurations have been proposed and simple prototype systems, with only a handful of phase elements, demonstrated, no full-scale implementations have yet been fielded. The primary obstacles are two-fold—from the microwave point of view, fiber-optic links, integrated optics, and electro-optical components are often quite lossy, with losses of several tens of decibels from microwave-in to microwave-out per element or per link not uncommon. This leads quickly to impractical requirements for total laser power to support real radars with thousands of phased-array elements. Approaches based on active integrated optics structures, with built-in gain to compensate for the losses in a distributed fashion, have been proposed and demonstrated to a limited extent, but these possibilities remain in the research stage at the moment. The second major obstacle lies with the analog nature of the photonics approaches—in particular, dynamic range. It is very difficult to find electro-optical components, e.g., modulators and OEICs, that have more than 30- or 40-dB dynamic range, and this just is not enough for most applications.

R&D efforts to improve and mature the technology of photonic components continues today. There is little doubt that photonics can implement such radar tasks as microwave signal distribution and true time delay. But whether this will lead to fieldable, high-performance radars is another question. The competition for the future is, of course, the digital radar. Ironically, the digital radar concept also exploits photonics, but in its most mature COTS form, i.e., fiber-optic digital communications. Just as with SAR coherent optical processing, the growth rate of digital technology so far exceeds that of photonics that it is inevitable that a digital approach will win sooner or later. Looking at the state of the art in these two technologies, it is clear that it will be sooner. Beyond the timing issue, digital radar offers a broader, more flexible capability. Although it may find some niche applications, the photonic radar, on the other hand, is unlikely to supply the general-purpose capability inherent in a digital approach.

Multidimensional Signatures

Collecting additional sensor information about a target from multiple points of view is an obvious technique to improve sensor recognition performance and is one of the important common technology trends identified earlier in this chapter. Multiple points of view can be interpreted in several different ways. It could mean observing from the same physical sensor on different spectral bands or, equally well, it could mean several sensors in different physical locations observing the target from varying geometric aspects. In radar technology, both forms are in practice. Certain missile seekers currently under development are exploiting the first concept—looking in several widely spaced spectral bands, e.g., RF and millimeter wave, or RF and IR, to generate complementary information. The second interpretation is already in use in the Navy's cooperative engagement capability (CEC), which combines multiple radars with modern communication to produce a result that is greater than the sum of its parts. By sharing the data from all of the existing radars in the vicinity and fusing the results to eliminate duplications or resolve apparent contradictions, a single radar view of the battle space can be constructed for all to use. The radar vision of a single participant is thus usefully extended well beyond the reach of his own radars, greatly enhancing his battlefield awareness.

Passive Radio Frequency

Passive systems do not generally receive much attention in the radar community, although they offer a number of attractive features. The most obvious is that since they do not emit radiation, they are difficult to find and thus are relatively covert. Of course, not emitting actively is not sufficient to be entirely undetectable. RF and optical receiving systems, which focus or image the received radiation, generally produce strong retro-reflections when actively illuminated within the field of view of the sensor, thus redirecting some of the incident radiation back along the same direction it came from. Careful attention to signature control is required to minimize these enhanced cross-section effects that can make a passive sensor less covert than might be thought.

Passive RF sensors exploit the fact that the targets or objects of interest can be illuminated from a number of sources of RF radiation that are not co-located with the receiver. These sources could be artificial and coordinated with the receiver, as in a bistatic radar configuration; they could be artificial and uncoordinated, as when the receiver detects the scattered radiation generated by independent active radars that happen to be operating at the same time; or they could be entirely natural, as a result of the direct, scattered, and reflected thermal radiation from all the objects in the environment, including the targets of interest themselves, the sun and the sky. Recently, as a result of a Navy study of multifunction phased-array applications, bistatic configurations have come under serious scrutiny as the only feasible approach to achieving the transmit-receive isolation required to permit multiple RF systems to operate simultaneously through the same aperture. Bistatic configuration issues will be receiving increased attention in the near future.

Bistatic and Multistatic Systems

The ability to provide U.S. military forces with new types of active radar sensors continues to stress the limited availability of interference-free radio-spectrum resources. However, many active emitters of radio energy in the battlefield have the capability to be used as illuminators in either cooperative or noncooperative bistatic or multistatic radar modes in addition to filling their primary function. Although a few successful bistatic cooperative and noncooperative radar systems have been developed for military purposes over the years, in general the U.S. radar community has considered bistatic radar to be less satisfactory than the more conventional, monostatic radar systems. However, radar systems in which the transmitter and receiver are co-located end up being excellent targets for enemy antiradar missile (ARM) attacks, whereas multistatic radars that make use of existing (even enemy) emitters that are present for other, nonradar purposes are not similarly vulnerable to enemy ARM attacks. The richness of radio emitters in conflict areas and the availability of highly capable processing electronics suggest that meeting the need for new radar capabilities involves exploration of the possibility that they can be provided by bistatic and/or multistatic radar configurations using noncooperative emissions from such emitters.

It is possible that the exploitation and further development of this branch of radar technology can be used to provide new radar capabilities for the naval forces—radar capabilities that do not require the expense of radar transmitters and that can function in an almost totally covert manner. Full exploitation of the rapidly evolving capability and capacity of digital data processing and the funds to carry out proof-of-principle exercises involving bistatic and multistatic radar technologies and concepts are worthy of detailed examination. Some applications of bistatic and multistatic radar could be started now; technology understanding is currently adequate to consider new radar solutions today. Certainly, the more applications considered, the more they can become part of the future radar solutions into the indefinite future.

Passive Imagery

Using the sources of natural noise in a scene to provide the illumination for generating a passive image is just what is happening when we use our eyes or a video camera to view a scene. Remarkably, this same concept applies to radar and millimeter-wave RF frequencies (and to acoustic sonar, discussed below) and may offer a breakthrough technology for viewing targets thought to be hidden by smoke, foliage, or camouflage.

Even in the RF spectral range, natural surfaces will emit different amounts of radiation depending on parameters such as temperature and emissivity. In addition, metals are strongly reflective in the RF, which reduces a metal surface's

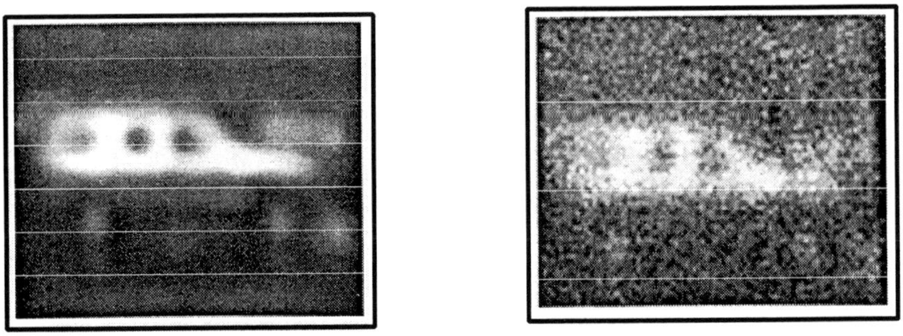

FIGURE 4.9 Millimeter-wave images of an automobile in a garage. The automobile is recognizable in both pictures, but the second picture was taken with the garage door closed. SOURCE: Smith, R.M. 1996. "The Passive MM-Wave Scenario," *Microwave Journal*, Wright Laboratories and Millitech Corp., March, p. 22.

emissivity and allows it to produce reflections of other sources in the scene—the most significant being the sky, which looks quite cold compared with Earth's surface environment. Because of these nonuniformities, with enough antenna aperture to achieve narrow beam resolution, passive radar imaging is possible.

For practical reasons involving antenna size, this kind of imaging is usually confined to the millimeter-wave spectral range and has already been applied successfully to targeting of weapons using the apparent cold spot associated with the image of the sky reflected off the top of a tank as the key discriminator. It has long been appreciated that such imaging millimeter-wave systems can see through fog and smoke as these frequencies are not attenuated very strongly, that is, neither scattered nor absorbed. What has not been as well appreciated is that such systems can see through other dielectric materials as well, such as wood, plastic, or tree leaves. Figure 4.9 shows a millimeter-wave image of a car taken through a garage door.

The ability to see through dielectric materials with millimeter-wave systems suggests that passive RF deserves further exploration for the detection and imaging of hidden targets. A combination of thinned arrays and digital techniques, by providing adequate spatial resolution, could extend the practicality of these techniques to lower RF frequencies, where the penetration is even better than at millimeter-wave frequencies.

Electro-optics

The optical portion of the electromagnetic spectrum has been growing in importance for many decades. In spite of the optical spectrum's lack of all-weather capability, its ability to generate very-high-spatial-resolution, two- or three-dimensional images from conveniently small apertures makes it extremely useful in situations that are not troubled by weather obscuration. As the trend in weapons concepts moves inexorably toward smarter, more precise surgical strikes with little or no collateral damage, optical systems are being called on to supply tactical capabilities in target designation, ATR, aim point identification and selection, terminal guidance, and so on, as well as the more conventional surveillance capabilities for sensing and detection.

Passive Imaging

Conventional Focal Plane Arrays. Modern optics has benefited greatly from the development of the computer and the semiconductor microelectronic technology that supports it. When computers first appeared in the 1960s, optical systems—the mirrors and lenses—took a great leap forward as the invention process that had characterized the development of good designs up to then was replaced by a computer-aided engineering discipline that routinely produced affordable, optimized designs. Later, as various semiconductor detector materials were developed to cover different portions of the optical spectral bands, microelectronic fabrication techniques were applied to create imaging detector focal plane arrays (FPAs), i.e., semiconductor chips containing multiple optical detector elements along with the electronic circuitry needed to read them out. FPAs have enabled a wide range of new imaging applications, for not only can they be made much more sensitive than the human eye in certain useful spectral ranges, e.g., the IR where the eye cannot see at all, but also the image information is directly available in electronic form for further enhancement via amplification and filtering or computer processing for sophisticated information extraction.

In recent years, FPA technology has advanced rapidly, following the same kind of exponential growth curves that characterize the rest of semiconductor and microelectronics technology. Figure 4.10 shows a history of the development of visible FPAs.

The growth rate of pixels per chip closely mirrors the exponential growth rate of conventional digital circuits. That the visible and near-IR ranges are the most advanced is no surprise, given that excellent, almost optimal detectors for these spectral regions can be fabricated from silicon. With the read-out electronics also in silicon, completely monolithic focal plane arrays can be fabricated and form the technology basis for the high-quality imagery found in the television industry, as well as camcorders and the digital cameras that are increasing in popularity at the moment. State-of-the-art visible FPAs are now available with as many as 5,000 × 5,000 pixels and all of the support electronics on a single chip.

For other spectral regions, the detector material systems are less mature than silicon and lag in their development accordingly. In the 1- to 5-μm region, monolithic platinum silicide (PtSi) FPAs are available, as well as hybrid arrays of

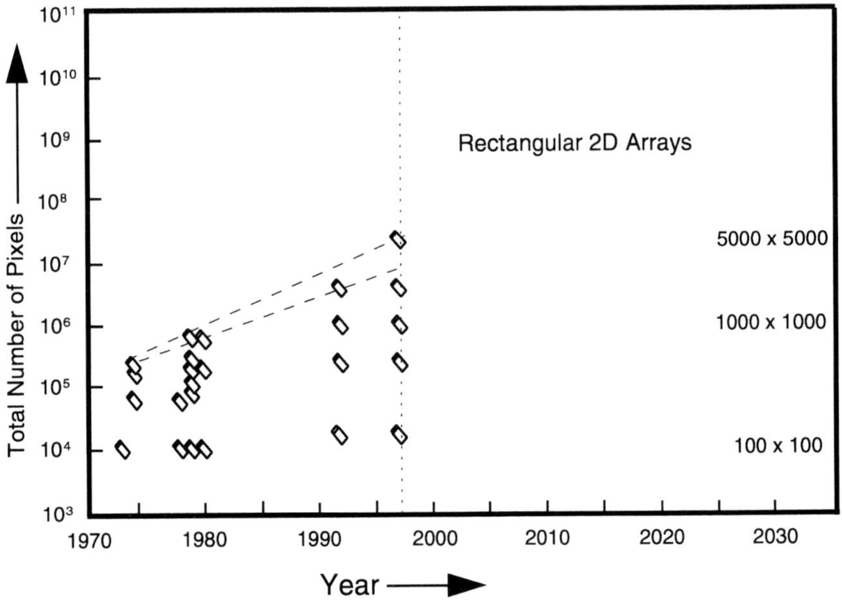

FIGURE 4.10 Visible focal plane array technology. SOURCE: Figure created with data from (1) Theuwissen, Albert J.P., 1995, *Solid-State Imaging with Charge-Coupled Devices*, Kluwer Academic Publishers, Dordrecht, Boston and London; (2) EG&G Reticon Product Line Catalogs for 1993/1994 and 1994/1995; and (3) Thompson CCD Products Catalog 1992/1993.

indium antimonide (InSb) and mercury cadmium telluride (HgCdTe). The current state of the art on these devices is near the 10^6 pixel level (1,000 × 1,000 pixels) and can be expected to follow Moore's Law for the near future. In the longer IR wavelengths (8 to 20+ μm), special forms of extrinsic silicon and HgCdTe are becoming available in sizes of 256 × 256 pixels, or a little larger, and can also be expected to grow by a factor of two about every 2 years. In addition to the level of technology maturity, IR FPAs, for reasons of semiconductor physics, must be cooled during operation—often to liquid nitrogen temperatures (77 K) or lower—an awkward, expensive, and often logistically complicated disadvantage.

From a fabrication point of view, there is nothing obvious that would prevent FPA technology from exploiting much of the growth in microelectronics in the near future—certainly with respect to the on-chip read-out electronics. Reducing the detection pixel sizes, however, reduces the number of photons collected by each pixel, thus increasing the statistical noise levels and reducing the available signal dynamic range. It is likely that a point will be reached in the not too distant future at which it no longer makes sense to attempt to further reduce FPA pixel dimensions. The imagery being detected by an FPA will always be limited by the collection optics and the interface to the physical world, and once the pixel-to-pixel sampling distance satisfies the Nyquist criteria for the primary spatial frequencies allowed by the aperture and the residual detector noise, further oversampling, i.e., reducing the pixel-to-pixel separation, adds little benefit. At the moment, current state-of-the-art FPA implementations are not yet well matched to their optics, and so there is room for improvement. After matching is achieved, further growth in FPA pixel count will evolve, not as a fabrication-limited process, but as an activity closely coupled with improvements in the collection optics—e.g., larger apertures, lower f-number designs, and wider fields of view.

Uncooled Focal Plane Arrays. Recently, a new class of IR FPAs has appeared, using simple thermal bolometer concepts rather than matched bandgap, exotic semiconductor materials. Microbolometers and the read-out electronics are integrated onto a silicon substrate by means of MEMS technology. Incident IR radiation heats microbolometer pixels and the resulting temperature rise is directly measured by the on-chip electrical circuitry. No cooling to low temperatures is needed and, as the effect utilized is a thermal phenomenon, these FPAs are inherently broadband, responding to a wide range of wavelengths.

Although less sensitive than conventional photodetectors, the uncooled MEMS-based FPAs have demonstrated performance capabilities measured in fractions of a degree—easily enough for a number of important imaging applications. Imaging implementations based on these MEMS focal planes are physically smaller, logistically simpler, and significantly less expensive than those based on conventional FPAs, promising breakthrough applications to many critical naval force applications such as affordable missiles and miniature UAV sensors.

Multi- and Hyperspectral Signatures. Another technology trend, evident in RF and radar technology, is also evident in optics—an interest in collecting multidimensional signatures for more robust image understanding. Dual-mode missile seekers using two different IR bands, carefully chosen to match the scenarios, have already demonstrated superior performance against IR countermeasures and a large reduction in false alarms resulting from such things as sun glints. In support of these applications, FPAs that image simultaneously in two IR spectral bands have been demonstrated in sizes up to 256 × 256 pixels. In some designs, the two arrays are stacked one on top of the other and precisely aligned, pixel to pixel, with the shorter-wavelength detectors substantially transparent to the longer-wavelength radiation that passes through and is detected in a smaller bandgap material below. Stacking in this way leads to convenient, very compact implementations of dual-spectrum FPAs, which have broad applicability to space-constrained applications such as missile seekers.

Extending this direct approach to many more than two spectral bands, however, is not practical for a variety of fabrication and performance reasons. Yet there is great interest, particularly in satellite ground-imaging applications, to collect many simultaneous images in as many different spectral bands as possible, from the UV, through the visible, and well into the IR. The extreme situation, known as hyperspectral imaging, collects hundreds to thousands of such simultaneous images in spectral bands, only a few nanometers wide. As computer technology continues to improve, it will become easier and easier to extract useful information from hyperspectral images in real time, and its use is certain to expand. The potential for subpixel detection offers great promise and should be explored further. Of particular interest to the Navy and Marine Corps is the potential for using spectral imaging to characterize the littoral regions.

Laser Sensors

Although long used successfully on the battlefield as range finders and target illuminators for precision-guided weapons and, most recently, as real-time wind sensors for gunship weapons targeting improvement, lasers have other sophisticated capabilities that have not yet been fully exploited.

Imaging LIDAR

After many decades of development, LIDAR or laser-based radar seems poised to enable a variety of important military imaging applications. Combining the range measurement capabilities of microwave radar with optical-quality two-dimensional imaging, imaging LIDAR offers unique sensor characteristics. Its range-to-pixel three-dimensional imagery, with high resolution in all three spatial dimensions, contains much more useful information than either ordinary radar results or a passive optical image of the same scene or target. For target detection and ATR applications, the geometric measurements generated by the LIDAR provide direct information about object dimensions and orientation without the need to search through ranges of possibilities and are completely insensitive to the ambient illumination and temperature effects of a scene that complicate the interpretation of IR imagery. Excellent detection and ATR results have been demonstrated for such imaging LIDAR systems operating from the ground and airborne against a variety of targets in realistic cluttered natural environments, but operational systems have yet to appear. No doubt, with increasing availability of computational resources, more effective algorithms will emerge.

Complementing these performance advantages has been a long-term trend away from the earlier 10-mm carbon dioxide (CO_2) gas laser technology, toward diode-pumped, solid-state lasers that operate in the 1- to 2-μm region, near IR spectral bands with good coherence properties, sometimes supported with fiber-optic optical power amplifiers. This has led to significant reductions in transmitter power and volume requirements, enhancing the feasibility and practicality of using LIDAR to detect missiles or other aerial vehicles.

Unfortunately, because the power available from these laser sources is typically measured in watts, rather than the kilowatts that are common in microwave radar, laser radars have limited range capabilities—a few to a few tens of kilometers are typical. And of course, operating at optical wavelengths brings other well-known penalties—lack of all-weather capability and limited search capabilities; that is, the power-aperture products, which are proportional to volume search rate, are typically very small.

In spite of these limitations, however, there is a broad class of applications for which the imaging LIDAR is well suited and for which it offers breakthrough capabilities. Effective target recognition, aim point selection, and precision guidance in the end game are critical for the success of surgical-strike weapons. LIDAR technology can supply these capabilities through its high-quality, three-dimensional, direct geometric imagery and the resulting robust algorithms. The weather-penetration and range limitations can be avoided by employing the LIDAR only in the end game where the ranges are small, whereas the search constraints can be addressed by using a coarse guidance technique, such as GPS or ordinary RF radar, to provide a final LIDAR search region that is guaranteed to contain the target.

Imaging LIDAR technology deserves serious consideration for these classes of precision-guidance applications.

Environmental Sensing LIDAR

The relatively strong interaction of laser beams with the atmosphere and its constituents presents serious performance limitations when the objective is to image solid objects through the atmosphere. However, this same interaction can be turned into an advantage by setting as the objective the measurement of the atmosphere itself, as described below:

- *Wind sensing.* The coherent backscatter of focused and scanned CO_2 laser beams, from the aerosols naturally found in the atmosphere, has permitted remote characterization of wind fields, i.e., velocity and magnitude, at distances up to several kilometers from the LIDAR. Although first demonstrated almost three decades ago, such a capability has been fielded as part of a weapons system only recently. The multiple rocket launch system (MRLS) employs such a LIDAR to determine, in real time, the winds in the intended path of the rockets, particularly those closest to the launcher where the rockets are moving slowly and are most susceptible to deflection. Algorithms convert these measurements into prelaunch aiming corrections. Altogether, this system has proven successful in enhancing missile accuracy by an order of magnitude. Other possible applications, such as the determination of helicopter local down drafts for similar weapons prelaunch corrections, are under consideration.

The CW CO_2 laser technology that supports the MRLS LIDAR is already well developed. It is a highly coherent and fairly compact laser with good power output and efficiency. Whether the shorter-wavelength, diode-pumped, solid-state laser technology currently under development can supplant CO_2 in this application, with further reductions in size, is a complex issue. The coherency and power requirements favor the longer-wavelength CO_2 laser, although the wavelength dependency of the optical scattering cross section of the aerosols, which increases strongly with wavelength because of Mie scattering, favors the shorter-wavelength, solid-state lasers. Since the current development trends emphasize the solid-state developments in preference to further refinements of gas laser technology, it would not be surprising someday to find solid-state technology winning in this arena also.

- *Remote sensing of chemical and biological agents.* Perhaps a more promising application of LIDAR sensing lies with its potential for the remote detection and localization of environmentally released clouds of chemical or biological agents. Given the ease with which these threats can be created, their timely and accurate detection at a distance will be of increasing importance in the near future. When absorbed by an atom or molecule, a high-peak-power pulse of laser radiation can cause a detectable reradiation (via Raman scattering, for example) that is spectrally shifted by amounts and in patterns characteristic of the scatterer. Because of their nonlinearity, effects like Raman scattering require high-peak-power pulsed sources to generate useful levels of response, which, when combined with a low density of scatterers, e.g., a dispersing cloud of pollutants, severely limits the range capabilities of such a system. Remote sensing LIDARs, based on effects like Raman scattering, have been demonstrated many times in the laboratory and in the field over modest distances, but the technology is a long way from being incorporated into an effective, affordable battlefield system. Tuning the laser source to enhance specific atomic or molecular resonances could greatly improve systems performance, and so research into tunable high-power pulsed solid-state laser sources, for these applications, should be of great interest.

Electronic Scan—Optical-phased Arrays

Optical systems have traditionally relied on analog techniques such as lenses and mirrors to provide beam control. Radar initially adopted the same approach with the use of steerable reflector antennas, but in recent years has largely abandoned this approach in favor of electronically steered phased-array systems that offer the capability for rapid, random beam placement, orders of magnitude faster than can be obtained mechanically.

Direct translation of radar phased-array concepts and techniques into the optical regime is difficult because of the tiny dimensions of the optical wavelengths. With improvements in high-resolution microelectronic fabrication, however, and the discovery and refinement of new classes of optical materials, it became possible about 10 years ago to envision a phased-array, optical beamsteering technology based on liquid crystals and other electro-optical materials. In the liquid crystal approach, voltages are applied to small volumes of liquid crystal by means of transparent electrodes, which cause internal realignment of the rod-shaped molecules and a consequent voltage-dependent change in the optical phase shift experienced by an optical beam passing through the cell. For visible and near-IR wavelengths, convenient voltages of only 10 or 20 V are sufficient to produce 2π phase shifts, and the transparent electrodes are easily deposited to the fine dimensions enabled by optical lithography. Spatially linear phase distributions produce beam deflections, whereas more complicated phase distributions can provide focusing and beam splitting.

Liquid crystal optical phased arrays, with thousands of pixels, already have been fabricated for two-dimensional steering of laser wavelengths from the visible out to 10 μm. Single incident beams can be steered to random positions, can be repositioned with high-precision repeatability, and can be split into complex patterns of spots that can be simultaneously focused and steered to all parts of the field of view. Currently limited to a response time of about 100 ms, new materials and new configurations will reduce this time to 10 ms in the near future.

Optical phased arrays seem to offer yet another possible breakthrough technology enabling electronically controlled modes and capabilities clearly not possible with conventional mirrors and lenses. Combined with advancing FPA technology and the predicted continuing growth of digital and computer capabilities, an all-digital LIDAR may become a reality in the future.

Computational Optics

Short of the revolutionary completely digital LIDAR envisioned above, many intermediate benefits could accrue to optical systems from a hybrid of analog and digital techniques. Fabricating optical imaging systems has always involved a challenging tradeoff between size and weight and the number of aberrations that must be minimized across the desired ranges of spectral wavelengths and physical field of view. As the experience with the Hubble telescope has so dramatically demonstrated, however, the optical system need not be perfect in order for the system to produce perfect images. It is enough to measure just what optical point response the actual optical configuration produces and use this information to correct the images by appropriate digital filtering.

Obviously, as digital and computer capabilities continue to improve at the rate of Moore's law, it will become more and more practical to consider this option. For example, optical assemblies for any application whatsoever could be designed and fabricated, at very low cost, to be nearly diffraction limited, the characteristics measured, and the corrections applied in real time to produce overall diffraction-limited performance at a fraction of the cost of diffraction-limited optics.

Another interesting possibility is to consider implementing sparse optical apertures, i.e., a number of small, coherently related optical subapertures, representing only a small fraction of a much larger aperture that might also be supporting one or more RF systems. With judicious placement of subapertures, and computer compensation of the resulting complicated point-response function, very-high-resolution optical imagery, simultaneous with RF aperture sharing, may be obtainable at affordable costs.

Acoustic Sensors

Sonar

Sonar technology has long been limited severely by the nature of the propagation medium and the complexity of the environment. The ocean medium is inhomogeneous, with a sound velocity of about 1,500 m/s plus or minus about 3 percent, depending on local conditions of temperature, salinity, and pressure that give rise to variable refractive deflections of propagating sound beams. Combined with reflections from the ocean bottom and the sea surface, these inhomogeneities give rise to anomalous propagation channels, in addition to the direct path, e.g., bottom-bounce convergence zones where the propagation paths refract down and up in a sinusoid-like pattern never hitting the bottom or the surface, and surface ducts that combine surface reflections and deep-water refraction. In addition, propagation in the ocean is highly absorptive, with attenuation increasing rapidly with frequency, more or less as frequency squared. As a result, long-range detection requires low frequencies of tens to a few kilohertz, which conflicts with the requirement for narrow-beam resolution. To get many acoustic wavelengths across the aperture, the sonar array often must be longer than the ship or submarine, often requiring towed arrays for good beam resolution. Finally, sonar signals, both active and passive, are often contaminated by multiple reflections from the sea surface and bottom and can be masked by many sources of natural noise such as waves, moving surface ships, and sea animals, thus requiring substantial clutter rejection processing for effective signal extraction.

Recent passive acoustics experiments at sea have demonstrated that signals are far more coherent than previously believed, both spatially and temporally, if accurate propagation models are used and sensor position and source/receiver motion are compensated for. As a consequence, significant gains in passive detection from large, densely populated arrays with both horizontal and vertical apertures and long-duration observations are achievable. Required technologies include miniaturized sensors; towed arrays with fiber-optic links and in situ digitization; fiber-optic connectivity; navigation; unmanned undersea vehicles; high-power wideband active sonar; and capable, intelligent displays.

A full description of this phenomenon is found in another report in this series, *Volume 7: Undersea Warfare.*

Because of the complexity of the medium and because A/D conversion was never a serious obstacle, sonar processing became digital long before radar. Very early, sonar adopted transducer arrays as a practical alternative to physically awkward acoustic lens and reflector configurations. Although originally processed with analog techniques, sonar arrays were converted to digital systems several decades ago, and adaptive beamforming was implemented using dedicated, special-purpose custom digital hardware. And a decade ago the use of arrays of COTS microprocessors for digital beamforming, clutter filtering, and ATR imaging processing began. As discussed above, radar is just now on the verge of becoming all-digital for these same functions.

With this early digital start, it is no surprise to find that the dominant trend in modern sonar is the application of more and more computer processing and the continued evolution of increasingly sophisticated algorithms. Two-dimensional image-processing concepts taken from the optical world, such as edge and textural feature extraction, wavelet processing, and neural net ATR, are being applied to sonar images for such tasks as mine location and identification. And the impressive sonar performance of bats and porpoises, long the object of academic research, has recently led to the implementation of promising, nonlinear, biologically motivated, signal-processing algorithms that have demonstrated greatly reduced rates of false alarm as compared with traditional approaches.

In terms of the transducer hardware, the extreme physical constraints of the medium and environment do not seem to leave much room for significant improvements over the already mature piezo-electric ceramic-based technology of today. On the other hand, there has been continuous slow progress in the application of polymer-based transducers that are easily patterned by photolithographic techniques to have low sidelobe characteristics. This class of transducer materials offers good broadband receive sensitivity but rather limited power-generation capabilities for transmit. Excellent fiber-optic acoustic sensors have been developed for receivers and have found limited application so far. Whether MEMS technology can be applied to provide very compact, inexpensive, sensitive acoustic sensors is yet to be determined, but it seems likely.

Another direction in transducer development involves attempts to achieve the broad-bandwidth high-power transmit waveforms assumed by the biologically motivated algorithms, through novel structures such as the slotted cylinder and multilayer stacks of polymer-based transducers. No doubt, efforts to increase the transducer bandwidths will continue in the future.

To achieve very-high-resolution sonar imagery, very large physical or synthetic apertures are required. This has led to the deployment of large fixed arrays for passive monitoring of critical regions of the world's oceans and to the development of large towed arrays for ships and submarines. Exploiting the obvious alternative of synthetic apertures, which has proven so successful for radar, has been greatly hindered by the fluctuations in the propagation medium and the limited coherence times that result. Although the feasibility of synthetic aperture sonar (SAS) has been demonstrated, no such device has yet been fielded. Mastery of this technology would seem to be a worthwhile investment for the future.

After many years of emphasis on the deep ocean, the Navy today finds itself increasingly concerned with close-to-shore, littoral operations—a terrible sonar environment with limited direct path opportunities, characterized by numerous reflection sources and other natural sources of noise. Extracting useful information from active or passive sonar returns in this environment is extremely challenging, requiring massive computational resources and clever algorithms. Since exponential growth in computational power is the direction in which technology is progressing, the ability to operate in littoral environments will certainly improve with time.

The multitude of sources of strong noise in the littoral environment can be used as sources of natural illumination in the acoustic spectrum. This may enable passive acoustic imaging, in direct analogy with conventional optical imaging (video) or the millimeter-wave passive RF imaging discussed in Chapter 5. So far, because of the practical difficulties of achieving high-enough resolution from convenient aperture at the frequencies tested, preliminary experiments along this line have not met with great success. The obvious expedient of moving to higher frequencies will improve resolution while simultaneously decreasing the useful range because of the frequency-squared dependence of acoustic absorption. But for littoral environments, with limited line-of-sight direct paths, this may be a reasonable tradeoff. This passive imaging option should be thoroughly explored, for it may offer the possibility of breakthroughs in littoral operations.

The limited direct path characteristic of littoral environments suggests yet another approach, which mirrors trends in land battlefield surveillance—a network of distributed, autonomous, communicating short-range sensors. Assuming progress in power sources, the limited range requirements and the expected continued progress in microelectronics and computers will lead to inexpensive, autonomous minisonar in the future that could be deployed in the suggested distributed fashion. The sticky point in implementing this concept is achieving the degree of communications needed between the individual sensors and back to the ultimate user. Among the possibilities are direct fiber-optic links, which would provide unlimited communication bandwidth if deployment conditions permit; indirect high-bandwidth satellite-mediated links, again if deployment permits; or low-bandwidth acoustic links, which offer serious challenges in terms of low detectability, antijam capability, information extraction, and data compression. It may be possible to disguise acoustic links by imitating the 1/f spectral distributions that characterize many sources of natural noise, including the communications of whales and porpoises. Wavelet mathematics can readily provide such spectra and may offer interesting communication coding possibilities for low-detectability transmissions.

Seismic and Vibration Sensors

Sensing acoustic or mechanical vibrations outside the context of sonar includes applications to monitoring of equipment and machinery for condition-based maintenance. For these applications, both fiber-optic and MEMS vibration sensors—or perhaps a MEMS-optical hybrid for sensitivity—seem ideal, as both approaches lead to very compact, sensitive, inexpensive sensors that have a natural affinity for interfacing with modern digital optical communication networks.

Inertial Sensors

The clear overall trend in inertial sensors is toward miniaturized, all-solid-state implementations, enabled by MEMS and optical technologies. Inertial sensors measure accelerations of the sensor—both linear and rotational—relative to local inertial space and thus always involve mechanical phenomena. Classically, inertial sensors consist simply of a static or a uniformly rotating mass, constrained by mechanical, electric, or magnetic forces. Accelerations are sensed by measurements of the small induced mechanical displacements allowed by the constraints or by measurements of the changes in the electrical currents required to hold the sense mass in its original position relative to the sensor's support structure or local frame of reference.

MEMS permits the sensing masses to be very small and to be fabricated as an integral part of a silicon chip, which can also contain monolithic sensing electronics along with self-test, calibration, and signal-conditioning functions. Air bag sensors, utilizing MEMS technology to measure linear acceleration, are already available commercially.

MEMS offers the possibility of implementing angular accelerometers by monitoring the displacements of miniature vibrating or rotating structures. Application of MEMS to these classes of measurements is still at an early stage, and the resulting measurement capabilities are not yet as refined as can be obtained from the much larger conventional inertial instruments; but improvement is certain to be rapid because the technology is so straightforward.

Aside from MEMS, optics has already made significant inroads into angular acceleration-sensing technology through the introduction of laser and fiber-optic gyros. Still not as capable as the best mechanical gyros, laser and fiber-optic gyros are far less expensive and often more compact than the mechanical alternatives and have found wide use in advanced avionics and missile-guidance applications where the ultimate in performance is not required. Because the physical phenomena on which laser and fiber-optic gyros are based require long optical path lengths for high accuracy, it is not clear that these concepts can exploit MEMS directly. However, integrated MEMS and micro-optomechanical assemblies have already been demonstrated, suggesting that optical on-chip measurements of small

MEMS sensing element displacements may some day be implementable with greatly improved sensitivity over the capacitance or resistance noise-limited measurement techniques currently employed.

Ultimately one can expect to see the A/D conversion function integrated onto the sensor chip, along with exponentially increasing digital signal-processing and computing capabilities, to produce a self-contained, miniature inertial navigation capability—full navigation, not just sensing—and, no doubt, aided by GPS information, when available.

Chemical and Biological Sensors

Although sensing chemical or biological substances remotely at a distance is possible through the LIDAR, most chemical and biological sensors rely on direct physical contact between the sensor interface and the unknown or sought-for substance. For detecting known (sought-for) substances, physical matching between the detector and the target substance in the form of atomic or molecular templates, specific responses, or selective chemical reactions is used. This approach can produce conveniently small, cost-effective, sensitive detectors of sought-for target substances but yields no information about a nontarget that may be present. A more fruitful technique for dealing with unknowns is to carry out basic physical measurements of the molecular structure, by optical or mass spectroscopic techniques, and identify the unknowns by pattern matching against the large databases of known material parameters that have been laboriously accumulated over the years. Until recently, it has been sufficient to make these spectroscopic measurements in the laboratory. As a result these classical, physics-based, chemical and biological sensors are often large (e.g., table-top to room-sized), slow (e.g., minutes to hours), and expensive.

Tiny Time-of-flight Micromass Spectrometer

The threat posed by chemical and biological weapons together with the opportunity afforded by modern optical and digital technology will drive the rapid development of new sensors. Portable, sensitive, fast, inexpensive sensors for chemical and biological sensing are needed for field use today, and development activities in this arena are increasing. The Defense Advanced Research Projects Agency, for one, has mounted a major thrust, emphasizing the detection and identification of biological agents. One of the most sensitive and selective detectors of biological and chemical agents known is the mass spectrometer. DARPA, with Johns Hopkins University's Applied Physics Laboratory, has initiated an ambitious program to develop, in about 5 years, a tiny time-of-flight (TOF) micromass spectrometer. This rugged, fieldable instrument would be small enough (< 5 lb, < 3,000 cm^3, < 50 watts) and inexpensive enough (< \$25,000) to fly on an unmanned aerial vehicle, would generate data at 10 ms per spectra, and would be sensitive down to the single-molecule level and flexible enough to work with solid particulates, chemical vapors, and aqueous materials. Combined with advanced sampling and signal processing, this could provide a major advance in chemical and biological sensing with broad applicability to biological and chemical defense, battle-space management, damage assessment, and intelligence collection.

Microelectromechanical Systems

Although the tiny TOF micromass spectrometer represents a quantum leap, so to speak, in the engineering of mass spectrometers and does exploit some modern technology in the form of digital and computer processing and a unique laser vaporization scheme for biological sample preparation, it is only one step toward what may be possible in the future. The ultimate device may be a MEMS mass spectrometer-on-a-chip. With limited prototyping and testing under way at the University of Minnesota, this project targets a 200-g, 0.5-W mass spectrometer, about the size of a penny and costing only $20. As digital technology continues to evolve, doubling in capabilities every 2 years, the signal and data processing eventually will migrate onto the chip to form a smart mass spectrometer-on-a-chip. This seems to be a natural and very promising direction for mass spectrometer sensor development.

MEMS technology enables a host of other creative possibilities for detecting and recognizing biological materials. Among the miniature mechanical components that can be fabricated in MEMS are microfluidic components, such as tiny valves and pumps, which can be configured to create an integrated DNA amplifier based on the PCR process. With microvolumes (1 or 2 μl) of sample material and integral on-chip heaters, the temperature cycles required by PCR for doubling the amounts of DNA each cycle can be carried out extremely rapidly—generating detectable amounts of DNA in seconds to minutes, rather than the hours that characterize conventional PCR equipment. Manipulated by microfluidics, the DNA-amplified samples can be inserted into an on-chip microfabricated capillary electrophoresis system with integrated sources and CCD detectors, for optical identification of the DNA. The result is a fast, accurate, and inexpensive DNA identification sensor. Active development of the required microfluidic MEMS technology and the DNA sensor itself is currently under way.

Another MEMS-based biological sensor, currently in development in a number of laboratories, uses an array of microcantilever beams on a MEMS chip to which biological materials can be selectively attracted. Whether something attaches to a specific microcantilever can readily be detected through measurable changes in its mechanical behavior, specifically, its vibration resonance frequency. Larger matrix arrays of cantilevers can readily be fabricated and individual beams selectively coated with specific materials such as oligonucleotides by applying electrical voltages only to the selected beam or beams while flooding the chip with the oligonucleotide. Washing and flooding with another oligonucleotide, and a different pattern of applied voltages, can be repeated until the whole array is loaded, so to speak. When exposed to an unknown (but anticipated) nuclear material, hybridization will take place only at the beam locations prepared specifically for that material, which can be determined by surveying the mechanical properties of the individual beams in the array. Of course this is a pattern-matching sensor, not a physical-properties sensor, and as such is ineffective when confronted with an unanticipated unknown substance.

In addition to MEMS, a second large class of fiber-optic-based biological and chemical sensors[17] seems quite promising in its sensitivity and breath of applicability. The potential of fiber optics for all kinds of sensing,[18] particularly remote sensing, was recognized in the 1960s, and the technology was applied to the measurement of the oxygen content of the blood as early as 1962.[19] Typically, the material to be identified is brought into contact with the end or a portion of an optical fiber, light from a laser source is sent down the fiber, and through selective absorption or scattering of the incident light by the unknown material or through the generation of fluorescence, the unknown is identified by pattern-matching techniques based on the returned optical signals.

Because of the poor detection sensitivity of absorption and scattering measurements, fluorescence is the most commonly used technique. Fluorescence can be generated directly by the sampled material in some cases or can be assisted by another fluorescent compound that selectively attaches (to tag or label) specific biological or chemical materials. One example of the latter approach is to coat the surface of the exposed core of an optical fiber with a particular antibody that remains attached to the surface of the fiber. When the fiber is exposed to various antigens, only those specific to the bound antibodies will attach. Introducing a free antibody that is fluorescently labeled and that is also specific for the targeted antigen then results in attached (antibody-antigen-labeled antibody) complexes that are detected by their characteristic fluorescence spectrum. The fiber can be prepared with multiple antibodies, which, when combined with differently tagged antibodies, gives the potential for simultaneous detection of multiple antigens.

Under DARPA sponsorship, such fiber-optic biological sensors have been tested against a number of battlefield threats, such as botulism, ricin, plague, and anthrax, and they promise near-real-time (~5 minutes) detection down to the level of 0.5 to 5.0 ng/ml. Efforts are now under way to miniaturize these sensors. In the long term, this class of sensor may also be compatible with a MEMS imple-

---

[17]Boisde, G., and A. Harmer. 1996. *Chemical and Biochemical Sensing with Optical Fibers and Waveguides,* Archtech House, Boston, Mass.

[18]Giallorenzi, Thomas G., Joseph A. Bucaro, Anthony Dandridge, and James H. Cole. 1996. "Optical-fiber Sensors Challenge the Competition," *IEEE Spectrum,* 23(9):44-49, September.

[19]Polyani, M.I., and R.M. Hehir. 1962. "In Vivo Eximeter with Fast Dynamic Response," *Rev. Sci. Instruments,* 33:1050.

mentation. Using on-chip lasers and detectors, with integrated optical waveguides in place of fibers, very small throwaway ($20) biological sensors could result.

A persistent problem with all types of biological sensors is the difficulty in cleaning or resetting the sensor so that it can be used multiple times, instead of being discarded after a single use. The antibody-antigen-labeled antibody fiber-optic biosensor just described is somewhat reusable in that it can be used a second time if the first trial gives a negative result, that is, when no antigens or labeled antibodies are detected. On the other hand the dog's nose is infinitely resettable (with saturation effects and time constants, to be sure), but ultimately recyclable an unlimited number of times. There is much to learn in this discipline in the future.

Other Sensors

General Local Sensors

Beyond the major sensor classes explicitly discussed above, there are numerous other sensors for measuring just about anything one can imagine, including temperature, humidity, position, stress, strain, speed of flow, shape, roughness, stiffness, compliance, viscosity, electrical resistivity, inductance, interatomic distances, and so on. In terms of future growth, all can be expected to benefit from the digital revolution and the significant microelectronic and digital technology trends that have been described. With few exceptions, the signal conditioning is already, or will soon become, solid state with the signal processing handled digitally. These sensors, except where constrained by the physical interface, will exploit the progress in microelectronics—digital VLSI as well as MEMS—to grow increasingly monolithic, smaller, cheaper, smarter, and distributed, with built-in microprocessor capabilities (i.e., a sensor on a chip) supporting test, calibration, communications, and sophisticated signal-processing algorithms for greatly enhanced performance.

Fiber-optic Sensors

Perhaps the most interesting and versatile of the general sensors are those based on fiber-optic sensor technology.[20] Using the same technologies that enable the optoelectronics (e.g., CDs and laser printers) and the telecommunication industries, fiber-optic sensors are rapidly displacing a number of traditional sensors, achieving both higher performance and lower cost. Fiber-optic sensors are discussed above in the context of chemical and biological applications and their role in inertial sensing in the form of the fiber-optic gyro. But fiber-optic sensors

[20]Giallorenzi, Thomas G., Joseph A. Bucaro, Anthony Dandridge, and James H. Cole. 1996. "Optical-fiber Sensors Challenge the Competition," *IEEE Spectrum*, 23(9):44-49, September.

are also capable of measuring electric and magnetic fields, temperature, humidity, viscosity, stress, strain, position, acoustics, and other parameters. Any physical phenomenon that can be used to affect a light beam can provide the basis for a fiber-optic sensor.

The fiber-optic sensor operates by sending a light beam through a fiber to a sensing element that, by an interaction with the physical phenomenon to be sensed, modulates some aspect of the light beam—e.g., amplitude, polarization, and frequency. The modulated return beam then propagates back through the fiber to an optical receiver and a signal/data processor where the information generated at the sensing element is extracted.

Fiber-optic sensors offer a range of inherent advantages including high sensitivity, small size and weight, low power requirements, reliability, and inexpensive components. Of particular value is its ability to remotely locate the passive sensing element in hostile environments or at a significant distance from the power and processing elements because of its use of simple inert materials, its immunity to EMI, and the broadband and low-loss communication capabilities of the fiber. This ability of fiber-optic sensors to sense remotely, along with the safety aspects associated with inert materials and a lack of electrical danger, has been a key element in encouraging their current widespread use in medicine.

One of the most promising new applications, where no equivalent sensors exist, is the concept of smart structures. Building on the capacity for remote operation, the concept envisions networks of fiber-optic sensors, embedded into or attached to structures as they are manufactured, to permit continuous monitoring of the status of the structure for nondestructive evaluation or real-time damage assessment or perhaps to permit control of the shape or strength of the structure via actuators. Applications to all sizes and kinds of structures are possible—structures as small as a single aircraft panel or as large as a bridge, a dam, or a whole ship.

Beyond its versatility, fiber-optic sensor technology possesses an inherent compatibility with the distributed communication networks that will likely dominate future sensor systems. In the near future, fiber-optic sensors of all kinds should continue to become less expensive, displacing more and more traditional sensors and finding new applications. In the long run, they will become increasingly capable along with every other sensor and will find broad use in distributed sensor networks where mobility and wireless communications are not mandatory.

FUTURE IMPACT ON NAVAL OPERATIONS

Expected Evolution of Sensor Technology

Based on the technology trends and historical growth patterns described, the panel anticipates that future sensor technology will be characterized by the following:

- Ever-decreasing size and cost as microelectronics evolves into nanoelectronics within the limits and constraints implied by the physics of the interfaces.
- Migration of the analog-to-digital conversion to the front end of the sensor, leaving only those analog elements absolutely necessary for interfacing with the physical phenomenon to be sensed—e.g., microwave LNA, filters and power amplifiers, fiber-optic transducers, MEMS transducers, and the like.
- Ever-increasing application of computer processing as gigaflops grow to teraflops and then to petaflops.
- Development of monolithic smart sensors, combining sensing transduction, ADC, digital signal processing, communication input and output, and perhaps power conditioning on a single chip. This offers interesting possibilities for very small, very smart weapons such as affordable smart bullets.

  Note that not all sensors can be small, even though the electronics can be. Size depends very much on the physics of the physical interface constraints. For example, propagation-based sensors such as RF radar and sonars typically require many wavelengths across the T/R aperture for good spatial resolution. Optical and millimeter-wave sensors, however, with their small wavelengths, and all MEMS-mediated sensors can and will become small and integrated.
- As increasingly capable sensors evolve, it will be natural to deploy collections of autonomous, mobile, communicating sensors that can cooperate to function as a single, higher-level metasensor.

In a sense, the Navy's CEC already functions as a metasensor but is not yet viewed as such. In CEC, the radars are thought of as individual, independent sensors that are cooperating. The meta-interpretation views the cooperating radars as a single sensor, which happens to have distributed and mobile components.

In the future, as individual sensors grow smaller and more capable, and perhaps become autonomously mobile, they will be deployed in environments where each can see only a small part of the scene and can communicate only in a limited sense with other close-by minisensors. Under these conditions it becomes natural to think of the individual sensors as members of a distributed ant-like society that, through only local communications and simple local protocols, manages to behave as a single purposeful entity—that is, a metasensor. Investigations of the dynamics and potentially chaotic behavior of such distributed systems (e.g., flocks of birds or schools of fish) have recently begun to appear in the physics literature. This direction of research should be carefully nurtured. Figure 4.11 illustrates this evolution.

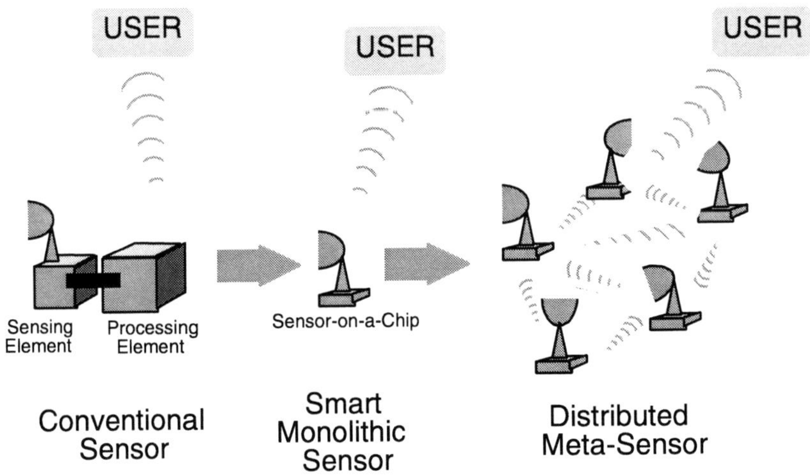

FIGURE 4.11 Sensor evolution.

What Will the Evolution of Sensor Technology Enable?

Broadly speaking, eventually sensor technology will allow us to know everything and to hit anything.

Know Everything—Situational Awareness

With unlimited computational power and affordable, micro- and nanoelectronic monolithic sensor implementations in hand, the battlefield environment can be thoroughly examined by distributions of smart sensors and metasensors from multiple points of view, in multiple spectral bands, with high spatial and temporal resolution, through natural and manmade obscurations to provide a continuous awareness of the current situation. Who and what are present? Precisely where are they? Which are friends and which enemies? Are there threats present? What kind? Precisely where? What are they doing? These are the kinds of questions for which sensor information will be supplied.

Hit Anything—Surgical Smart Weapons

With high-resolution imaging sensors of many kinds, effective ATR and aim point selection algorithms, and large amounts of affordable and compact computational power, ever smaller and smarter weapons can be envisioned. Such inexpensive weapons, able to identify and precisely strike intended targets with high probability, can have great cost and logistic advantages. As the sensors get smaller and smarter, one can even imagine shrinking smart bombs to the level of smart bullets—a small bit of explosive, guided by a miniature monolithic smart sensor and controlled by means of MEMS boundary layer actuators. The panel believes that an adequate aerospace industry manufacturing infrastructure remains to produce smart weapons at a reasonable cost if the production runs are large enough.

Caveats

The panel sets forth the following caveats:

- The transition from microelectronics to nanoelectronics or to superconductor RSFQ logic implies significant reductions in operating voltages along with the threat of increasing EMI vulnerability.
- As the number of observing, communicating sensors on the battlefield increases, the threat of data overload also increases. At the very least, efficient and effective information-extraction procedures must be developed.
- Ever-increasing trends toward small, smart sensor and weapon combinations with various degrees of autonomy conjure up the possibility that the weapon will somehow turn on its masters. As autonomy is a matter of degree rather than a binary issue, confidence in the systems will evolve gradually.
- As mobile autonomous smart sensors and cooperating metasensors evolve, communications integrity will become an increasingly important issue.
- In addition, the information-networking aspects inherent in these distributed concepts suggest increasing vulnerabilities to the threat of conscious attack by means of the techniques of information warfare.

DEVELOPMENTS NEEDED—MILITARY VERSUS COMMERCIAL

The panel sees the five steps listed below as promising an overall sensor development strategy:

1. *Application of existing currently available state-of-the-art technologies.* COTS products will cover much of this need, perhaps with some special packaging to make them suitable for the military environment.
2. *Basic research and development to keep the technology growth curves growing.* DOD is the primary support for these critical activities. This includes such topics as wide-bandgap semiconductors, quantum wires and dots such as single-electron transistors, superconductors such as HTS, RFSQ logic, quantum computing, molecular computing, passive imaging RF/acoustics, meta-sensor community dynamics, and so forth.
3. *Realization that this technology growth does not imply that only 6.1 ex-*

penditures are necessary. For many technologies, superconductor RSFQ logic, for example, considerable 6.2 and 6.3 engineering developments as well as investments in manufacturing technologies are required to attain the desired economical performance levels.

4. *Attention to DOD-specific needs that must be addressed (applicable to the Department of the Navy also).* This category includes such topics as digital radar, condition-based maintenance, and common-aperture electromagnetic sensor combinations, and so forth.

5. *Attention to Navy Department specific needs that must be addressed (ocean issues that will not be addressed by anyone else).* This category includes such topics as littoral environments, synthetic aperture sonar, communications in the ocean, and so forth.

FOREIGN TECHNOLOGY STATUS AND TRENDS

The technologies underlying sensors, including semiconductor and superconductor technology, digital microelectronics, computers and software, microwave, optics, and acoustic technology, and biotechnology are available worldwide. In many of these areas, the United States is not necessarily the leader, although we are not seriously deficient in any of them. Generally, the U.S. advantage lies more with the large amount of resources it is willing to invest in vigorously applying the technology to military problems, rather than with an inherent dominance of the individual technologies.

Some of this is suggested by a critical technologies plan[21] crafted by the DOD in 1990, when the Soviet Union was still thought to be the prime threat. Table 4.3 is a summary assessment of foreign technical capabilities vis-à-vis those of the United States adapted from the DOD report. Interestingly enough, this study concluded that in the all-important area of microelectronics, Japan led the United States in all areas except radiation-hardened electronics, for which there is only a limited commercial market, principally satellite and nuclear power plants. Japan also excelled in photonics, superconductivity, and biotechnology, with considerable strength in machine intelligence and robotics, which are particularly relevant to the development of metasensors. The manufacturing of microelectronics has now spread throughout Southeast Asia. And China, with its vast human resources and unquestionably competent scientists and engineers, is rapidly coming up to speed. The 1990 DOD report also suggested that although North Atlantic Treaty Organization (NATO) allies were lagging at that time in most microelectronic technologies, this situation could change drastically in the near term if the capabilities of individual European countries were integrated.

[21]Department of Defense. 1990. *Critical Technologies,* Department of Defense, Washington, D.C., March 15.

TABLE 4.3 Summary of Foreign Technical Capabilities

Critical Technologies	USSR	NATO Allies	Japan
Semiconductor materials and microelectronic circuits	♦	◊◊	◊◊◊◊
Software producibility	♦	◊◊	◊◊
Parallel computer architectures	♦	◊◊	◊◊
Machine intelligence and robotics	♦	◊◊◊	◊◊◊◊
Simulation and modeling	♦	◊◊◊	◊◊◊
Photonics	♦♦	◊◊	◊◊◊◊
Sensitive radar	♦	◊◊	◊◊
Passive sensors	♦♦	◊◊	◊◊
Signal processing	♦♦	◊◊	◊◊
Signature control	♦♦	◊◊	◊◊
Weapon system environment	♦♦♦	◊◊◊	◊◊
Data fusion	♦♦	◊◊	◊◊
Computational fluid dynamics	♦	◊◊	◊◊
Air-breathing propulsion	♦♦	◊◊◊	◊◊
Pulsed power	♦♦♦♦	◊◊	◊◊
Hypervelocity projectiles	♦♦♦	◊◊	◊◊
High-energy-density material	♦♦♦	◊◊◊	◊◊◊
Composite materials	♦♦	◊◊◊	◊◊◊
Superconductivity	♦♦	◊◊	◊◊◊◊
Biotechnology materials and processes	♦♦	◊◊◊	◊◊◊◊

Position of USSR relative to the United States		Capabilities of others to contribute to the technology	
♦♦♦♦	Significant leads in some niches of technology	◊◊◊◊	Significantly ahead in some niches of technology
♦♦♦	Generally on a par with the United States	◊◊◊	Capable of making major contributions
♦♦	Generally lagging except in some areas	◊◊	Capable of making some contributions
♦	Lagging in all important aspects	◊	Unlikely to make any immediate contribution

SOURCE: Adapted from Department of Defense, 1990, *Critical Technologies,* Department of Defense, Washington, D.C., March 15.

That other nations excel in various sensor areas is well known. In fact, the Navy is currently deploying a shipboard infrared video system made by the French rather than a U.S. version. In another sensor application, it is known that the former Soviet Union has already deployed dual-band optical missile seekers with antijam performance superior to that of U.S. systems.

In other words, just about everybody has the requisite technology to compete in sensor technology. The race will go to the diligent.

TIME SCALE FOR DEVELOPMENT AND DEPLOYMENT

In a general way, the time scale has been addressed through the identification of the exponential growth characteristics of various technology components and their projected performance capabilities as a function of time out to 2035. Over the next four decades, sensors will continue to evolve, and advances can be expected to be deployed within about 5 years of the time of the projected state-of-the-art capability.

Note that all of the technology growth curves presented are conservative, in that they represent affordable, obtainable capabilities, representative of the near state of the art—not the best, one-of-a-kind achievements. For example, the clock speed of microprocessors in desktop personal computers is indicated as 200 MHz today, which is available as a COTS Pentium or Cyrix 6 × 86 from multiple suppliers. Processors with speeds exceeding this have been demonstrated already, with the best performance currently reaching the 400- to 600-MHz range. With this in mind, desktop computers with clock speeds of 1 GHz are expected to be commercially available by about 2005, and military applications of 1-GHz computers might reach the field over the next 5 years from 2005 to 2010.

RECOMMENDATION

Naval operations are increasingly dependent on enhanced sensor data to provide situational awareness, target designation, weapon guidance, condition-based maintenance, platform automation, personnel health and safety monitoring, and logistic management. The Department of the Navy should provide continuing support of sensor technology for areas critical to future naval operations. Special attention should be given to applications of microelectromechanical systems technology because it offers the advantage of low-cost, high-capability systems-on-a-chip that will enable future cooperative sensor networks.

5

Automation

INTRODUCTION

In the future, affordable automated systems and equipment will supplement or supplant many current operator functions in routine operations both aboard ships and on the shore. Automation technologies will also be employed in many steps in the chain from target search and acquisition to weapon impact in an effort to reduce platform size, manpower, and budgets and to minimize the number of personnel exposed to hostile action. Key areas involving automation technologies include (1) unmanned underwater vehicles (UUVs), (2) unmanned aerial vehicles (UAVs), (3) ship automation, (4) platforms and weapons guided by integrated inertial navigation systems (INSs) and the global positioning system (GPS), (5) robotics, and (6) automatic target recognition (ATR).

Opportunities for automating operator functions range from mundane housekeeping and monitoring tasks to intelligent robots for performing hazardous or superhuman tasks. In the time frame of reference, the panel believes that technology will permit semiautonomous, remotely operated combatants as well as fully autonomous low-observable combatants with a variety of tactical capabilities, themselves controlling smart UUV and UAV off-board sensor networks for an extended spatial and temporal awareness of the surrounding space. These unmanned, independent platforms will benefit from all-weather continuous communications and autonomous sea-keeping and logistics support.

The reduction of operators will allow naval and expeditionary platforms to become significantly smaller, more robust, and survivable for the same sensor and weapons load. Many current operators will be replaced with internal smart sensors teamed with robotic devices and highly reliable, fault-tolerant hardware and software systems to deal with casualty and damage detection, control, correction, and recovery.

UUVs and UAVs will go beyond just serving as off-board sensor platforms. Some will be capable of long-term independent operations. Others will combine sophisticated target detection, classification, and tracking capability with maneuvering and weapons-launch capability to serve as semiautonomous and possibly autonomous strike or fleet defense platforms. Those serving as off-board sensors will possess all-weather autonomous launch-and-recovery capability, with the mother platform able to provide automated logistics support.

Platforms, UUVs and UAVs, and weapons will enjoy the benefits of continuously available position information to 1 m or less accuracy from small, inexpensive integrated GPS/INS navigators. Sophisticated adaptive mission planning/replanning processors and ATR software on board these semiautonomous vehicles and weapons will support obstacle avoidance and mission restructuring in real time.

UNMANNED UNDERWATER VEHICLES

Description of the Technology

Unmanned underwater vehicle technology is actually a combination of a number of emerging technologies that can provide the naval forces with new operational capabilities in undersea warfare, reconnaissance, and surveillance. During the latter days of the Cold War, UUVs were envisioned as force multipliers for the blue-water antisubmarine warfare (ASW) mission, including mobile acoustic barrier patrol, off-board active acoustic sources, deep-water mine avoidance for submarines, and tactical aids for port egress. Although considerable research was undertaken and a number of operational demonstrations were conducted in the late 1980s and early 1990s, the blue-water ASW problem was perceived to disappear before any operational UUVs were deployed. In the post-Cold War era, potential UUV missions have centered on undersea littoral warfare, including shallow-water mine reconnaissance, diesel submarine ASW, surveillance, and counterproliferation. The Department of the Navy has just started an acquisition program for the first operational UUV, the long-term mine reconnaissance system (LMRS), which is a submarine-launched and submarine-recovered countermine system.

Relevance to the Naval Forces

UUVs can conduct a series of primary and support missions to extend the fleet's warfighting capability, provide surveillance and reconnaissance functions, and conduct tactical oceanographic missions. UUVs not only provide increased mission capability but also should not add significantly to operational costs, since no crew is required and, for the most part, maintenance and support can be conducted by existing ship's force.

Technology Status and Trends

There is no current tactical application of UUVs. The various technologies are just becoming mature and operational confidence is growing. The Department of the Navy has just started its first UUV acquisition program, LMRS.

Future Impact of Technology Trends on U.S. Naval Operations

UUVs will be operated by both submarines and surface combatants, and variants may well be launched from fixed-wing and rotary aircraft. Autonomous, independent UUV operations, without a host platform, are also likely for certain missions. UUVs will provide economical force multiplication, increased battle-space awareness, reduction in exposure to hostile action, and extend operational capability.

Future UUV applications will include the following:

- *Off-board sensing.* Platforms for acoustic and nonacoustic detection and localization and tracking of targets. With multiple vehicles operating as nodes in a far-ranging, cooperative, smart-sensor network, detection thresholds will be significantly improved.
- *Intelligence collection.* Covert platforms for gathering tactical and strategic intelligence, including minefield surveillance, harbor and waterway traffic surveillance, surveys of uncharted bodies of water, plus the ability to carry a limited range of weapons for attacking detected targets.
- *Mission support.* Autonomous, covert, and stealthy electronic warfare (EW) and information warfare (IW) platforms for decoy operations in coastal areas, for example.

Developments Needed

The most significant UUV technology needed for long-duration (30 days or longer) autonomous operation is safe, reliable, high-energy-density and high-power-density batteries and fuel cells operating from seawater or equivalent sources of energy. Of almost equal importance is the need for compact, cost-effective, reliable UUV hardware and software systems.

On-board intelligent navigation and control and decision planning and re-planning are essential technologies for reliable, long-term unattended operations. Other important technologies include low-power, high-performance acoustic and nonacoustic sensors; high-bandwidth, continuous, covert communications between cooperating UUVs, between UUVs and hosts, and between UUVs and global-reaching operators; and automated resupply, launch, and recovery from unmanned hosts.

Low observability to avoid acoustic and nonacoustic detection is essential to the survivability of autonomous UUVs. This requires vehicle shaping, acoustic and optically absorbing material coatings, equipment quieting, and low-noise propulsion, such as direct electric drive.

Current commercial applications focus more on remotely operated vehicles (ROVs) rather than on autonomous vehicles. Because these vehicles share some of the same needs as the autonomous UUVs envisioned for naval force operations, a small portion of the technical advances in the above technologies can be expected from the commercial sector. Technologies for long-duration, highly reliable autonomous surveillance and tactical missions will need to be developed by the Department of the Navy.

Foreign Technology Status and Trends

In general, foreign developments in UUV technology have lagged significantly behind the work conducted in the United States. The United States has made a significant technology investment in the military application of unmanned undersea vehicles, whereas most foreign UUV research has been at the university level, and much has been oriented toward scientific and oceanographic uses of the technology. The only exception is the former Soviet Union, which did invest in UUV research and development for the military.

Time Scale for System Insertion

The first operational UUV, the LMRS, is scheduled for initial operation in 2003 to 2004. It is likely that other UUV systems would be easily inserted within the same time frame. More sophisticated systems with higher degrees of autonomy enabled by intelligent software systems that can plan and replan to deal with failures and contingencies will follow in subsequent years. Toward the 2035 time frame, the panel can envision highly autonomous UUVs that operate in cooperative engagements. These societies of UUVs will be capable of sensing their environments and communicating with each other to optimize underwater missions.

UNMANNED AERIAL VEHICLES

Description of the Technology

Unmanned aerial vehicles are small to large unmanned aircraft that can be launched from ships and shore to conduct reconnaissance and surveillance, to coordinate and participate in cooperative engagements against targets, and to act as weapons in some applications.

Relevance to the Naval Forces

UAVs have the potential for becoming pervasive solutions for many future Navy and Marine Corps needs. The rapid evolution of mission-payload technologies, including sensors and sensors combined with weapon systems, and aeronautical technologies, including navigation, autonomous control, and propulsion, will enable UAV missions heretofore not possible.

A future vision of the UAV for naval operations might be as follows: Navy and Marine Corps forces will control a host of UAVs sweeping the combat areas around the battle group and onshore with electro-optical/infrared (EO/IR) and synthetic aperture radar (SAR) reconnaissance sensors, collecting enemy radar and communications signals, coordinating friendly communications, detecting mine fields, enemy submarines, and cruise missiles, jamming enemy radars and communications, and performing targeting and control missions for precision weapons.

Linking several UAVs with precision timing into a society of agents would create a signals-collecting dragnet, not allowing a single RF signal to escape collection and geolocation. Similarly, UAVs of the future will be able to deliver the appropriate response, whether it is jamming, performing as an antiradiation weapon, generating directed-energy weapon (DEW) disruption signals for information warfare purposes, or delivering precision-guided munitions.

Although most UAV systems envisioned today involve remote piloting and require some takeoff-and-recovery real estate, future UAVs can well use short takeoff and landing (STOL) or vertical takeoff and landing (VTOL). They will be totally autonomous with sufficient on-board intelligence to change their operational behavior on the basis of observations in the conflict areas. Strategic intelligence will be acquired with high-altitude, long-endurance (many days), low-observable UAVs. They will be refueled in flight by tankers or perhaps by beams of microwave energy from ships or shore.

Technology Status

Most current UAV technology development is being carried out by the Joint DOD Cruise Missile and UAV program office, administered by the Department of the Navy. This office was established at the request of Congress to ensure that UAV development programs were coordinated and nonduplicative between military departments. In addition, DARPA is funding the development of high-altitude, long-endurance UAVs.

Developments Needed

The panel believes that the highest priority should be given to the development of gas turbine engines for propelling UAVs. Gas turbine engines have long life and low weight-to-thrust ratios, use heavy fuels, and possess low acoustic signatures.

High-altitude and long-endurance operation would enable UAVs to continually observe the entire sphere of importance around the battle group and the expeditionary forces on shore. Payloads equipped with sensors sensitive over the spectral range from EO/IR to millimeter waves with radar and SAR and with electronic surveillance devices would observe all enemy activities and report this intelligence immediately to all friendly users. These UAVs would participate in cooperative engagements against difficult targets, such as low-observable missiles. The ability to time-integrate weak ocean-surface signatures from ships, low-flying missiles, and submarines would greatly increase the detection of these threats. With the future availability of affordable and reliable solid-state blue-green lasers, direct detection of mines and enemy submarines in the littorals and direct communications with friendly attack submarines and UUVs will become possible.

Because of the importance of these UAVs to surveillance and detection of hostile threats and targets, the panel strongly encourages the continual support of technology development in high-altitude, long-duration flight and in low-cost, small-size sensor payloads. Low-cost communications between UAVs and the user community will also need to be developed.

An interesting new airframe technology that should be considered for UAVs is the so-called free-wing. This wing configuration is free to rotate in pitch, providing for a completely passive means to adapt to the instantaneous direction of air flow. Thus, as the wing encounters turbulence, the attitude reconfigures without the necessity of active control. Not only does this technology allow for much higher platform stability because the entire airframe is not buffeted by turbulence as with a fixed wing, but also the wing is by its design stall-free, thus mitigating one of the major sources of aircraft failure.

The free-wing combined with another new concept called the tilt body, which adds trim tabs to the fuselage, provides a capability for small-area launch and recovery as well as a simple transition from helicopter-like flight to conventional flight. UAVs with VTOL capability would bring a new dimension to Navy and Marine Corps missions. The basic free-wing concept has been flight demonstrated both as a UAV airframe and as a manned ultralight vehicle. The free-wing tilt body has been built and flown successfully but has not yet been commercially or militarily applied. The use of this technology could lead to UAVs that can operate more effectively from a variety of naval ships. Both analytical and hardware designs are required, prototypes need to be built, and in-fleet experience needs to be gained.

UAVs with very low observability (VLO) features will be required for a variety of covert missions, and all UAVs must possess some degree of low observability for survivability. The panel believes that attention is needed to developing the low-cost materials and active signature management technologies that will enable these capabilities.

Efforts should also be undertaken to develop a family of small UAVs that are significantly less expensive, so inexpensive that they could be considered as expendable. Such UAVs would open up entire new operations. Equipped with the low-cost, smart sensor-on-a-chip systems, as described in Chapter 4, and modest munitions, these micro-UAVs could perform functions such as search and find, followed by homing and kill.

SHIP AUTOMATION

Description of the Technology

The surface combatants, submarines, and amphibious and auxiliary ships of the future naval forces will have a much different configuration than the ships of today. Ships will be highly automated and conduct extensive semiautonomous (limited need for remote operator intervention or decision) operations. Many of the functions currently conducted by personnel on board today's ships will be performed by autonomous systems on board or in some cases teleoperated by humans via communication links. The reduced or nonexistent manning that results from the broad application of automation will allow smaller, more damage-resistant and survivable ships, with fewer personnel being exposed to hostile action.

The set of technologies required for full automation of a naval ship is diverse in scope and includes the following:

- Over-the-horizon (OTH) communications that are jam resistant, damage tolerant, and all-weather capable;
- Distributed computer systems with high-performance capabilities and architectures that are fault and damage tolerant;
- Advanced computer algorithms and programs for operation, monitoring, control, maintenance, and logistics of all combat, weapon, hull, mechanical, and electrical systems;
- Robotics for damage control and special functions such as under way replenishment and launch and recovery of UAVs and UUVs;
- Reconfigurable power-generation and auxiliary systems that are highly reliable, fault tolerant, and damage tolerant;
- Navigation technologies that are continuously all-weather capable and jam resistant and provide passive knowledge of the geographic location of a ship to 1-m accuracy; and
- Security technologies for physical, communications, and control systems.

Relevance to the Naval Forces

The implementation of ship automation will significantly reduce manning, currently the highest cost element of naval ship operations. The reduction of personnel aboard ships will also reduce casualties in future hostile actions.

Applications Today and in the Future

The fiscal reality of the high costs associated with people on ships has forced efforts within the Department of the Navy to reduce manning. The Navy's new surface combatant has a goal of 95 people compared with a manning level of about 320 on an equivalent ship today. The arsenal ship has established a goal of 50 people maximum. The Navy's next aircraft carrier will attempt to reduce the manning to one-half the present manning levels. The smart-ship initiative and some advanced technology demonstrations will make some progress toward removing people from ships, but much further work is needed. Significant reduction of people from ships requires not only definition and development of the technologies required but also acceptance of the manning philosophy and doctrine to make autonomous ship operations a reality.

Technology Status and Developments Needed

The technology status and the developments necessary for autonomous ship operations are as follows:

- OTH communication capability exists today; however, the issues of being noninterruptable, jam resistant, damage tolerant, and all-weather capable must be addressed. The criticality and reliability of the communication link are overall system issues that will determine the extent and reliability of autonomy.
- The computer-processing power required to enable ship automation is expected to continually increase over the next 40 years and is not envisioned to be a limitation. The development of computer architectures and systems that are fault tolerant and damage tolerant is required, however.
- Advance computer programs and algorithms will require a large amount of development work. The philosophy and methods developed for the control of small autonomous systems, such as present-day UAVs and UUVs, will have to be greatly expanded and enhanced to address all the complex issues required for a ship to carry out extremely complex and lengthy missions.
- The rudimentary diagnostics used today to implement condition-based maintenance will have to be further developed into sophisticated prognostics in the future. Such prognostics must deal not only with the condition of machinery and equipment but also with the condition of the overall system, the impact of this condition on the warfighting capability of the ship, and the ability to reason, reach decisions, and take appropriate action.
- Robotics systems for damage control will require some development. Some of the technology development utilized for such things as factory automation may be adaptable to many of the damage-control and special applications on a ship.
- The ability for a ship to navigate anyplace on the globe in a nearly hands-off mode exists today. Expected advancements in antijam capability in GPS

receivers and integration with inexpensive inertial navigation systems will permit the ship, the off-board sensors, and the weapon systems to precisely locate themselves at all times and under all circumstances. Automated mission planning and on-the-fly replanning using the certain knowledge of current geographic location will enable autonomous obstacle avoidance and mission reconfiguration in the future.

GPS/INS SYSTEMS FOR NAVAL PLATFORMS AND WEAPONS

Description of the Technology

The Precise Positioning Service (PPS) of the Global Positioning Satellite system currently provides spherical error of position (SEP) accuracy to 16 m and CEP accuracy between 8 and 10 m over most parts of Earth. This impressive technology enables exceptional worldwide navigation, especially when multiple GPS measurements are combined in a Kalman filter to update an inertial navigation system on a platform or a weapon. The Kalman filter provides an opportunity to calibrate some of the GPS errors, such as clock and ephemeris errors, as well as several of the inertial system errors, and when properly implemented, CEPs of better than 8 m have been observed.

Relevance to the Naval Forces

The ability to provide precise navigation for Navy and Marine Corps platforms anywhere in the world and to deliver precision weapons on target through combinations of external GPS and internal INS navigation is essential to future naval operations.

Technology Status and Trends

The path to the projected 1-m accuracy in the 2010 time frame includes the use of additional GPS ground monitor stations and advanced ground-based software to generate and uplink more accurate and frequent corrections to satellite clock and ephemeris data. Interranging and communications between satellites will be used to further refine the navigation information. Platform receivers will also use all-in-view tracking and will be navigating using 8 satellites and perhaps up to 12, as opposed to the 4 satellites in many current implementations. Advanced atmospheric models and algorithms will also be implemented to compensate for tropospheric and ionospheric effects, as well as multipath signals to the receiver.

In military applications, users are also concerned with the integrity and reliability of the system in the presence of spoofing, jamming, and interference. In the future, communication crosslinks between satellites and improved ground monitoring of satellite health status will allow almost instantaneous information to users about failing satellites whose signals should be ignored. The widespread use of all-in-view receivers will allow continued high-accuracy navigation even with fewer satellites. In addition, the tight-coupling between the inertial navigation systems and the GPS data in future platforms will allow continuous inertial navigation with tracking from just a single satellite. This capability is particularly important for the Marine Corps ground forces and in low-level aerial applications. Low-level flight creates severe signal masking because of foliage, terrain, and urban structures. The continued use of encryption and the ability to rapidly change the military code will minimize false satellite navigation signals.

Jamming and intentional interference will remain serious issues for military as well as civilian operations. The GPS satellites are typically in high-altitude 12-hour orbits and broadcast at low power (on the order of 20 W). Thus, very low power jammers on or near Earth's surface can jam the weak received signal. Inertial navigation systems, however, are not jammable, and so there is real synergy in combining satellite and inertial navigation. Through the use of inertial system information, the tracking loop bandwidth in the GPS reception can be narrowed by an order of magnitude, thus increasing the antijam margin by 10 dB. In designing an integrated GPS/INS system, tradeoffs must be made between the accuracy and cost of the inertial system and the cost associated with an antijam GPS receiver/antenna capability.

Proponents of inertial systems argue that a high-antijam GPS receiver is not required because of the inherent high accuracy of the inertial information, whereas GPS proponents argue that use of a high-antijam GPS receiver will substantially reduce inertial system accuracy requirements. Both arguments are strongly dependent on the usually poorly defined mission and jamming scenarios. However, what has generally become accepted is that GPS is remarkably vulnerable to jamming during the signal acquisition phase with conventional receivers. For example, a broadband CW jammer with 1-W effective radiated power located at 100-km line-of-sight distance from a GPS receiver antenna would prevent acquisition of any satellites using the civilian coarse/acquisition (C/A) code. A 1-W jammer is cheap and is about the size of a hockey puck. Furthermore, the current civilian code can be spoofed by an even smaller power jammer. In general, a GPS receiver cannot be expected to acquire the C/A signal in a hostile environment. For long-range navigation applications, the C/A signal acquisition could be accomplished outside hostile territory, with a transition inside hostile territory to military P(Y) code lock that has a higher level of jamming immunity. A 1,000-W jammer at about 100 km would now be required to break receiver lock. If the vehicle approaches within 10 km of the jammer, power levels of about 10 W would be effective in breaking military code lock.

The military P(Y) code (an encrypted precision code) has more antijam protection than the civilian code because of its 10-times-larger spread-spectrum bandwidth. Therefore, it is important to develop receivers that can acquire the military code without having to first acquire the civilian code. Because the military code is very long, however, many seconds of time and many correlators are needed for a two-dimensional search over code timing and Doppler frequency. It would be faster if satellite ephemerides and accurate code timing were available to perform a "hot" start. For a GPS-aided weapon launched from an aircraft, accurate timing and satellite position could be transferred from the aircraft to the weapon. This transfer normally requires a wideband data bus, but few aircraft are currently so equipped. However, in the future, most aircraft will be so equipped and weapons will normally be hot-started directly into the military code.

As new receiver technology with massively parallel correlators, improved algorithms, and adaptive or nulling antenna technologies are incorporated into weapon and platform systems, antijam performance will improve significantly. If antijam performance is doubled (in decibels), then the jammer in the previous cases would have to be five orders of magnitude more powerful to be effective. Such powerful jammers would present inviting targets to antiradiation homing missiles. In the terminal flight area, the GPS receiver will probably be jammed and the weapon or platform will have to depend on inertial-only guidance or the use of a target sensor. Thus, it is important to make sure that adequate backup vehicle guidance and navigation capability is provided to meet military mission requirements against adversaries who are willing to invest in electronic countermeasures. This is true today and is expected to remain so in the foreseeable future.

To prevent the use of GPS by hostile forces, it may be necessary to jam the C/A code in the battlefield area, thus giving more emphasis for our own forces to directly acquire the P(Y) code. Additionally, a hostile platform using a combination of correction signals transmitted from a fixed ground station and the direct GPS signal to improve accuracy, a technique known as differential GPS, is vulnerable to jamming of the data links and destruction of the ground station. Many of these notions have been described in the 1995 report by the DSB Task Force on GPS.[1] The Russian GLONASS system currently provides accuracy similar to GPS-PPS, and hence some requirements could also be satisfied by using GLONASS under jamming conditions.

Time Scale for Development and Insertion

In the next few decades, accuracy of the integrated navigation solution will improve from the current 8-m to the 3-m range (for precision strike applications) and then to the 1-m range. Planned programs by the GPS Joint Program Office

[1]Defense Science Board. 1995. *Report of the Defense Science Board Task Force on Global Positioning System,* Office of the Under Secretary of Defense for Acquisition and Technology, Washington, D.C.

(JPO) are in place to accomplish the latter. Complementary techniques to provide geo-location targeting information at comparable accuracy to the GPS/INS solution will be developed. Techniques to deny the use of accurate GPS (or the Russian equivalent GLONASS system) to hostile forces will also be developed in the next decade. In parallel, lower-cost inertial components with improved accuracy will be developed using MEMS technology and other approaches. Tightly integrated architectures for high-antijam GPS/INS systems will become common, replacing avionics architectures based on functional black boxes where receivers and inertial systems are treated as stand-alone systems. Techniques will be developed to deny the use of GPS or GLONASS by hostile forces.

ROBOTICS

Description of the Technology

A robot can be defined as a general-purpose machine system that, like a human, can perform a variety of different tasks under conditions that may not be known a priori. Robots are intelligent, taskable mechanical entities that can assist or replace sailors and marines performing high-risk operations in hazardous environments. Over the last decade, robotics technology has made significant inroads into high-volume manufacturing operations. Recent improvements in sensing, artificial intelligence (AI), and control technology, coupled with a significant improvement in low-cost, low-power computational devices, have made possible intelligent robots capable of operating in unstructured environments.

The key elements of robot systems are (1) *effectors*, which corresponds to the arms, hands, legs, and feet of humans; (2) *sensors*, which detect the conditions around the robot; (3) *computers*, which bring intelligence to the performance of the robot tasks; and (4) *communications systems*, which allow robots to communicate with other robots and with humans.

Robot systems are deployed and operated in different ways depending on the application as described below:

1. *Teleoperated systems* are devices that are controlled directly by a human operator. Such systems are also called remotely operated. The operator is usually connected to the device over dedicated or shared telecommunication links that allow control signals to be sent to the robot and allow sensor signals from the robot to be displayed to the operator.

2. *Telepresent systems* are specialized teleoperated systems in which the operator's perception of the remote environment is identical to what it might have been if the operator were present at the robot site. Achieving telepresence requires the use of immersive displays, high-resolution sensors, and much higher communication bandwidths.

3. *Telerobotic systems* involve the transfer of human control to autonomous

robot control sometime during the course of activities. For example, the operator may position the robot at the start of a task and then turn over control to perform the task to the robot.

4. *Robot systems* are the pinnacle of robot complexity. These systems can perform tasks autonomously under task guidance by human operators.

Robotic systems have been used extensively in repetitive manufacturing environments where their speed and precision offer significant advantages over human performance. Autonomous and teleoperated vehicles (ground vehicles, underwater vehicles, and aerial vehicles) allow extension of human presence into environments that are dangerous, e.g., battlefields, or inaccessible by other means, e.g., deep oceans.

Robotics Technologies Status and Trends

Actuator and Locomotion Technology

Robot systems manipulate and navigate through the environment using actuators. Most robot manipulators are built as a collection of articulated links with a firm stable base and ending in a specialized end effector. End effectors can be specialized tools, such as a welding torch, or generic articulated grasping devices. Most ground-based mobile robots use tracked or wheeled platforms, although legged devices have significant advantages in harsh terrain, e.g., on beaches or in urban debris-filled buildings.

Until recently, design efforts were targeted at building large actuators with very high power-to-weight ratios, resulting in heating and vibration problems encountered during operation. Recently, however, there has been a shift toward building small, cheap, dispensable systems. Current actuation technology, however, is far from what might be required for the realization of small machines. This has resulted in a significant interest in the miniaturization of electric motors, as well as alternate means of actuation that might be applicable to small systems. The latter has resulted in the development of materials that might serve as artificial muscles, expanding gels, and electrostrictive polymers. And finally, there is a growing body of results related to microactuation, and a small number of actuator technologies have actually been scaled down to the micrometer scale. These technologies have been applied to a small number of problems, but the true challenge will be to integrate them into a centimeter-scale system using MEMS and other approaches to demonstrate coordinated micro/macro actuation.

Sensing Technology

Sensing is an integral part of most machines. It is used for taking the measurements required for controlling the machine. Currently, in addition to various position and velocity sensors used for providing internal state information, the use of IR or ultrasound sensors for the detection of obstacles is quite common. Sophisticated systems also use one or more visual sensors for acquiring information about the nature of a robot's environment. And finally, a wide array of domain-specific sensors, such as subsurface imaging sensors or magnetic sensors for mine detection, or thermal-imaging sensors for detection and classification of humans, have been suggested for successfully tasking the robots within various scenarios.

Recent advances in stereo-vision algorithms coupled by low-cost, high-speed digital signal-processing chips have resulted in the ability to build extremely cheap vision sensors that can provide future robots with a very-high-resolution sense of its environment.

Control Technology

Control typically refers to the choice of actuation signals that might be required for forcing a machine to exhibit some prescribed behavior. Although traditional control has extensively studied problems related to set-point and trajectory control of linear systems, few formal concepts exist for the synthesis of control methods that have the ability to integrate information from a range of sensors, such as vision and touch. This has resulted in the investigation of alternate methods such as intelligent control, fuzzy control, neural-network control, and rule-based control for autonomous systems. Intelligent control refers to the ability of a control system to make decisions online and to reconfigure control structures as necessary. Rule-based control attempts to express decisions as heuristically derived rules driven by sensory information. Fuzzy control applies fuzzy membership functions for the determination of control structures, and neural-network control uses neural networks for functional approximations in the identification and control processes.

With the advances in the enabling technologies outlined above, current robotics research is focused on the development of taskable, cooperative, autonomous devices in a variety of environments. The Department of the Navy is funding the development of a variety of semiautonomous underwater vehicles. In collaboration with the Army Unmanned Ground Vehicle office, robot vehicles are being investigated.

In the future, collaborative systems consisting of specialized robots at multiple scales, for example, miniature robots controlled by a larger mother robot, communicating with each other and with external controllers in very complex environments, will be needed in many military scenarios being discussed today. Research into computationally tractable representations of the environment, reasoning with uncertain and partial knowledge, cooperative tasking of heterogeneous devices, techniques for manipulation and locomotion in hazardous terrain, sensing systems, robot power systems, and system-integration techniques will be required to bring these technologies into fieldable systems.

Robot Applications Relevant to the Naval Forces

Shipbuilding

Robots are used extensively in manufacturing operations that are repetitive and require precision motions. The automobile industry has been the leader in the application of such devices. Shipbuilding, however, presents a very different set of manufacturing problems. Most ships are fairly dissimilar and built in very small quantities. Many of the structures within a ship hull are modified and changed in the field.

Sensor-based robot systems are therefore indispensable for automating the shipbuilding processes. Several advanced technology programs, such as DARPA's Intelligent Design program and Simulation-Based Acquisition program, have focused on automating the design of ships. The next step in the evolution of this process is to use automated design information to control robots and other devices to perform the manufacturing tasks. Such manufacturing technology has been demonstrated in the past in specialized areas like robotic arc welding. Similar systems for grinding and deburring are being actively researched in the manufacturing community.

Another application of intelligent, sensor-based robots in shipbuilding is in creating as-built drawings of ships by actively surveying ship interiors. This brings together the problems of mobility in restricted spaces, sensing, and interpretation of sensed data into a single application.

Munitions Handling

One critical and dangerous military task is the handling of munitions, especially on the flight deck of aircraft carriers. Heavy munitions have to be moved from storage to aircraft through the clutter of pipes, hoses, and other obstacles typically found on an active flight deck. These movements have to be coordinated with the other time-critical activity that surrounds the landing, refueling, preparation, and launching of aircraft. Errors in this process can be fatal.

Combining dexterous manipulation and sensor-based navigation and locomotion, munitions-handling systems capable of operating in the flight deck environment (wheeled or tracked vehicles will not be suitable) will provide significant benefits. Further, such a system would be coupled with intelligent processing of logistics and operational-planning information to achieve end-to-end control over the matching of mission-specific munitions to aircraft. The core technologies required to build such a system are already available or are maturing at a rapid rate (see Chapter 10, "Technologies for Enterprise Processes").

The Naval Explosive Ordnance Disposal Technology Division is tasked with the mission of disposal of unexploded ordnance as well as defusing and disposal of improvised explosive devices. This is a fundamentally risky task because it involves putting humans into close proximity to unknown explosive devices. Robotics can play a key part in these tasks.

Robotics technology can be applied to the explosives disposal problem in many ways. The simplest is the use of low-cost, expendable, teleoperated devices to approach, handle, and move explosives to a safe location. This keeps the operator at a safe distance from the explosive device. Teleoperated devices could also be used to neutralize the explosives by using a variety of conventional manual disrupters.

More challenging is the use of robot devices that can make intelligent decisions about the nature of the environment and can position sensors in close proximity to, or even inside, the shell of explosive devices. Such robots require significant dexterity to manipulate the explosives, penetrate the outer shells, and position sensors inside the shell with high precision. Snake-like manipulators being developed today hold the promise of providing such capabilities. In addition, telepresence that conveys high-resolution visual, tactile, auditory, and haptic sense to the operator is crucial because the smallest miscalculation or uncontrolled movement can lead to a catastrophic outcome. Telepresense systems being developed for remote surgery demonstrate many of the key characteristics necessary for explosives neutralization.

Undersea Search and Rescue

The use of unmanned ROVs and their autonomous equivalents (UUVs) provides the naval forces with critical tools for deep-sea search and rescue. High-profile applications such as the search for the TWA Flight 800 debris and photographing of the *Titanic* have raised the visibility of the technology. This technology is being actively pursued by Department of the Navy laboratories, such as NRaD (e.g., with the Advanced Unmanned Search System capable of searching at depths of 20,000 ft), and by others (MITI in Japan, for example).

In the future, these systems are likely to evolve toward cooperative collections of vehicles. Significant research in core technologies is necessary to achieve this goal. To build an effective collection of collaborative automatons, two things are necessary: each device must be able to (1) model and understand its environment, and (2) communicate this understanding and its intentions to other automatons. Both of these abilities are difficult in an undersea environment because visibility is very limited and communication bandwidth between untethered vehicles is very low, acoustics being the primary medium of effective communications. Both of these limitations place significant stress on the ability of the undersea vehicles to interpret the condition of their environment. Collaboration between UAVs presents far fewer problems and will probably occur earlier.

Robot Assistants in Small Unit Operations

The Marine Corps is in the process of evaluating a fundamental change in its operational doctrine. The Sea Dragon doctrine dictates a large collection of small, dismounted units ranging across the target area, controlled from protected cells on offshore ships. Unlike conventional warfare where units are in close proximity to each other and can provide mutual support, the Sea Dragon concept places a significant stress on the small units. Robot assistants can play a significant role in extending the surveillance and offensive capabilities of these small units.

One approach is to develop small, portable, and rapidly deployable robots with a collection of tactically relevant sensors, such as video cameras, audio sensors, and chemical and biological sensors, that can be carried into the battle space by the expeditionary forces. When required, these devices would be deployed to autonomously perform a range of tasks such as watching the perimeter, protecting the rear, and looking around corners or up stairwells in an urban environment. The devices must be autonomous because the marines cannot take on the additional task of teleoperating the devices for any significant periods of time. These challenges are exacerbated by the poor RF communications and difficulties common in urban environments where the Marine Corps will likely operate.

Littoral and Surf Zone Operations

Mines that are originally placed during various internal and border conflicts can remain in place for a number of years. Mines continue to be a very cost-effective defensive weapon and hence are deployed by many adversaries. Of particular interest to the Navy and Marine Corps, therefore, are the detection and clearance of various mines in surf zones, in shallow waters, and in deep waters. Most littoral mine clearance strategies involve the dispatch of Navy Seal units. The consensus emerging in the robotics community is that taskable, expendable machines could significantly reduce the risks involved in such operations for two reasons. First, in view of the inherent wear and tear on the trigger mechanisms of older mines, as well as the movement of these mines resulting from water currents, waves, and other natural phenomena, they could potentially exhibit unreliable behavior; and second, there are areas in the world where sophisticated smart mines have been deployed. Such mines could detect the approach of Navy Seals, explode, and cause lethal damage.

In the past two decades, various programs have been directed at building small, lightweight, low-cost UUVs that can be easily deployed and tasked to deal with various littoral scenarios in deep waters, shallow waters, and surf zones.

The technologies that must be developed in support of these missions are described above.

AUTOMATIC TARGET RECOGNITION

Description of the Technology

Automatic target recognition is the use of computers to process outputs of one or more sensors to locate or identify specific objects. The problem is made difficult by the presence of sensor noise, degradation of sensor responses resulting from adverse sensing conditions or countermeasures, and scene clutter. Typical sensors used in ATR systems include forward-looking infrared radars (FLIRs), millimeter-wave imaging sensors, synthetic aperture radars (SARs), and laser radars (LIDARs)

ATR methods themselves fall very roughly into three, heavily overlapped categories as follows:

- *Statistical pattern recognition.* Statistics regarding some aspect of an object's appearance are used to discriminate targets from nontargets. Many approaches, varying in mathematical rigor and the manner in which features are chosen, fall into the pattern-recognition category. Familiar techniques include Bayesian probability calculus as well as clustering methods.
- *Neural networks.* Although neural networks are also statistical classifiers, they often differ from typical pattern-recognition approaches in the style of computation and in the means by which features are chosen. Both model-based and statistical approaches require the system designer to choose discriminating features, whereas many types of neural networks can learn features on their own. Some researchers believe that this advantage indicates that the neural networks are the future of ATR and sensor fusion.
- *Model-based recognition.* Thought to have originated when artificial intelligence techniques were applied to computer vision, the model-based approach relies on a hypothesis-and-test paradigm that attempts to reconcile predicted features based on a target model with extracted features from real sensor data. The extent to which predicted features match sensed data is evidence that a target is present.

When considering factors affecting ATR, one cannot ignore the impact of the following supporting technologies:

- *Sensors.* Better sensors are generally expected to afford better ATR performance.
- *Signal processors.* Faster processors mean that more targets, models, and

situations can be processed. Fuller search and fewer shortcuts will yield better performance.

- *Intelligence gathering.* The better the a priori knowledge of the target, the better one can design recognition algorithms.
- *Target modeling and simulation.* Simulation with high-fidelity phenomenology will allow discriminating features to be developed and thoroughly tested.

Relevance to the Naval Forces

Naval forces need to hit more targets with higher accuracy and from greater distances using fewer humans. The Department of the Navy is faced with an optimization problem, since it is trying to make the best use of increasingly limited resources. Automatic target recognition plays a key role in this problem, as it essentially replaces human pilots during some or all phases of tracking and homing on both fixed and mobile targets. Improvement in recognition algorithms means that human operators are required to do less and less, freeing them to attend to other problems. Hence, ATR gives the Navy and Marine Corps leverage in solving their increasingly complex resource allocation problems.

Technology Status and Trends

Because ATR encompasses a large number of applications and because of the general lack of standardized testing procedures, the current state of the art in ATR is not easily quantifiable. However, some general observations can be made. For instance, there is a widespread reliance on correlation techniques. In addition, variants of correlation, such as the Hausdorf distance—a measure of shape similarity—are also used. An increasingly popular method of integrating fragments recognized by correlation or other means is the elastic template. Regardless of the particular processing steps employed, ATR techniques generally encounter difficulty in highly cluttered scenes. Hence, there is an increasing interest in multisensor fusion technologies to more easily distinguish targets from clutter.

To fully understand future directions in ATR, one must consider the larger context in which these algorithms function. In particular, the success or failure of ATR is dependent on the sensors employed and the a priori modeling used to process the sensor information.

Trends in Sensors

Listed below are trends in sensors:

- Increasing sensitivity, higher resolution, lower noise;
- Large increases in data rates and processing capability;

- Increased use of multiple distributive sensor systems to combat weather and countermeasures and to provide multiple viewing perspectives;
- Increased use of multiple wavelength and multimode sensing to see through foliage and camouflage and to detect low-observable targets; and
- Increased employment of LIDAR and imaging millimeter waves to provide two- and three-dimensional images of targets.

Trends in ATR Algorithms

Trends in ATR algorithms are as follows:

- Use of adaptive systems that are insensitive to component failure and changes in target appearance;
- Use of uncertainty reasoning in the representation of ambiguous sensor reports and in the handling of modeling limitations; and
- Increased use of context in processing of models and sensor data.

The increasing sophistication of targets and countermeasures cannot be overlooked. Failure to deal with new countermeasures and stealth techniques may nullify gains resulting from these trends.

Future Impact of Technology Trends on Naval Operations

The following vision illustrates how ATR could work in the future if the supporting technologies are developed:

- Surveillance missions to supplement data found to be incomplete are automatically performed.
- Target-and-scene models, replete with phenomenology and weather reports, are automatically extracted from existing and newly gathered intelligence databases.
- Relevant target models are communicated to smart weapons and submunitions.
- Intelligence data archives are automatically polled.
- ATR algorithms for on-board smart weapons and submunitions specialize themselves by uploading models, off-board cues, and context information.
- Once approaching the target, smart weapons automatically adjust for differences between the real and hypothesized scene attributes.
- Intelligence databases and simulations are updated from the information sent back by weapons approaching the target and from platforms assessing damage.
- Updated scene-and-target models are unloaded into subsequent rounds of smart weapons and submunitions.

Developments Needed

Although ATR can be used in commercial applications, such as automated assembly, typical industrial problems are vastly simpler than those encountered in battle. Hence, the main drivers for improvement of ATR are likely to come from the military rather than the commercial world.

The following developments in supporting technologies and ATR algorithms are needed to make significant improvements in the future:

- *Standardized testing through simulations.* The establishment of standardized tests and testbeds would allow different approaches and progress in the field of ATR to be objectively measured. No single set of sensor readings can adequately capture the variation encountered in the real world. Hence, rather than evaluating candidate ATR systems based on fixed sets of sensor outputs, simulations incorporating sensor noise and realistic target and countermeasures phenomenology should be used. In simulation-based testing, candidate ATR systems could be judged based on performance in a number of standardized scenarios.

  The fragmented nature of existing modeling and simulation packages is an impediment to simulation-based testing. The need is clear for an accessible, standardized set of multimode target models. Once available, these models can be incorporated into simulation-based testing procedures for candidate ATR systems.
- *Ability to assess information availability.* The need to rapidly furnish smart weapons with accurate target-and-scene models is a requirement of the future naval forces' global surveillance and communications mission. In any area of conflict, the naval forces must know what information exists and what information must be collected to create accurate models. The processes of updating intelligence repositories are recurrent, in that the appearance and position of targets change over the course of a conflict. Intelligence repositories must be transparently connected to battle managers and battle managers to weapons and surveillance systems. In the future, ATR will be part of a feedback loop. Precursor surveillance missions having smart-weapon capabilities will seek targets, given the best information of the targets' appearance. Target statistics will be communicated back to the battle manager, and subsequent smart weapons will capitalize on the information.
- *Use of flexible recognition schemes.* As the use of multiple on- and off-board sensors becomes more common, processing will have to become more flexible in that some sources of information may not always be available. A related problem is that information sources have differing degrees of reliability, particularly in the presence of countermeasures. Future systems must be flexible enough to recognize when the real world is departing from assumptions made during training. ATR systems must be able to measure the quality of various

information sources and adapt accordingly. For a more in-depth discussion of recognition schemes, see Chapter 3 in *Volume 3: Information in Warfare,* another of the reports in this series.

- *Phenomenology supplanting ad hoc development.* In the past, development of many ATR approaches was ad hoc, relying on the experience and intuition of designers for the selection and implementation of discriminators. As countermeasures become more sophisticated and low observables more common, the success of such systems will decline. Instead, a phenomenology-driven approach is needed in which features that can discriminate targets in the presence of countermeasures are automatically or semiautomatically determined. More research in feature discovery and in methods of navigating the mazes of target and sensor physics is needed to accomplish this goal.

Time Scale for Developments and Insertion

Standardized, simulation-based testing could be carried out in parallel within the next 1 to 2 years. Creation of realistic scenarios might require an additional 3 to 4 years.

The information availability assessment task is quite comprehensive in scope. Efforts should build on work done in projects such as DARPA's Battle Assessment and Data Dissemination program, which attempts to provide warfighters with needed information in a timely fashion. After a language is defined to describe what is needed by way of target-and-scene information, an assessment system must be devised to check which pieces of information are already available. Finally, a planning system must be developed to suggest and schedule intelligence-gathering missions. Early versions could simply point out what information is needed and where it can be found. The initial definition phase might require 2 years; a system that relies on heterogeneous databases and inference engines to detect missing and outdated information might require an additional 2 to 3 years. An advanced planning and scheduling system to identify missing information would take far longer—perhaps 5 to 7 years.

Maturation of flexible recognition systems is expected to be a continuing process, taking advantage of more capable sensors and fusion calculi as they become available. Funding should be increased in this area to stimulate continued development and insertion over the next 20 years. Periodic reassessment may be warranted in order to identify and encourage particularly promising areas. Insertion would be a continuing process over the next 2 to 20 years.

Increased use of phenomenology-driven feature selection techniques requires development of tools that simplify the visualization and manipulation of target and countermeasure phenomenology. Funding for such tools and feature-discovery techniques should be increased and development combined with flexible recognition schemes encouraged. Insertion would continue periodically over the next 5 to 20 years as deep research insights mature.

RECOMMENDATION

Automation increases manpower effectiveness and warfighting capability by performing routine functions, conducting superhuman and hazardous operations, and minimizing casualties. The Department of the Navy should field a vigorous program in the technologies for ship automation that will realize these benefits. Unmanned aerial vehicles and unmanned underwater vehicles will play a major role in future naval warfare as surveillance, communication, targeting, and weapon-guidance platforms. The Department of the Navy should support technology developments to increase mission duration and operational capability, enhance sensor payloads, and increase survivability.

6

Technology for Human Peformance

INTRODUCTION

During the next three decades, revolutionary developments are expected in several areas of science and technology that will enhance human performance and significantly benefit the naval forces. The challenges associated with providing the manpower required to conduct naval missions include recruiting competent people, matching talent to tasks, training for utmost proficiency, obtaining optimal performance on the job, maintaining individuals at peak competence, protecting personnel in combat environments, enriching the quality of the Service experience, and retaining highly trained personnel. The following sections discuss anticipated technological developments and how these might be employed to accomplish these tasks.

Among technologies of particular significance are (1) communications, including global access to people and information sources, distance learning, and interactive multimedia information transfer; (2) information processing, including human-machine interfaces, computer-processing speed and memory, intelligent software, modeling and simulation, and embedded information (e.g., smart card); (3) health care, including telemedicine, drug delivery systems, surgical techniques, and disease prevention and treatment; (4) biotechnology and genetics, including gene therapy, human genome mapping, DNA identification and categorization, susceptibility testing, disease diagnosis, and vaccine and drug development; and (5) cognitive processes, including knowledge of brain functions, optimal learning modes, memory expanding drugs, and fatigue management. These areas will likely experience extraordinary leaps in knowledge and applicability within the next few years.

The forthcoming technological advances could have a profound impact on many aspects of life in the Navy and Marine Corps, especially (1) education and training, (2) operational performance, (3) health and safety, and (4) quality of life. The Navy has placed high priority on reducing crew sizes both aboard ships and on shore. By adapting technology improvements, the Navy will achieve a higher level of manpower readiness than ever before and thereby accomplish its mission with fewer, better-prepared people. Because the most consequential future technology advances will be in computers and the accompanying ease of access and assimilation of information, it is notable that future recruits will be better prepared to employ these advances by being uniformly computer literate and accustomed to learning via computer-aided, self-paced instruction.

The aspects of the technologies described in this chapter are limited to those that appear to be relevant to enhancing human performance in the context of the naval force requirements. The intent of this discussion is merely to highlight certain anticipated technology developments and their potential applications. Included is a concise description of the technology, with status and trends; relevance to the Navy and Marine Corps, with likely impact on future naval operations; needed developmental activities in the military and/or commercial sector, with possible time scale for use. The material is grouped into two segments: technologies and applications.

TECHNOLOGIES THAT WILL HAVE AN IMPACT

Among the many technologies undergoing rapid and continuous change, the following are thought to be particularly important to enhancing human performance of naval personnel over the next three decades. The intent of this discussion is to illustrate some of the gains that these technologies can provide the Navy and Marine Corps if properly developed and integrated into naval operations. It is noteworthy that the Department of the Navy is adept at keeping pace with technology developments and has continuously applied technology, within the limits of its resources, in a rational and timely manner to improve its effectiveness. This is especially true in regard to systems; it is less so for personnel, although this area has not been neglected.

Changes in technology are usually nonlinear and nonperiodic, whether evolutionary, revolutionary, or serendipitous; efforts to predict future developments are generally suspect. Trends can be identified and predictions made based on the knowledge that similar outcomes have occurred in the past. It cannot be known, for example, if gene therapy for hereditary disorders will be perfected as a result of insight gained in human genome research, but it is a reasonable expectation. What cannot be anticipated, of course, are scientific and technological breakthroughs that accelerate the pace of change and redirect the course of future events.

Communications

Within a decade, most of the world's population will have access to the Internet, and the Internet will access most of the major public databases in the world. This explosion of global information and rapid information availability is already altering, in very fundamental ways, the education process at the elementary, secondary, and college level. Although global station-to-station telephone service has existed for many years, the addition of cellular telephone and paging systems sets the stage for a time when virtually everyone will be instantly available via voice communication, day or night, anywhere in the world. Television transmissions via satellites now cover the globe, and soon interactive audio-video signals will facilitate essentially instantaneous interactive multimedia information transfer worldwide. Each of these modes of communication will continuously improve in access speed, data content, and information value over the next three decades. Today, the idea of creating hundreds of communication satellites in a low Earth-orbit network is only a concept; within 30 years it will be fully operational and readily available to all, friend and foe alike. These developments will have a significant impact on the performance of naval personnel, including improved operational readiness, on-site/on-demand training, and enhanced quality of life. These advances, however, will be driven by commercial applications, independent of military interests, and it will be the responsibility of the Navy and Marine Corps to exploit the benefits of these developments and adapt them to suit their unique needs.

The coming age of global connectivity and accessibility will cause a reevaluation of traditional organizational structures that are shaped in part by the traditional means of communication among its members. People will become increasingly able to obtain and act on information individually without need of supervision or organizational infrastructure. Further, the information an individual obtains and the actions he or she takes, or is preparing to take, can be instantaneously shared with many others. This environment lends itself to greater reliance on individual decisionmaking, aided when necessary by human and/or artificial (computer-based) experts located elsewhere. With the accompanying advances in information processing, the interrelationships and combined effects of multiple individual decisions can be evaluated and assimilated into a comprehensible view of events, thereby facilitating the effective management of extremely complex situations and increasing the value and utility of every individual participating in the process. This is the path to more efficient utilization of naval personnel and eventual reduction of their total numbers.

Information Processing

The staggering pace of improvement in computer-processing capabilities has been the dominant theme in information processing over the past 40 years, and hardware advances will continue at a similar pace for the foreseeable future. For many years, software lagged far behind hardware, but the advent of user-friendly, menu-driven procedures has given software development increased momentum. There is now an avalanche of new computer software covering an ever-increasing expanse of topics and relevant to an ever-broadening range of applications. This trend will accelerate in the future to eventually provide user-friendly intelligent software systems and a wide variety of expert systems.

These improved information-processing capabilities will continue to be important to the Navy and Marine Corps in upgrading the performance of Service personnel, but the current methods of implementing the human-machine interface are a barrier to realizing the ultimate value of this technology. Access via the mouse opened computer usage to many individuals who had difficulty overcoming the keyboard barrier, but the current processes of both computer input and output are inefficient and fail to utilize many human comprehension skills. They are barriers to achieving the maximum value of information-processing technology.

Significant advances in human-machine interface speed and efficiency will be forthcoming during the next three decades. The ultimate objective is to transfer human thoughts into digital information, and vice versa, with the least number of transformations and conversions. This problem is being attacked at both the input and output stages. Computer output has been enhanced using audio, graphics, images, and animated presentations, all in readily accessible, menu-driven formats. The next stage will include three-dimensional holographic displays to support virtual-reality simulation and modeling. To date, progress at the input end has been less spectacular. Some technologists forecast a time when there will be direct coupling between the human brain and machines, but that scenario is not a requirement for achieving greatly improved interface efficiency. The development of interface devices using handwriting and/or the spoken word will grow beyond the present notebook stage to provide a new class of input options for manual and vocal data entry. Although speech is generally a less accurate and less efficient means of computer interface, it will soon become a common mode of human-machine communication, particularly for command of relatively simple machine operations. This will open even more applications of digital information-processing technology to enhance the performance of Service personnel, for example, embedded training, intelligent tutors, and robotics.

The smart card containing a high-density read-only memory chip is a promising new form of machine output that will serve many personal and commercial functions, including compact personnel records, positive identification, secure money/phone/credit card, and emergency medical records (possibly implanted).

The smart card will become a common accessory in all industrialized nations within the next decade because of its convenience, security, and high-information content.

As computer speed and capacity have increased, and image and graphical representations have improved, modeling and simulation techniques have expanded into a wide variety of applications. In addition to the obvious value of these techniques for system design and optimization, they will be increasingly employed in decisionmaking, especially for complex situations. Much has been written about the potential of modeling and simulation techniques for evaluating and managing wartime situations, but the most promising application will be more efficient, cost-effective personnel training that exposes students to up-to-date, accurate, and realistic warfighting scenarios, equipment- and systems-maintenance information, and operation and safety procedures.

Health Care

Advances in health care during the past decade have been exceeded in magnitude and significance only by advances in information processing. Most notable have been improvements in pharmaceuticals, surgical techniques, and rehabilitation procedures. Future developments will benefit the Navy and Marine Corps not only in improved combat medicine, but also in the operational readiness and proficiency of sailors and marines, reduced health care costs for naval personnel and dependents, and enhanced quality of life aboard ship and on shore.

The enhanced effectiveness and applicability of recent pharmaceutical products for treatment of physiological, neurological, and psychological disorders are impressive even for drug development using conventional chemistry. With the advent of drug development using combinatorial chemistry and genetic-engineering methods, entirely new pharmaceutical products will cascade onto the market beginning within the next decade. The intense concentration on viral infections as a result of the AIDS epidemic has greatly increased knowledge of cell biology and facilitated customized drug-design procedures that reduce the drug-development task and improve efficiency and efficacy. The development process is further enhanced when modern molecular design and synthesis techniques are employed. Proceeding in parallel are innovative drug-delivery techniques such as microencapsulation, implants, and other controlled-release mechanisms. Currently, these techniques are being applied to deliver existing drugs, but eventually they will be an integral part of the initial drug-development process, thereby reducing the frequency of application, improving the effectiveness of the dose, and facilitating use of an expanded range of chemicals. These techniques will also enable the development of multicomponent, multivalent vaccination systems, including oral vaccines.

The rapid progress in drug and vaccine development, with the increased understanding of cell biology and genetics, suggests that many common diseases will be cured or controlled in the near future. HIV and other sexually transmitted diseases will pose far less of a threat. Improved diagnostic tests and treatments will sharply reduce the risk of cancer and other life-threatening diseases. Ongoing research indicates that gene-therapy techniques will be discovered within the next three decades to correct certain genetic disorders and/or reduce the onset of hereditary problems. Regardless of the ultimate technical success and social acceptance of gene-therapy techniques, the research is already producing methods to screen for genetic defects that pose health hazards and thereby facilitate early intervention and possible control.

Recent advances in surgical techniques have greatly reduced the recovery period associated with traditional surgery. Laparoscopic procedures are finding wide application in internal medicine to lessen the damage and risks involved in abdominal surgery. Similar procedures are becoming routine in orthopedic and cardiovascular examinations and treatment. These techniques, coupled with robotics technology and reliable global communications, pave the way for sophisticated telemedicine to provide higher-quality medical treatment aboard ships using shipboard personnel with less extensive education and training. Organ transplants are becoming routine, and new procedures are under development to reduce the risk of organ rejection with less hardship for the recipient. The problem of obtaining suitable transplant organs will soon be overcome through the use of animal surrogates, artificial organs, and eventually, a cloning or replication technique. Recovery from surgery and other injuries will be quicker and less debilitating because of continual improvements in rehabilitation methods and rejuvenating drugs. An added benefit of these new techniques will be the development of better methods for obtaining and maintaining peak physical conditioning.

The ongoing national search for more efficient, cost-effective health care is producing many new procedures and approaches to medical treatment that will ultimately benefit the Navy and Marine Corps. Further refinement of noninvasive examination and diagnostics techniques, e.g., ultrasound imaging and magnetic resonance imaging (MRI), will be particularly important, especially when smaller, portable instruments are developed for use aboard ships. Accompanying these developments will be real-time disease and injury surveillance, more rapid and accurate field diagnostic techniques, and more effective on-site medical treatment (artificial blood, smart splints, and the like).

Biotechnology and Genetics

"Genetics is the future of medicine." This was a key finding of a 3-year study conducted for the Surgeon General of the U.S. Army.[1] Collection of

[1]Surgeon General of the U.S. Army. 1996. *Consolidated Military Genetics/DNA Program,* U.S. Army, Washington, D.C., March.

genetic information is integral to providing health care today, and the demand for genetic services will greatly increase in the future. Within the next decade, every gene in the human genome will have been mapped and sequenced. Nearly 2,000 genetic disorders have already been defined. The report to the Surgeon General stated: "The opportunities stemming from the new-found knowledge in molecular biology . . . present military medicine with the chance to affect operational readiness of the armed forces in ways never before possible."

At present, clinical genetic laboratories can provide a series of highly specific and sensitive analyses for genetic and infectious diseases using techniques such as the polymerase chain reaction or ligase chain reaction. Civilian medicine has also introduced large-scale efforts for genetic testing for both breast and colon cancer. Other common genetic tests will soon emerge, including tests for hypertension, atherosclerosis, and many types of cancer. Testing for conditions such as cystic fibrosis, factor V (Leiden) mutation, familial hypercholesterolemia, certain metabolic disorders, and diseases of premature neurologic degeneration (Alzheimer's and Huntington's disease) are relevant to the determination of the fitness of military recruits. Conventional approaches toward these tests, however, are labor intensive and expensive. Recent developments in miniaturizing bioanalytical instrumentation will become important in reducing the cost of these tests and therefore allow for the widespread adoption of multiple genetic testing.

Knowledge derived from human genome research is particularly important to military medicine with enormous implications for resource allocation, military medical practices, and preventive care. The information and technologies derived from this research are rapidly being developed for health care applications, and they will have an impact on all of medicine in ways not anticipated at the present time. Of special significance will be the development of human gene maps, the discovery of genetic disorders, the ability to conduct rapid and inexpensive genetic tests, the development of inexpensive and portable analytical instruments, and the perfection of technologies to permanently preserve the DNA content of human tissue.

The rapid advance of new biotechnology techniques will accelerate during the next three decades. Despite the fact that the level of knowledge in this field is still quite low, there are now several hundred biotechnology companies developing commercial products, and more than 100 genetically engineered products are currently undergoing clinical trials. The recent intense concentration on HIV has, coincidentally, yielded a much greater appreciation of the structure and operation of the human cell. This knowledge, coupled with rapidly improving tools such as computational chemistry, molecular modeling, crystallography, and high-throughput screening, has opened the way to more efficient development of drugs and drug-delivery systems that are customized to specific diseases. The concept of gene therapy remains a promising future possibility at this time, but ongoing research suggests that a proven treatment methodology will exist within a few years and that it will be put into practice for a wide variety of genetic disorders during the next three decades.

Cognitive Processes

New imaging techniques, analytical methods, and scientific insights have opened the way to fresh paradigms in the study of the human mind and its substrate, the brain. Images of the brain taken from positron emission tomography (PET) scans and MRI have enabled scientists to observe, for the first time, certain brain functions. Enhanced electroencephalogram (EEG) sensor arrays can now detect brain-wave patterns at specific frequencies and locations within the brain. These data are being analyzed using increasingly powerful computational tools to model the workings of the mind to gain understanding of the learning process, memory storage, and many other aspects of the mental process. Rapidly emerging areas of scientific study, such as cognitive neuroscience, could have considerable potential value in several military applications.

The U.S. Air Force study *New World Vistas*[2] expresses considerable enthusiasm about the potential for increased understanding of brain function, so much so that it urges a commitment to a consistent level of investment in cognitive science research. The report notes that several recent developments make possible a vastly improved understanding of how people process information. This understanding can be used to better design information and control systems and to select, classify, and train people more effectively. Among these developments are improved methods of observing brain activity by monitoring electrical activity; improved understanding of brain biochemistry and neurology; development of computer models, such as neural networks; and the emergence of interdisciplinary collaboration in the study of the human mind.

Within the next three decades, a computational model will be available that accurately simulates many brain functions, particularly the functions of memory storage and retrieval. Some believe this model could become the basis for a new generation of computers that either employ an electronic equivalent to the human neurological process or employ biological systems as computer processors. With this higher level of comprehension of brain functions, the learning process will be better understood and optimum learning modes will be developed. In addition, there will be a substantially increased understanding of the root causes of mental illness, abnormal behavior, mood changes, motivation factors, sleep cycles, and many other personality and attitudinal traits. From this knowledge will flow a

[2]U.S. Air Force Scientific Advisory Board. 1995. *New World Vistas, Air and Space Power for the 21st Century,* U.S. Air Force, Washington, D.C., December.

wide range of new drugs and psychotherapeutic agents designed to correct, enhance, subdue, and modulate human performance and attitude, including enhancing an individual's ability and desire to learn, achieve, and relate positively to fellow human beings. These drugs will provide means to enhance information retention, memory recall, attention span, and mental endurance. These technological developments will be especially important to the Department of the Navy's efforts to optimize the effectiveness and efficiency of its human resources, especially learning ability, decisionmaking skills, and memory capacity.

Other Relevant Technologies

The recent history of rapid technology evolution in areas relating to human performance, comfort, entertainment, education, and training makes it obvious that numerous advances in these fields will occur during the next three decades. Of particular significance to the Navy and Marine Corps will be food and nutrition technologies, environmental-pollution control and waste-elimination technologies, and air- and water-quality technologies. These will be driven by consumer demand and commercial opportunities, and the Department of the Navy will benefit from them by remaining attentive to developments and trends. Robotics encompasses technologies that have considerable potential to aid human performance, but because the cost-benefit ratio remains unfavorable for most commercial applications, the full value of robotics will not be realized in the near future. The opportunity exists, however, to employ robotics in specific, high-value naval applications. The Department of the Navy will have to fund the R&D to develop such devices.

Technologies required to better prepare the Navy and Marine Corps to deal with chemical and biological warfare threats will remain dependent on DOD investments. There have been predictions of significant improvements in technologies for personal protection, including the possibility of vaccines. These will likely occur, particularly as a result of advances in biotechnology; however, it is equally likely that the severity of the threat will accelerate at about the same pace as the quality of the protective measures, if not faster. As knowledge of biotechnology and genetics increases, the potential for use of this knowledge to design even more sinister weapons will also increase. This situation will place demands on DOD for continued vigilance and long-term research support. Protective clothing and decontamination methods are the most fruitful areas for research investments.

APPLICATIONS

Enhancing the performance of military personnel involves, for example, improving the match of personnel to the job at hand, the competence of an individual to conduct a task quickly and accurately, the quality and reliability of decisions under stress, the survivability of sailors and marines in adverse combat situations, the ease and cost-effectiveness of maintenance functions, and recruiting and retaining competent people. The Navy and Marine Corps have in place an extensive infrastructure to enhance the performance of the people in the naval forces. Future technology developments can be incorporated into this system with relative ease; however, in many instances the full value of emerging technologies will not be realized without a restructuring of the relevant management and delivery methods. The following sections endeavor to forecast the impact of anticipated technology advances on the processes used to enhance human performance. The most significant opportunities will likely be in the area of education and training, but substantial gains are possible in several other areas as well. A supplementary discussion of the application of technology to the performance of Navy and Marine Corps personnel is provided in *Volume 4: Human Resources,* of the nine-volume series *Technology for the United States Navy and Marine Corps, 2000-2035: Becoming a 21st-Century Force.*

Education and Training

Traditional training techniques have proven satisfactory for preparing individuals to perform specific functions in a manner adequate to their assigned tasks. They are less satisfactory for preparing individuals to function interactively in a group setting and promptly render intelligent decisions, independently or collectively, when confronted with unfamiliar multimodal and complex information. As systems, equipment, and operating procedures become more information intensive, traditional training techniques will become increasingly inadequate.

Future training should be organized in a manner that exploits information-processing and cognitive-process technologies to maximize the learning experience. It must be more flexible, more portable, more user-friendly, and more cost-effective than present methods. It must place greater emphasis on computer-based, self-paced, on-site training using modeling and simulation techniques, embedded training, and intelligent tutors. Education and training modules must be tailored to the individual or groups by making use of selection, classification, and characterization methods to better match the people to the processes.

The advantages provided by the fast, accurate, reliable global communications systems of the future will be employed commercially to make distance learning a routine occurrence. Financial pressures on colleges and universities, plus the increasing scarcity of technically trained teachers, will contribute substantially to enabling distance learning to achieve its long-sought potential. As these methods are developed and refined such that the lecture format is augmented or replaced by a tutorial, interactive, computer-aided instructional format, the educational experience will become more flexible and better able to provide the level of education and training needed to supply Navy and Marine Corps needs.

Trends in public education, including extensive development of computer-aided, self-paced instruction and interactive distance learning, will not only provide new and improved training tools relevant to naval force tasks, but also will provide a valuable resource for use on an outsourcing basis. Access to formal education in science and technology via interactive distance learning from colleges and universities will ease the problem of preparing and maintaining competent officers. Similar procedures will provide timely, on-site training of technicians by equipment vendors and other specialists.

Operational Performance

Traditional organizational structures, both industrial and military, were formulated to a large extent on the basis of the need to communicate rapidly and accurately throughout the organization and ensure a timely, reliable response. As communications and information-processing technologies continually improve, and the work force becomes increasingly more adept at understanding and independently applying information to carry out individual and collective responsibilities, industrial organizational structures are changing. Because workers are better equipped to participate directly in decisionmaking processes, layers of middle management are being eliminated, and productivity and worker satisfaction are increasing. This trend will continue to spur fundamental changes in organizational concepts that will have an impact on all institutions seeking to optimize individual performance.

The future benefits to the Navy and Marine Corps from the emerging technologies that enhance human performance can best be realized by creating "smart" sailors and marines and placing them in a work environment that enables them to function at maximum capacity. Industry is finding that achieving that end involves less supervision, fewer layers of management, and the placement of greater responsibility further down in the organization. If the Navy and Marine Corps increasingly incorporate this approach in the future, the operational performance of Service personnel will increase and the number of people required to conduct specific functions will decline.

The term "smart sailor" is a convenient descriptor for the ultimate goal of applying technologies that enhance human performance. The smart sailor of 2035 will possess many desirable attributes, including being (1) compatible with work assignments as a result of matching with aptitude and native ability and customizing education and training through sophisticated selection and classification techniques, included genetic markers; (2) adequately prepared to handle new assignments as a result of efficient education and training processes using realistic modeling and simulation techniques, comprehensive self-paced instruction, and multimedia, interactive distance learning, plus the availability of on-site intelligent tutors and embedded training; (3) competent to execute complicated assignments and arrive at intelligent decisions as a result of an abundance of quality information, multimedia presentation and automated analysis methods, expert decision aids, and heightened human cognitive processes resulting from improved training and use of performance-enhancing pharmaceuticals that expand memory, improve sleep efficiency, and reduce stress; (4) equipped to undertake arduous assignments as a result of good health care and excellent physical conditioning, as well as extensive use of work aids and robotic devices; (5) stimulated to conduct routine assignments as a result of having a heightened sense of participation in the decisionmaking processes, ready access to relevant information, and a broad range of individual responsibilities; and (6) able and available to perform warfighting assignments as a result of effective personal protection and safety equipment, good damage control systems, and the accessibility of high-quality field medicine, including use of a smart-card personal health history, telemedicine, multiple vaccines, and effective pharmaceuticals.

Health and Safety

Both civilian health organizations and the military will be profoundly affected by advances in genetics. Some believe genetics to be the next major paradigm shift in public health with an impact as significant as the impacts of introduction of sanitary measures, aseptic anesthetic surgery, and antibiotics. Genetic therapy represents the potential to cure, not simply to provide symptomatic treatment of human suffering.

Military personnel face exposure to a broad range of physiological situations determined by the nature of military missions and operations. The physiological responses of individuals exposed to different environmental conditions are determined at least partially on a genetic basis, and military personnel could be better protected and trained if they were known to be at risk in certain hazardous environments. For instance, military personnel might be selected for specialized training based on their physiological tolerance for such activities as work at high altitudes, deep-sea diving, or exposure to loud noises such as high-blast overpressures generated by weapons detonation. Similarly, screening can be carried out based on individual differences in susceptibility to threat agents, toxicants, and other environmental variables. As another example, in job settings requiring exposure to inhalation of dusts, toxic chemical vapors, or allergens, testing for alpha-1-antitrypsin deficiency could be used to assess whether the genotype of certain people predisposes them to emphysema or reactive airway diseases. Future military applications include fitness screening, tissue typing, gene therapy, vaccine production, environmental damage assessment, and diagnosis of disease and susceptibility to disease.

Given validated genetic markers derived from DNA samples first obtained on entry to active duty, military personnel might be screened for certain demanding types of work or disqualified based on both physical condition and genetic predisposition. Screening based on certain disqualifying military conditions has historically been done by all DOD services and is governed by specific and changing service regulations. Disqualifying conditions with a hereditary component include seizure disorders, heart disease, diabetes, asthma, color blindness, and cancer. Genetic tests for several types of cancer are available and are becoming more widespread. Genetic tests for diabetes and asthma are likely to emerge within the next 2 years. Technologies for performing these genetic tests are continually improving. It is anticipated that the advent of microchip-based molecular diagnosis and inexpensive microanalytical instrumentation will substantially reduce the cost of genetic tests.

The future use of genetic screening methods relies on the present authority of DOD to define medically disqualifying conditions for military service for various specialized duties. It is reasonable to assume that future knowledge of genetic markers associated with genetic predisposition will become more important for preinduction and other military physical examinations. Of course, the prospect of rejecting individuals who are genetically at risk raises ethical issues, plus opening the issue of the possible use of genetic information outside the military. Society's views on the use of genetic information, however, will change significantly as the value of this knowledge becomes widely recognized and understood. Within the next 40 years, it would not be surprising if the personal health record of virtually every U.S. citizen were to contain a comprehensive DNA analysis highlighting health risks and genetic predisposition.

Quality of Life

Several anticipated technological advancements have the potential to significantly improve conditions that determine the quality of life of military personnel. Most notable among these are communications, health care, distance learning, and personnel selection. The ability to rapidly, reliably, and routinely communicate with family and friends from anyplace on the globe at any time of the day or night will greatly reduce the anxiety, stress, and loneliness of military life. The ready availability of excellent health care, including a wider range of vaccines and more effective pharmaceuticals, will ensure more physically fit, mentally alert, and contented personnel. The expanded use of computer-based, self-paced instruction and distance learning will open new opportunities for personal development and career preparation and serve to create a constructive, self-satisfying environment. Improvement in the selection and classification processes will increase compatibility, decrease stress, and improve job satisfaction.

In addition to the application of these technologies, Service personnel will benefit from changes in organizational concepts that place greater responsibility on each individual. Increased use of decentralized decisionmaking, participatory management, and distributed information will flatten organizational structures, reduce personnel, and require a higher level of performance and a greater degree of responsibility at every Service grade. This will produce self-satisfaction and a sense of accomplishment that will enrich the Service experience and enhance the quality of life at every level.

RECOMMENDATIONS

Economic and social conditions will force the Navy and Marine Corps to conduct future missions with fewer people and lower overall manpower costs. To accomplish this, the Department of the Navy must achieve higher levels of personnel performance in an increasingly more complex, information-dependent environment. The Department of the Navy should exploit the technology advances in communications, information, health care, biotechnology and genetics, and cognitive processes to enhance human performance through expanded education and training, improved personal health and safety, and enhanced quality of life throughout the Navy and Marine Corps.

This effort will require special attention to the following actions:

- Enhance the capability and competence of personnel through improved education and training using computer-aided, interactive distance learning, intelligent tutors, embedded training, and better selectivity and classification.
- Upgrade the operational performance of personnel by providing higher-level information and support to all personnel grades, increase participation in decisionmaking processes, flatten organizational structures, and develop smart sailors serving on smart ships.
- Improve the health and safety of personnel by ensuring rapid access to personal medical history and exploiting advances in biotechnology and genetics to provide chemical and biological warfare protection, fitness screening, tissue typing, gene therapy, multicomponent vaccines, disease diagnostics, and genetic predisposition classifications.
- Increase job satisfaction and personnel retention through enhanced quality-of-life factors by providing ready access to loved ones via global communications, supplying comprehensive personal-development opportunities via interactive distance learning, and building self-esteem by granting higher degrees of individual responsibility.

7

Materials Technologies

INTRODUCTION

Materials science is an interdisciplinary field, and the national research and development effort in this area is robust and broadly based. In 1989, the National Research Council published a report, *Materials Science and Engineering for the 1990s*,[1] which was the culmination of a comprehensive survey of materials science and engineering in the late 1980s. The technology-development trends identified in that report are likely to continue to be relevant over the next few decades. Rather than conduct its own independent survey of such an extensive field, the Panel on Technology chose instead to highlight a few materials technologies that may be useful for developing improved capabilities for future naval forces. The panel's review is not meant to be comprehensive. For example, semiconductor technology is expected to continue its rapid pace of development[2] and contribute significantly to the capability of future Navy and Marine Corps forces, but it is not discussed in detail here.

That being said, the opportunities are legion for developing enhanced capabilities for future naval forces through the application of materials technologies. Materials affect the operations, maintenance, and safety of all naval systems. The

[1]National Research Council. 1989. *Materials Science and Engineering for the 1990s: Maintaining Competitiveness in the Age of Materials*, Board on Physics and Astronomy and the National Materials Advisory Board, National Academy Press, Washington, D.C.

[2]Semiconductor Industry Association. 1996. *National Technology Roadmap for Semiconductors*, San Jose, Calif.

development of advanced materials with improved properties can lead to enhanced capability and reduced life-cycle costs. The Department of the Navy can ensure the development of new materials required for future naval systems by continuing to fund basic research in materials science.

DEVELOPMENT DRIVERS

The panel envisions a systems approach to materials development in the future, whereby physical and mechanical properties are understood at the atomic scale and the design of new materials is carried out computationally based on this knowledge. Capitalizing on the opportunities presented by an atomistic simulation-based design approach may enable breakthroughs in, for example, ferrous alloys, titanium matrix composites, polymer composites, high-temperature ceramics, wide-bandgap semiconductors, optical materials and coatings, and smart materials based on ferroelectrics, ferromagnetic materials, and ferroelastic materials. Advances in materials technology are expected to enable the development of improved microelectromechanical systems, leading to more capable systems-on-a-chip. Sensors that are mechanically and electrically adaptive will likely be part of this improved systems-on-a-chip capability.

Advances in materials processing and synthesis will be needed in order to meet the requirements of future Navy and Marine Corps missions. The processing of nanophase materials could be scientifically based on high-frequency cluster sputter techniques. The processing of diamond, based on chemical vapor deposition and flame techniques, could be extended so that low-defect-density materials can be achieved.

In the panel's view, the technology drivers for future naval requirements will be as follows:

- Computationally designed materials and processing;
- Unique nanophase materials systems for new applications;
- Smart materials and systems based on shape memory alloys, ferroelectrics, and ferromagnets adaptive for mechanical and electrical applications; and
- Wide-bandgap semiconductors, materials for microelectronics, photonics, and RF and microwave applications.

TECHNOLOGY TRENDS AND FUTURE MATERIALS DEVELOPMENTS

Computationally Driven Materials Development

The panel believes that a key development in the materials area will be the emergence of a computational design approach based on knowledge of the physical and mechanical properties of materials at the atomic level. This approach will lead to the development and manufacturing at low cost of smart materials and nanophase materials that will, in turn, be enablers of new structures and devices. Design will follow from first principles, giving rise to a systems approach in materials development. Simulation of materials characteristics using this approach will include modeling of microstructure, defects, surface structure, interface properties, transport properties, reaction kinetics, prediction of adhesion and bonding, thermodynamic properties, and general mechanical behavior.

Process-simulation modeling will be another materials technology enabler. Processes based on thin-film growth techniques, such as advanced chemical vapor deposition (CVD) and molecular beam epitaxial growth, could be first simulated in the computer before actual growth takes place. Other processing techniques for molding, extrusion, solidification, and physical spray techniques for nanophase materials could also be simulated.

Materials Simulation

Materials simulations can be classified into three domains: (1) simulations based on quantum mechanics and statistical mechanics, (2) atomistic simulations based on classical equations of motion, and (3) simulations based on phenomenological models. These three domains will be interfaced with the appropriate process models and finally extended to the simulation of materials systems. Such an approach will likely be cost effective, since it will allow the development of materials systems specifically designed to meet the required specifications.

Simulation of Materials Component and Materials System Integration

Integrated simulations of materials systems will require greater understanding of thermomechanical phenomena for naval components. A continuously upgradable materials database would greatly facilitate the development of computational materials science and engineering, and the development of such a database is worthy of the Department of the Navy's attention. The database should provide the material parameters required for system simulation and for prediction of bulk properties.

Modeling of Heterogeneous Materials

Current models typically focus on homogeneous materials, but future performance requirements will most likely be met with heterogeneous materials that have properties controlled by surfaces and interfaces and behavior based on phase transitions and material degradation. Therefore, simulation models for heterogeneous materials that incorporate the role of defects and dynamic behavior must be developed. Further development of molecular dynamics for organic molecular polymers and inorganic crystals will be needed.

Processing and Synthesis

Modeling of reaction paths for intermetallic compounds, ceramics, metal alloys, high-strength steels, and titanium alloys will be an enabling path for the development of these materials. Likewise, the processing and synthesis of photonic materials and phonon-engineered materials will be required to meet the system requirements utilizing these materials.

Smart Materials and Sensors

Smart materials technology consists of the application of ferromagnetic, ferroelectric, and ferroelastic materials, better known as shape-memory alloys, as mechanical actuators and/or sensors to improve the performance of components, structures, and systems. The panel envisions the integration of smart materials with nanoscale electronic processors resulting in mechanically and electrically adaptive elements.

Many proposed systems will benefit from the utilization of smart sensors. For example, smart sensors could increase the performance and efficiency of personnel and equipment in areas such as condition-based maintenance. Overall, a full assessment of smart materials and MEMS materials will need to be carried out. System integration including data sampling, networking, and communication issues will have to be addressed. Smart materials on the microscale will be combined with electronics on the nanoscale to form smart sensors, all as part of a micro-nano-electronic technology thrust.

Nanophase Materials

A new emphasis in materials science centers on the nanometer (10^{-9} meter) size regime, which is intermediate between the well-studied macroscopic and atomic size regimes. The understanding of structural and compositional features in the nanometer size range will facilitate the control of the magnetic, electrical, and optical properties of materials. Nanophase or nanomaterials is an area of prime importance for future naval applications, especially with the expected conversion of most ships to integrated electric power and propulsion systems. Magnetic nanomaterials may offer dramatically improved performance for magnetic-storage applications. The enhanced strength of nanophase coatings and the potential for improved mechanical behavior of consolidated nanocrystalline materials have obvious applications in the area of structural materials. One important example is the superplasticity of nanocrystalline materials, a property appropriate for missile nose cones and armor.

Nanophase materials could be combined with nanoscale electronics to produce a new class of sensors able to achieve ultrahigh speed and low-power dissipation. The capacity to carry out high-resolution lithography capable of manufacturing devices with critical dimensions on the order of a nanometer is required before nanophase materials technology can become practical for naval applications. Other related technologies that will require further development before nanophase materials can be widely deployed include plasma-etch technologies and interconnects for quantum electronics. In photonic systems, nanophase structures will enable the development of nonlinear optical systems or possibly smart nanosensors that are optically interconnected to form a highly capable metasensor.

Structural Materials

Structural materials are widely used in naval systems, and some applications, such as engine components and ship structures that are exposed to salt water, are quite demanding. The future trend in the development of new structural materials will be to integrate functionality into the structure. An example of this type of integration would be the development of a submarine hull that contains embedded MEMS devices to maintain laminar flow around the hull and embedded networked conformal sensor arrays for both acoustic (sonar) and nonacoustic sensing.

Improved strength and stiffness are always desirable in structural materials, and such improvements may become available through the engineering of covalently bonded materials or, alternatively, though the use of thin lamellar structures that combine high strength and high modulus with a designed-in anisotropy to fit the specific application. The panel anticipates that materials will be available by 2035 with a strength-to-density ratio that is double that available today. This projection, which is illustrated in Figure 7.1, is based on the development of high-strength single-crystal graphite fibers.

High-temperature Structural Materials and Coatings

High-temperature materials and coatings, including metal composites, ceramic-metal composites, intermetallic alloys, and carbon-carbon composites, which are useful in a number of applications including aircraft engine components, may be amenable to low-cost synthesis through the application of computational materials design. Metal-matrix composites will likely address most requirements for materials that can withstand temperatures up to 500 °C. For systems requiring components that can withstand 1,000 to 2,000 °C, oxide materials will be needed, such as the yttrium aluminum oxides.

Metallic and ceramic surface coatings are currently used to improve the performance and prolong the life of advanced turbine materials. Protective coatings used in aircraft, marine, and power generation turbines to increase operating temperatures extend component life by providing protection from high-temperature oxidation and high-temperature corrosion. This dependence on coatings to

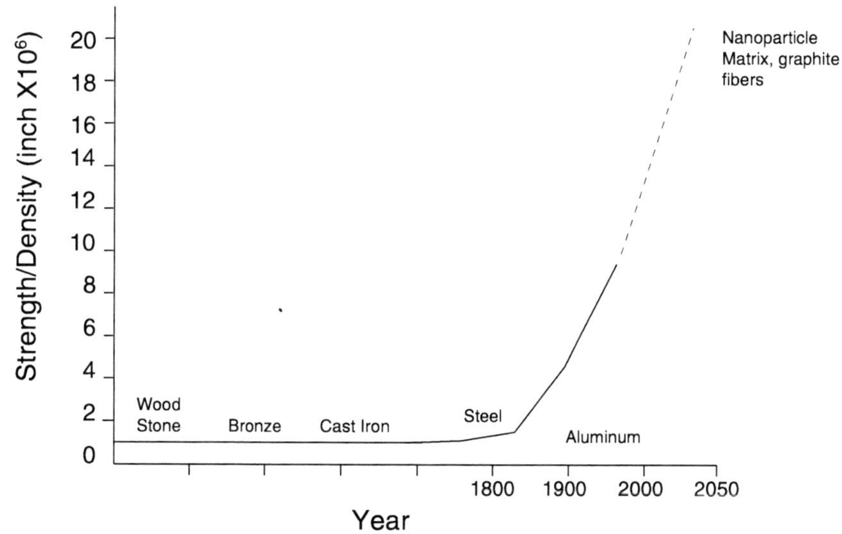

FIGURE 7.1 Tensile strength-to-density ratio of structural materials projected through 2035. The extrapolated values are based on a calculation of the maximum strength-to-density ratio of high-density single-crystal graphite fiber. SOURCE: Adapted from National Research Council, 1989, *Materials Science and Engineering for the 1990s: Maintaining Competitiveness in the Age of Materials,* National Academy Press, Washington, D.C., Figure 1.1, p. 20.

improve performance, extend service life, and reduce maintenance will be increasingly important as military budgets decrease. Advances in ceramic coatings will be required for future naval systems.

In the temperature range of 1,500 to 2,100 °C, materials such as silicon carbide, silicon nitride, and other systems able to limit oxidation will be needed. Candidate processing technologies for these difficult-to-shape materials include microwave and laser processing.

For systems above 2,000 °C, carbon-carbon composites, diamond-like coatings, synthetic-diamond thick films, and carbides such as boron tetracarbides and titanium carbides will be needed.

Processing and Synthesis of High-temperature Structural Materials

The technologies that may enable the manufacture of high-temperature structural materials are rapid solidification (splat cooling) and electron-beam evaporation. These techniques will allow the development of lamellar composition and functionally graded materials. Methods of processing of fiber with a polymer matrix that combine joining processes with material synthesis will be needed.

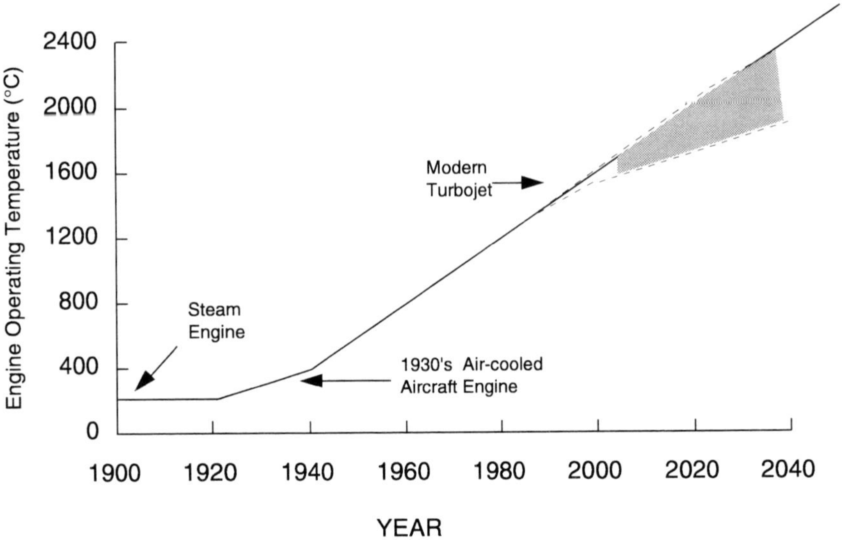

FIGURE 7.2 Projected high-temperature capability of NiAl superalloys. SOURCE: Adapted from National Research Council, 1989, *Materials Science and Engineering for the 1990s: Maintaining Competitiveness in the Age of Materials,* National Academy Press, Washington, D.C., Figure 1.2, p. 21.

Engine Materials

Key materials-related goals for future naval engines are to reduce weight, increase temperature capabilities, improve mechanical properties, and improve corrosion and oxidation resistance. Candidate high-performance materials include organic matrix composites, titanium alloys, and intermetallic compounds. For turbine components, nickel aluminum (NiAl) polycrystalline materials could be extended so that they are available in a single-crystal form. Intermetallic compounds, along with titanium-based metal-matrix composites such as TiAlNb with SiC fibers, may be useful for compressors. Static engine components will require high-modulus intermetallic compounds such as g-TiAl. The high-temperature capability of superalloys based on NiAl is expected to exceed 2,000 °C by the year 2035, as shown in Figure 7.2. Progress will be based on advances in synthesis and microstructure control.

Marine Materials

There is a continuing need for stronger, easily weldable, and less expensive steels for marine structures such as ship and submarine hulls. Steel-alloy designs based on first principles with controlled microstructures and predicted mechanical properties should be available in the near future. An achievable goal is 130 kips per square inch (ksi) yield-strength steel with high-fracture toughness that is able to withstand stress-corrosion cracking and fatigue-crack propagation. By using basic atomistic principles to model stress-corrosion cracking, greater understanding of stress-corrosion effects can be achieved. This knowledge can be extended to the design of new steels. A combination of new materials such as ultralow-carbon bainite (ULCB) and high-strength, low-alloy (HSLA) steels will yield significant advances in strength and corrosion resistance.

Titanium and titanium alloys exhibit good fracture toughness, corrosion resistance, high-temperature strength, and low magnetic signature. Titanium alloys in aircraft offer as much as a 50 percent weight savings as compared with aluminum parts. Ti-Al-V is used extensively in air frames today, but higher-temperature titanium alloys such as alpha-2, g, and orthorhombic titanium aluminides are under development and offer improved temperature capability beyond the 700 °C limit of current production alloys. The new alloys, which should be available in about 10 years, possess ductility in the range of 2 to 4 percent, which is adequate for most manufacturing processes. Titanium matrix composites (TMCs) consisting of titanium alloys reinforced with silicon carbide fibers may provide significant performance improvements, particularly for use in high-temperature engines.

Specialty Materials

Future naval systems for 2035 will require technology advances in the following specialty areas: superconductors and magnetic materials, organic materials and coatings, energetic materials, and high-temperature semiconductors. These material systems affect a number of areas and are summarized in the following sections.

High-temperature Superconductivity

Naval applications for superconductivity include (1) superconducting magnets for electrical motors and ship propulsion, (2) superconducting magnetic sensors for mine detection, (3) superconducting magnetic systems that store energy for burst power, (4) high-Q cavities for high-resolution radar systems, and (5) low-power analog and digital circuits. In order to realize the benefits of superconductivity in fleet applications, further technology development will be needed in materials engineering, manufacturing, and systems integration.

Since the discovery of HTS in 1986, numerous applications have emerged, including superconducting cables, transformers, motors, and energy-storage devices. HTS conductors are typically fabricated as a multifilamentary flat tape. These conductors use a ceramic precursor powder placed in a silver billet. The billet is then formed into a thin filament using commercial deformation processes, and multiple filaments are then placed into a silver tube and deformed

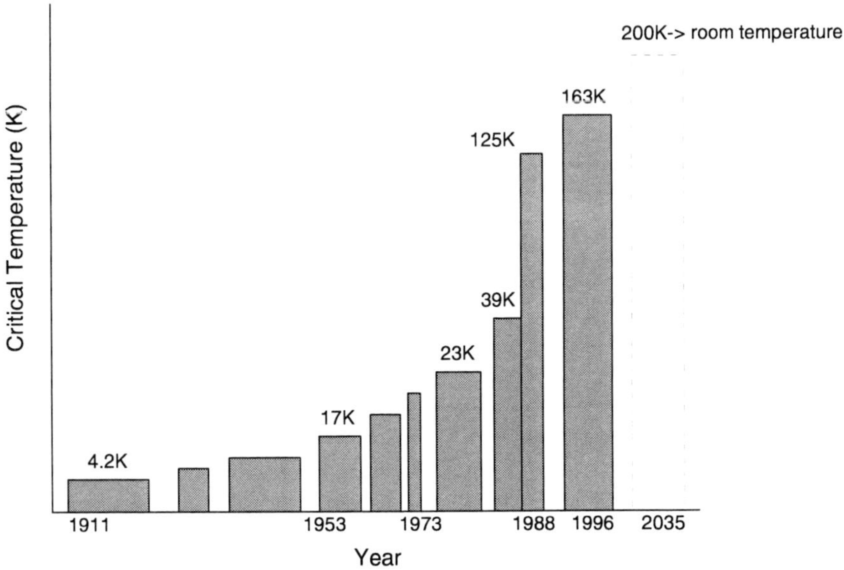

FIGURE 7.3 The increase over time in the critical superconducting temperature of HTS materials. SOURCE: Adapted from National Research Council, 1989, *Materials Science and Engineering for the 1990s: Maintaining Competitiveness in the Age of Materials,* National Academy Press, Washington, D.C., Figure 1.4, p. 23.

again into a bundle of filaments. These steps are repeated until the conductor contains the appropriate number of filaments. The conductor is then rolled into a flat configuration and heat treated to transform the ceramic precursor into a superconductor. This process is referred to as oxide powder in tube (OPIT).

OPIT conductors have shown linear performance improvements over the last 10 years, and manufacturing costs have steadily declined. HTS manufacturers are working with U.S. national laboratories to develop the next generation of HTS-coated conductors. Coated conductors use a thin film of HTS deposited onto a substrate; they exhibit significant performance gains as compared with OPIT conductors and can be significantly less expensive to manufacture.

The projected future critical superconducting temperature for high-temperature superconductors is shown in Figure 7.3.

Magnetic Materials

Improved magnetic materials will be required for magneto-optic devices and high-sensitivity, low-cost magnetic sensors to be utilized as magnetometers, radio-frequency antennas, and biological and chemical sensors. Improvements in

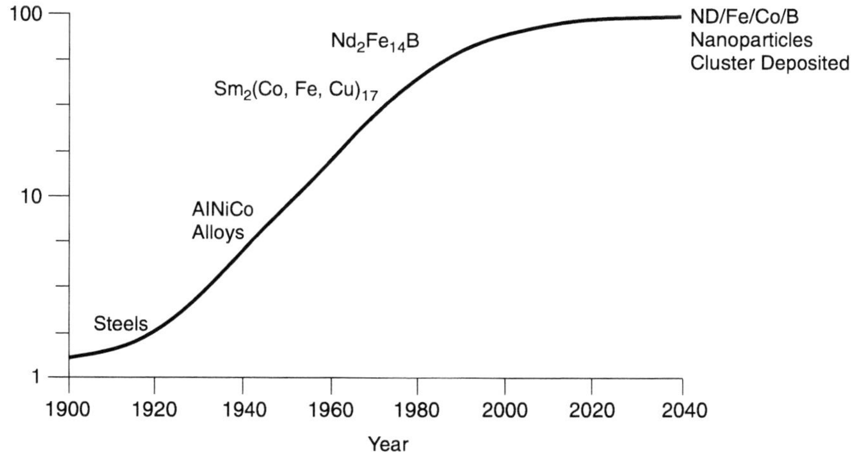

FIGURE 7.4 The projection for magnetic materials, with the flux-magnetization product as a figure of merit. SOURCE: Adapted from National Research Council, 1989, *Materials Science and Engineering for the 1990s: Maintaining Competitiveness in the Age of Materials,* National Academy Press, Washington, D.C., Figure 1.3, p. 22.

material properties through enhanced processing techniques and modeling will enable these applications. The projection for magnetic materials is shown in Figure 7.4 using the flux magnetization product as a figure of merit. Nanoparticle materials are expected to achieve the ultimate projected performance illustrated in the figure.

Advanced Energetic Materials

For the naval forces of 2035, a competitive edge in the power and range of missiles is required, in addition to improved warhead explosion devices. Advances in techniques for the synthesis of very dense organic compounds that are highly substituted with energetic groups will be required. The approach will be computationally based initially, followed by a synthesis simulation and prototype production. Continued development of new chemical processes to produce novel energetic materials and improvements of initial chemical processes to produce novel structures economically and environmentally are essential.

Coating Materials

The issue of life extension will be a critical one for the Navy and Marine Corps of the future. In addition, coatings reduce drag, increasing speed and range. The development of silicon-based coatings with natural antifouling agents will be required. In most instances, these coatings will be self-cleaning through the action of water flow across the hull.

Organic Materials—Flame-resistant, High-temperature Organic Composites

Polymers and polymeric composites are required for superior flame-resistant and high-temperature properties. These proposed materials are phthalonitrile-based composites with thermo-oxidative stability up to 500 °C. These novel flame-resistant materials will enhance ship and submarine safety.

ANTICIPATED BENEFITS OF THE APPLICATION OF MATERIALS TECHNOLOGIES

A competitive edge in the power, range, and performance of Navy and Marine Corps missiles will be acquired from advanced energetic materials. UAVs with high-performance sensors will be available. Naval craft will be protected with flame- and corrosion-resistant coatings able to extend the life of systems while achieving enhanced performance. New materials will enable new sensors, including those with the capability to detect and assess damage in microseconds, allowing for a reaction time in the range of seconds. Advanced sensor and actuator systems will combine damage detection with damage control.

The further advances in superconducting materials will lead to both new and improved military capabilities. Motors and generators will be smaller and lighter and more efficient. These will permit ship-design changes resulting in greater payload capacity or ships with smaller radar cross sections. Superconducting magnets will allow for high-sensitivity mine sweepers. Superconducting filters will allow for low-loss, in-band communications in areas of high electronic noise and will facilitate the development of high-bandwidth underwater communications.

High-temperature materials and coatings will result in more reliable, more efficient gas turbine engines with prolonged service life and lower fabrication costs. The development of new ferrous and titanium alloys will result in more capable ship and submarine hulls and structures with lower life-cycle costs.

Developments in wide-bandgap semiconductors and photonic materials will result in optimized performance of microwave and millimeter-wave detectors and transmitters. The development of diamond as a manufacturable semiconductor will allow it to be used in both analog and digital systems as well as photonic applications with smart sensors. Large-area silicon wafers and epitaxial layers (200 to 300 mm) will affect all microelectronic VLSI systems from computers to signal processing. Silicon quality and defect control continue to be important areas for Navy investment.

RECOMMENDATION

Materials are used in every aspect of naval operations. In the future, entirely new and enhanced existing materials will be designed and manufactured using a computational approach in which the physical and mechanical properties of materials are understood on an atomic scale. The nanophase materials engineered in this way will be tailored to meet specific requirements and to be reliable and robust at lower life-cycle cost. The Department of the Navy should strongly support the development of this materials design and processing approach.

8

Electric Power and Propulsion

INTRODUCTION

The panel focused on those electric power generation, storage, and propulsion technologies that, when applied as a system, will support the electrification of ships, submarines, and land-based vehicles. It surveyed the status of key elements of power systems such as energy storage devices, electric power recovery subsystems, and battery technologies. Industrial development is ongoing in most of these areas, and current trends suggest steady improvement in capacity, density, modularity, and reliability. The naval forces can take advantage of commercial developments by actively monitoring the commercial marine power and propulsion sector and by applying a top-down systems engineering process to fleet capability upgrades.

A recent naval planning document[1] outlined a development path for modernizing the power and propulsion capabilities of future surface ships and submarines. The plan does the following:

1. Takes a systems approach, allowing appropriate trades for optimization of performance within the constraint of affordability;
2. Takes a design approach in requiring modularity, commonality, and use of commercial technology for economy of scale wherever appropriate;

[1]Advanced Surface Machinery Programs Division, Ship Research Development and Standards Group, Engineering Directorate (SEA 03R2). 1996. "A Strategy Paper on Power for U.S. Navy Surface Ships," Naval Sea Systems Command, Arlington, Va., April 8.

3. Has initiated, and is following, a development path for those basic power generation, distribution and control, and propulsion technologies that are critical for shipboard implementation;

4. Has initiated a path whereby this capability can be used to backfit existing hulls in a hybrid configuration and be fully electric in new construction platforms; and

5. Provides a path for developing those critical technologies and beginning the integration of both hybrid and fully electric systems into the fleet during the next decade.

The concept appears to be sound, the enabling technologies are achievable, and the opportunity is now available to initiate a major change in the generation and use of electric power aboard ship. The benefits to the naval forces could be significant and far reaching. Advances in materials used in electromagnetic machinery, such as high-field permanent magnets and high-temperature superconductors, present opportunities for substantial increases in power density and efficiency. The electric ship approach is enabled by the ongoing revolution in electronics where advancements in solid-state semiconductor technology are being applied to power semiconductor devices. The emergence of direct electric conversion technologies, such as fuel cells, offers potential for fuel-efficient, low-emission, and low-noise sources of electrical power.

Within the plan, the Department of the Navy's integrated propulsion system (IPS) architecture will allow incorporation of developing technologies such as permanent magnet (PM) electric machines, fuel cells, and the power electronic building block (PEBB) architecture into future ship designs as programmed, preplanned replacements for the core technology in the first-generation IPS modules. Use of technology-independent module interface standards will facilitate technology insertion with minimum impact on the ship design and construction process.

IPS with electric propulsion motors allows for consideration of integrated motor and propulsor concepts. The steerable podded propulsor provides the greatest potential for meeting naval hydrodynamic and hydroacoustic requirements while also being competitive with conventional propulsors on commercial ships.

The benefits of ship electrification are as follows:

- *Combat systems effectiveness.* Integrated electric power and propulsion systems enable design flexibility that, in turn, will facilitate the optimization of topside arrangements for maximum combat system effectiveness.
- *Survivability.* The elimination of gear trains and propeller shafts, together with the flexibility and modularity of electric systems, will enable graceful degradation and rapid reconfiguration of vital systems, thus enhancing survivability.
- *Signature reduction and quieting.* By eliminating the mechanical link between the power plant and the propulsor (i.e., propeller), electric drive will

enable reduction of noise and vibration by allowing acoustic isolation of the engine generators.

- *Improved operational flexibility and reliability.* With the power station concept, all power will be supplied by a set of prime movers that provide power to propulsion, ship service, and other designated loads. This approach will provide the flexibility to shift power between propulsion, ship service, and other electrical loads, which cannot be done with a mechanical drive ship, and can enable improved speed control and steaming efficiency and the elimination of less efficient controllable pitch propellers in favor of fixed pitch types.
- *Increased flexibility and adaptability.* Integrating electric power and propulsion systems will provide flexibility in servicing other loads such as environmental controls (air conditioning) and launch and recovery systems, electric armor, high-power advanced electronics, and electric weapons.
- *More space available.* Eliminating gears and shafts from the propulsion system will make more space available for other uses. Also, the prime mover will no longer be tied to the propeller shaft line, and the power sources can be distributed throughout the ship as necessary.
- *Reduced manning.* Digital control and automation, which will be an integral part of ship electrification, will reduce the requirements for human operators (in machinery spaces, for example).
- *Reduced logistics.* Common power and propulsion modules can be used across the fleet.
- *Reduced costs.* Commercial technology appears likely to be available for many of the system elements.
- *Life-cycle cost and fuel-consumption savings.* Overall fuel efficiency can be improved if a ship is able to operate at varying speeds over a substantial part of its operational profile, and variable speed propulsion will be enabled by electric drive.

In general, the same line of reasoning that applies to surface ship power and propulsion systems also applies to submarines. Optimization of submarine systems will be driven by stealth, safety, power density, and other requirements that differentiate surface and undersea vehicles. Submarine power and propulsion system technologies of the future will include the following:

- Very low harmonic motor controllers;
- High-power-density and high-performance solid-state inverters and converters and the components thereof;
- Electrical substitutes for systems and components that now rely on fluid transport for energy and actuation, e.g., electric actuators, electromagnetic launchers, and thermoelectric coolers;
- Motors and generators with very low acoustic and magnetic signatures, including versions that can operate submerged in seawater at high pressure; and

- New technologies for motors and generators such as superconducting magnets, cryogenic coolers, current collectors, high-field permanent magnets, liquid cooling, and active noise control.

A HTS high-torque drive motor can be one-fourth the size of a permanent magnet motor, and its efficiency can be high enough to compete with existing reduction gear/steam turbine systems. HTS motors and generators can also be quieter because the high-density magnetic fields produced allow elimination of iron cores, simplification of armature designs, and elimination of other components that create noise.

The Marine Corps relies heavily on power and propulsion technology for its vehicle power systems. The requirements for higher water speed amphibious vehicles, signature reduction, adaptability to electric power, future electric weapons, and reduction of vehicle weight all lead to the need for an integrated electric power system for future land and amphibious vehicles.

Key technologies for land and amphibious vehicles mirror those described for surface ship systems but are driven by a different set of requirements. These include the following:

- Pulsed-power systems including energy storage pulse-forming networks and very high power solid-state switching;
- Flywheel energy storage;
- High-power-density engines, including electric drive systems with a unit size of several hundred horsepower; and
- Electrical steering, suspension, and actuators.

In conclusion, the panel believes that electric power science and technology holds great promise for major technical progress and payoff in the future. The remainder of this chapter is organized to first discuss electric power generation and storage technologies and then focus on the electric ship concept and associated technologies.

ELECTRIC POWER GENERATION AND STORAGE TECHNOLOGY

Introduction

Warfare in the early 21st century will involve the use of advanced technologies and combat systems with ranges, lethality, and detection capabilities surpassing anything known in contemporary warfare. In the battle zone surrounding and including the battle group and expeditionary forces, there will be requirements for electrical power at levels from tens of watts for surveillance and communication, to kilowatts for radar and for a variety of electric motors, to hundreds of kilowatts for field-base power demand and multi-megawatts for advanced weaponry, countermeasures, and propulsion systems. Mobility of all system elements will be essential for survival and success. Therefore, although fuel supply may dominate the total system operational weight for nonnuclear power, mobile weapons and sensors should be powered by supplies that have high specific power, are compact and quiet, and have minimal signatures in the full electromagnetic and acoustic spectra.

Nonpropulsive power generation requirements and implementations will be integrated into a single system with propulsive power generation. This, and the continued deployment of special electric loads, will determine the requirements for power distribution, storage, and conditioning.

The concept of an all-electric, or more electric, ship is gaining momentum in both U.S. and foreign navies. Simply stated, this concept embodies the use of electrical means for all power needs in lieu of other means such as mechanical, pneumatic, and hydraulic. This then provides an opportunity to standardize components and systems, reduce signatures, and better integrate the power and propulsion systems for all vehicles. The move toward increased reliance on electricity is also being applied to air vehicles and to Marine Corps amphibious and land vehicles.

Versatility in electric power availability will enable new varieties of weapons and ship systems. Some proposed new weapons present especially rigorous power supply requirements. Electrothermal, chemical, and electromagnetic guns and high-powered laser or microwave directed-energy weapons, for example, require large amounts of power over very short time periods, as do electric rail aircraft launchers (electric catapults), which are technically similar to electromagnetic guns. Sophisticated power generation, distribution, and control systems will enable local-zone power distribution and conditioning, which will enable lower platform signatures and localized platform damage control. An order-of-magnitude evolution in size and power requirements for specialized sensors and unmanned systems presents new application requirements for power storage systems.

A general schematic for the type of power system addressed by the panel is presented in Figure 8.1. A prime electric power source feeds directly, or via a load-leveling or storage stage, into a conditioning stage to alter the electrical parameters for optimum reliable power distribution to a load.

The panel believes that the following three technologies will require special attention and support if the full potential of integrated electric power and propulsion systems is to be realized:

- Power generation, including advanced fuel-efficient turbogenerator power units for continuous power and for pulsed or short-duration power generation applications and direct electrochemical and electrothermal generators such as fuel cells;
- Power conditioning and distribution, including passive components, such as capacitors, inductors, and transformers, and active devices, such as power

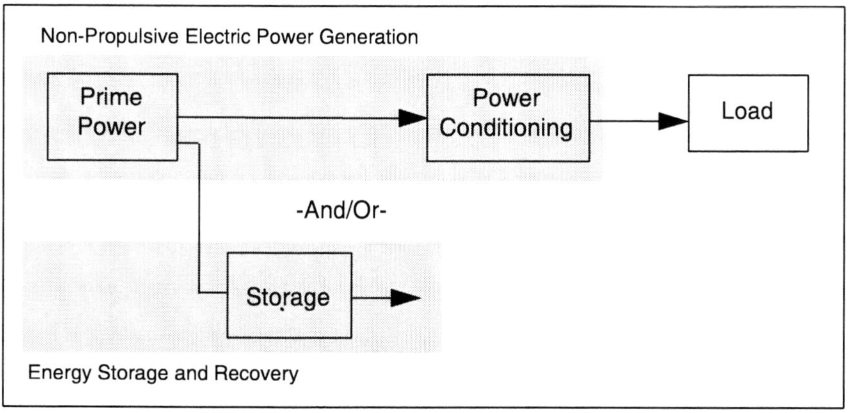

FIGURE 8.1 Schematic of a generic power and propulsion system.

electronic switches and inverters for continuous as well as pulsed or short-duration, high-power conditioning and distribution; and

- Energy storage, including primary and secondary electrical energy storage such as batteries, flywheels, and generator combinations, and magnetic energy storage devices.

Table 8.1 lists the technology areas covered by the panel. Technology forecasts are provided in the sections that follow.

TABLE 8.1 Electric Power Technology Areas

Electric Power Generation	Energy Storage and Recovery as Electric Power
Nuclear-electric generators	Primary batteries
Turboshaft engine generators	Rechargeable batteries
Piston-engine generators	Superconducting magnets
Fuel cells	Flywheels
H2	Pumped liquids
Other	Compressed gases
Explosive/magneto-hydrodynamic	Thermal storage
Power conditioning	
Bus power	
Slow power	
Fast power	

SOURCE: Adapted from the Board on Army Science and Technology, 1993, "Electric Power Technology for Battle Zones," *STAR 21: Strategic Technologies for the Army of the Twenty-First Century*, National Academy Press, Washington, D.C., Table 43-1, pp. 539-540.

Power Generation

The concept of the electric ship is a powerful enabler for new concepts in ship design and operation. Independent placement of propulsion and power generation elements and the integrated approach to electric power distribution will rely for their efficiency on new developments in prime power generation, conditioning, and distribution.

Electric propulsion assumes continuous generation capacity that will also provide electric power for ship auxiliary functions from the most mundane application to the more exotic ones, such as catapults and electric weapons. It will often be necessary to provide auxiliary power generation for backup, or for specific subsystems, or for zonal requirements. Aircraft and launch and recovery systems, electric armor, high-power advanced electronics, and electric weapons all require pulsed-power systems whose average power for the duration of output ranges to hundreds of megawatts. The mass and volume of generators in this class are half power unit and half power conditioning. The physical decoupling of power and propulsion subsystems afforded by electric drive will be a powerful enabler for compact, versatile, and powerful future ships.

Continuous Power Generators

For nonnuclear operation even the most efficient internal combustion engines consume their own weight in fuel in about 10 hours. Therefore, fuel storage determines overall operational system weight, and fuel supply will dominate logistics considerations. For mobile electrical generators and power conditioners, substantial weight reduction can be achieved by increasing the generating and distribution frequency from the present 60-Hz standard up to the 400-Hz range.[2]

Of the continuous power generation systems listed in Table 8.1, some are already in use by the naval forces. Nuclear-electric generators have potential for significant advances in specific power output. Although the pressurized water nuclear power plants currently in use do not hold much promise for greatly improved power density, a new generation of high-power-density nuclear power plants could be developed but would require substantial R&D investment. A development program could build on past R&D programs that have focused on compact, high-output reactor designs.

Turboshaft engine-generators, which represent another important class of continuous power generation systems, are currently used in military helicopters and some specialized boats. They are also being incorporated into mobile electric

---

[2]Board on Army Science and Technology. 1993. *STAR 21: Strategic Technologies for the Army of the Twenty-First Century*, National Academy Press, Washington, D.C.

power units, primarily by the Army. They burn jet fuel and have rotational speeds in excess of 20,000 rpm. The high rotational speed could be used to advantage in a high-frequency, lightweight, direct-coupled electric generator operating at 24,000 rpm (400 Hz) or higher. Typically, however, they are geared down to 60-Hz alternators, and these gears increase weight and noise.

The panel evaluated the various continuous and pulsed/short-duration power generation, power-conditioning, and energy-storage technologies by creating figures of merit in the near term (year 2000) and far term (year 2020). The selection of certain technology areas for inclusion in this report was based in part on the potential to achieve high values for figures of merit in the future. The figures of merit include functionality, mobility, supportability, manpower requirements, and cost. A significant improvement in any of these figures of merit could justify adoption of the technology by the Department of the Navy.

The panel believes that, of the various technologies evaluated, turboshaft-engine-driven alternator systems have the best potential for practical, continuous improvement in all figure of merit areas. To realize this potential, the Department of the Navy should support the development of advanced turboshaft engines such as in the DOD/NASA Integrated High-Performance Turbine Engine Technology (IHPTET) program and carry out an aggressive development effort in advanced lightweight alternators, power-conditioning control systems, and unit integration and modularization.

The electric power systems driven by turboshaft engines have the best potential for evolving into compact, lightweight units with reasonable noise control. Systems with these characteristics will be required in the future, particularly for the intermediate power ranges from 50 kW up to the megawatt level. The most significant factors controlling the specific power of an integrated unit are engine rotating speeds, generator frequency, and generated voltages. To a first approximation, the physical size of an engine is proportional to torque, and increasing the speed increases the shaft power proportionately with only a modest change in weight. The weight advantage gained by high-speed operation can be lost, however, by coupling through a gear box to heavy low-frequency generators and power conditioners. The present standard of 60-Hz generators requires that either engines with speeds greater than 3,600 rpm must be geared down to 3,600 rpm or the high-frequency alternating current (ac) that is directly generated must be rectified to direct current (dc) and then converted back to 60-Hz ac with an inverter.[3] Both approaches introduce weight and cost into the system.

Design optimization studies performed for the international space station suggest a different overall approach: Let the alternator rotate at the turbine speed and produce power at that frequency. Turbine speeds of 24,000 rpm or higher are

---

[3]Board on Army Science and Technology. 1993. *STAR 21: Strategic Technologies for the Army of the Twenty-First Century,* National Academy Press, Washington, D.C.

achievable and efficient. When directly coupled, 24,000 rpm yields ac power at 400 Hz, which is the current aviation standard. Using these higher frequencies for power generation can reduce the weight of the active components in the system by a factor of about six, compared with a 60-Hz system. Solid-state converters could then be used to shift the frequency to 60 Hz for distribution. Because of the ease of converting dc to any desired frequency, it is feasible to consider dc distribution in many cases.

The voltages used for power generation and distribution also have leverage with respect to high specific power. The required conductor cross section, and hence the weight, decreases as the square of the voltage, so operating voltage is an important parameter.[4] The practical voltage limit (currently about 1 kV) for mobile systems is set primarily by the reliability limits of semiconductor devices. Improvements in high-voltage semiconductor devices could permit more system-optimized voltages to be used up to the Navy standard bus transmission level of 5 kV.

As the technologies evolve over the next 20 years, the specific power of high-rpm engines should approach 10 kW/kg. Similar progress is anticipated for high-rpm alternators with stationary armatures and separately excited rotating-magnetic-flux generating units. The weights of these generators are almost inversely proportional to their rotating speed (rpm). Thus, they could be designed for 400-Hz operation and directly coupled to the gearbox output shaft at 24,000 rpm. Although some special generator units have been operated at specific powers as high as 20 kilowatts of electric power output per kilogram (kWe/kg), generators built to military standards will more generally achieve 10 kWe/kg.

At high-frequency operation, power-conditioning components are small, lightweight, and highly efficient, particularly when series resonant conversion is used in converters. Advanced technology should yield systems with power densities in the range of 15 to 30 kW/kg. Power densities as high as 100 kW/kg are conceivable for power conditioners when integrated-circuit fabrication techniques are used to manufacture integrated converters.

Figure 8.2 shows how the operating time affects the total specific power (power/weight ratio) for existing and estimated future mobile electric power units. The diagonal limiting line in the figure represents the maximum specific energy available from combustion of the JP-8 fuel that will be used in the Navy's diesel engines or turbine engines. An overall efficiency of 39 percent is assumed from maximum fuel heating value to conditioned electric power. The ordinate gives the specific power based on the total system weight: kilowatt-hours divided by the weight of the total integrated system, including engine, generator, power conditioner, integration factor, and fuel consumed. For many engage-

[4]Power (P) = V^2/R, where V is the voltage and R is the resistance. In a cylindrical cable of cross-section area A, R is inversely proportional to A. Therefore P is proportional to V^2A. For a given power, the required area A is inversely proportional to V^2.

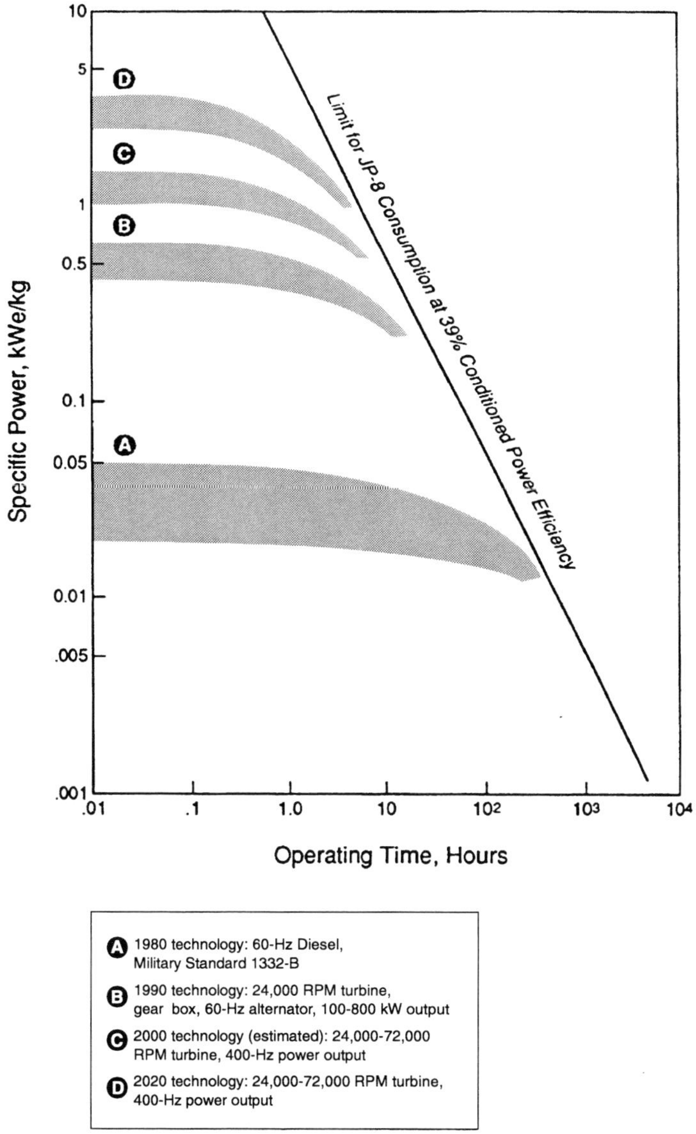

FIGURE 8.2 Specific power versus operating time for advanced turbogenerators. SOURCE: Adapted from the Board on Army Science and Technology, 1993, "Electric Power Technology for Battle Zones," *STAR 21: Strategic Technologies for the Army of the Twenty-First Century*, National Academy Press, Washington, D.C., Figure 43-1, p. 547.

ments in the future, actual operating times may be relatively short (hours to hundreds of hours). Therefore the specific power of the hardware units will be critical to mobility and initial operational capability.

Fuel cells hold promise for zonal power generation. They are a particularly efficient and clean method of utilizing hydrocarbon fuels and are appropriate for medium-sized power generation units (1 to 4 kW). Hydrogen-based fuel cells electrochemically oxidize hydrogen with oxygen (or air) and directly convert chemical energy to electric energy. The product of this combustion is water, and this combined with the extremely low signature of the cell make it a very attractive power alternative for Navy and Marine Corps operations. Because they are essentially isothermal devices and therefore not limited by heat engine thermodynamics, they can in principle have very high energy densities (watt-hours per kilogram), limited only by the specific free energy change in converting hydrogen and oxygen to water. The great obstacle to hydrogen-based fuel cells for mobile, battle-zone applications is the volumetric problem of storing hydrogen. Although hydrogen produces a great deal of energy on a weight basis, it produces very little energy on a volume basis, even when stored at high pressure.

Today hydrogen is the only fuel that can be electrochemically oxidized in a fuel-cell system at practical rates. The ideal system would directly convert the free energy of combustion of other fuels, such as hydrocarbons, into electric energy. Other fuels are used in prototype fuel-cell systems by either externally or internally reforming them to a mixture of hydrogen and carbon dioxide, then using the hydrogen electrochemically in the fuel cell. A major breakthrough in electrocatalysis would be required before other fuels could be electrochemically oxidized at useful rates.

Hydrogen fuel cells are impractical for at-sea applications because of the space required to store hydrogen, but hydrocarbon fuel conversion designs would be directly applicable to Navy use. Most units under development (primarily in the automotive industry) require higher-quality fuels than standard Navy distillates. Although there is no current evidence of an impending technical breakthrough in this area, the promise is such that near-term programs to develop converters for standard Navy fuels should be encouraged.

Pulsed-power Generators

Pulsed-power generation systems can be divided into two major subsystems as illustrated in Figure 8.3: (1) energy storage or prime-power generation, followed by (2) power conditioning. The panel considered five classes of pulsed-power generation as listed below:

1. *Open-cycle, combustion-driven gas turbine generators.* These systems burn fuel in excess air to produce hot combustion gases, which are then expanded through a turbine to drive a generator, producing the prime power. They have

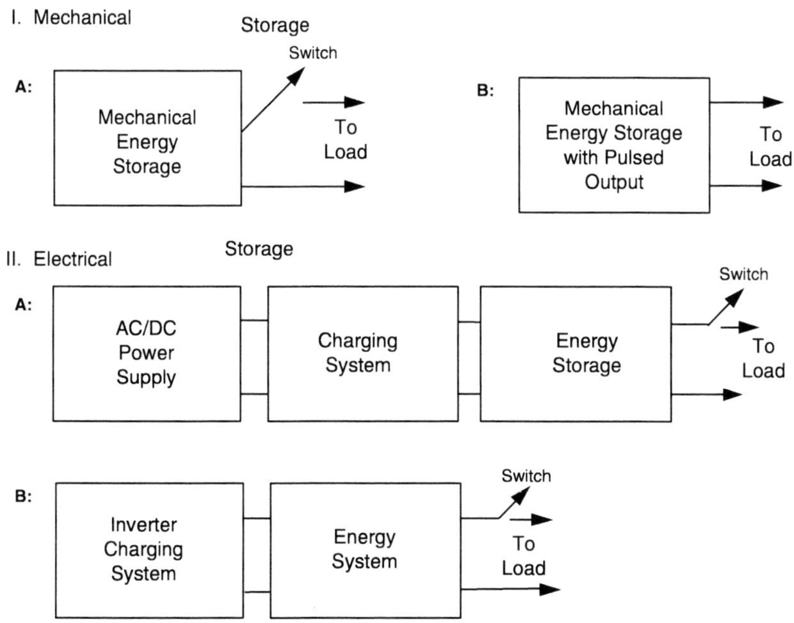

FIGURE 8.3 Alternative techniques to provide multi-megawatt prime pulse power and power conditioning. Which is chosen depends on the load to be served. SOURCE: Adapted from the Board on Army Science and Technology, 1993, "Electric Power Technology for Battle Zones," *STAR 21: Strategic Technologies for the Army of the Twenty-First Century*, National Academy Press, Washington, D.C., Figure 43-3, p. 552.

very attractive power densities, particularly for high power levels, and they are good candidates for future pulsed-power generation sources.

2. *Magnetohydrodynamic generator driven by solid or liquid propellants.* A magnetohydrodynamic channel consists of a linear array of electrodes on either side of the inside of a rectangular duct. A magnetic field is created by permanent or electromagnets oriented parallel to the electrode faces. Ionized gas is forced through this channel, and the velocity of the gas interacting with the magnetic field creates a voltage differential at the electrodes. To produce the ionized gas in these systems, solid-propellant grains or a bipropellant liquid combination burns at elevated pressure and extremely high temperature, and the propellant is seeded with an ionizable material such as potassium and then passed through a magnetohydrodynamic channel where part of the energy in the ionized gas is directly converted to electric power.

3. *Explosive-driven magnetohydrodynamic generator.* In this system, an explosive charge is used to drive a single-shot magnetohydrodynamic power generator. The advantage of this system is that a pulse with a very high power

level is produced, thus avoiding much of the power-conditioning requirements of other systems.

4. *Flywheel energy storage and power generation.* Large amounts of power can be stored in a flywheel as it is spun up. By coupling the flywheel to a generator, that power can be tapped rapidly. New composite materials have greatly increased the efficiency (inertia) of flywheel systems.

5. *Superconducting magnet energy storage and power generation.* Superconducting magnet energy storage (SMES) systems make it possible to store electrical energy directly with little loss other than the power required to cool the superconductor. Direct current can be fed into a superconducting coil to create a powerful magnetic field that stores the electrical energy. The field can then be discharged through a solid-state inverter to produce pulsed power as needed. The switching times between discharge cycles can be milliseconds, and charge and discharge cycles can differ by orders of magnitude. New superconducting materials and manufacturing technologies will provide rapid development for these devices in the future.

The panel believes that three of the five categories of energy storage or prime-power generation discussed above have broad-based applicability to naval force needs for high-energy pulsed power by the year 2020. These are (1) open-cycle, combustion-driven gas turbine systems, (2) flywheel energy storage and power generation, (3) fuel cells, (4) other direct conversion technologies, and (5) superconducting magnet storage and power generation. The panel anticipates major improvements in these systems over the next 30 years.

Open-cycle Combustion

Continuous development and improvement of open-cycle, combustion-driven gas turbine systems is expected, including systems that can use a broader range of fuel types, thus easing the logistics problem. These new systems will be much more rugged than today's turbogenerators because of the introduction of advanced materials and processing methods. Also anticipated are approaches to reducing signature of the effluent stream, thus enhancing survivability.

Flywheels

Flywheels have potential for improvement in the areas of functionality, mobility, supportability, manpower, and cost. Most of the anticipated improvements in flywheels will come from major advances in composite materials with ultra-high strength-to-weight ratios. These materials will enable stored energy density to increase by an order of magnitude over high-strength steels because they can sustain higher angular velocities. The cost of fabricating composite flywheels with optimally tailored structural properties will decline dramatically over the next 30 years, as more military and industrial applications are found.

Fuel Cells

Unlike turbogenerator technology, practical fuel cells that run on liquid hydrocarbons do not exist today. They are certainly a possibility, however, particularly in light of the advances made in the previous decade in understanding the catalytic reforming of hydrocarbons. Research in this area with the specific goal of developing a fuel cell that would run on liquid hydrocarbons (Navy distillate) and air would be worth supporting. If a breakthrough were to occur, the implications for the Navy of the future could be enormous.

Other Direct Conversion Technologies

There are other direct electrical-conversion technologies besides fuel cells, including thermionics and thermophotovoltaics, that may become viable in the future, but probably beyond the 2010 time frame. In thermionic conversion, heat is used to create electrical flow between electrodes. In the thermophotovoltaic converter, a two-step conversion process is used with heat causing an emitter to create photons that are captured by a photovoltaic cell to produce electrical energy. These technologies will likely be introduced in cogeneration schemes such as topping cycles for gas turbines or direct heat sources such as nuclear reactors or thermoelectric nuclear devices.

Superconducting Magnet Energy Storage and Generation

Superconducting magnet energy storage (SMES) is a scalable technology, and it is reasonable to consider storage capacities ranging from less than 0.001 MW/h to more than 10,000 MW/h. The technology has received government support since the mid-1980s—for example, the Air Force has conducted a program that demonstrates small, economically viable SMES systems for uninterruptable power and power-quality management, and the Los Alamos National Laboratory has run a program for high-discharge-capacity storage—commercial applications have focused on the high end of the scale (such as an operating load-leveling unit at the Bonneville Power Administration in Tacoma, Washington).

To achieve the improvements in capacity and reductions in size and weight required by naval applications, continued development of materials and manufacturing methods for SMES systems is required.

Power Conditioning and Distribution

Power conditioning currently represents between 50 and 80 percent of the total mass of high-power pulsed generator systems. For future Navy and Marine Corps applications, it will be essential to greatly increase the specific power density of these power conditioners. A reduction of at least an order of magnitude in specific weight is possible before intrinsic physical, chemical, or mechanical limits are reached.[5] The advances will come about primarily through new, molecularly tailored materials for passive components, advances in solid-state power electronic devices, and improved techniques for heat rejection and thermal management, such as heat pipes. High-temperature superconducting cables could increase current carrying capacity by factors of three to five times that of conventional copper cables. This factor includes parasitic cooling loads. The liquid nitrogen coolant required for HTS systems is nonflammable and would tend to extinguish any electrical fire that might arise from battle damage to the cable.

Power-conditioning systems that evolve over the next 30 years will be designed to degrade gracefully. In other words, through zonal power generation and distribution and intelligent management, single-point failures will result in only a small decrement to platform performance because the systems will be designed in a way that enables the automatic bypass of the degraded element.

Current and Near-term (Out to 10 Years) Characteristics

The three main elements of all power-conditioning systems are bus power conditioning, slow power conditioning, and fast power conditioning, as described below:

1. *Bus power conditioning.* This technology takes the electrical feed from one of the prime power sources listed above into a conditioning stage, where the voltage level is altered for optimum, reliable power distribution to a slow power-conditioning stage. The major components of this first stage of power conditioning are inverters, transformers, and rectifiers. The application of HTS technology could improve transformer efficiency and yield safer equipment by eliminating the dielectric oil that sometimes poses fire and toxicity hazards.
2. *Slow power conditioning.* The slow power-conditioning stage produces repetitive electrical pulses in the millisecond time frame. Switches, capacitors, and transformers are the major components used in this stage of power conditioning.
3. *Fast power conditioning.* If required for the particular application, a fast

[5]Board on Army Science and Science and Technology. 1993. *STAR 21: Strategic Technologies for the Army of the Twenty-First Century,* National Academy Press, Washington, D.C.

TABLE 8.2 Forecast of Power Conditioning Module Power and Power Density

Parameter	Today (per system)	Tomorrow (10 years) (per module)	Future (30 years) (per module)
Average power			
kWe/kg	0.05	10	>100
kWe	100	100	1,000
Pulsed power (as pulsed energy)			
kJ/kg	0.3	1	>10
kJ	50	500	5,000

SOURCE: Adapted from the Board on Army Science and Technology, 1993, "Electric Power Technology for Battle Zones," *STAR 21: Strategic Technologies for the Army of the Twenty-First Century,* National Academy Press, Washington, D.C., Table 43-3, p. 543.

power-conditioning stage can be used to convert the pulses with a millisecond duration from the slow power-conditioning stage to pulses of a microsecond or less. This stage uses transmission lines, capacitors, switches, and magnetic devices as the primary components.

The anticipated progress in key parameters for power conditioning subsystems is indicated in Table 8.2.

Technology Forecasts

Capacitors

As naval force requirements for compact electrical power systems grow substantially over the next 30 years, the development of capacitor technology will be a major enabling technology element. For microsecond to fractional-second energy storage and discharge, capacitor technology is unequaled in the flexibility and adaptability needed to meet a broad range of future requirements.[6]

At present, energy conditioning at pulse repetition rates of less than 1 Hz and 10 to 30 MJ per pulse has been achieved for pulse durations from 0.05 to more than 1,000 μs. The maximum voltages decrease from megavolt levels at shorter pulse durations to tens of kilovolts at the longer pulse durations. Voltage levels are determined by the nature of the load. Innovative capacitor systems, operating at near their ultimate voltage breakdown limits, which are 10 to 15 times today's operational levels, would enable the development of lightweight systems. The

[6]Board on Army Science and Technology. 1993. *STAR 21: Strategic Technologies for the Army of the Twenty-First Century,* National Academy Press, Washington, D.C.

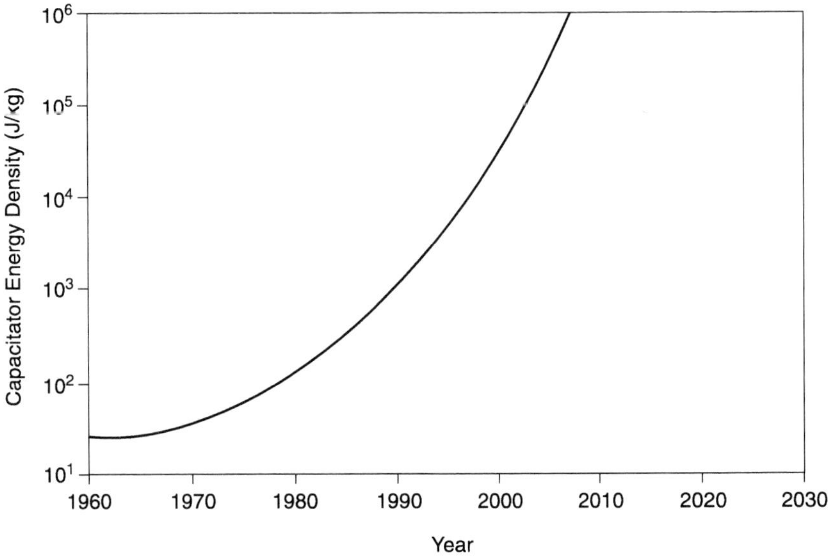

FIGURE 8.4 Projection of specific energy density for new capacitor technology. SOURCE: Adapted from the Board on Army Science and Technology, 1993, "Electric Power Technology for Battle Zones," *STAR 21: Strategic Technologies for the Army of the Twenty-First Century*, National Academy Press, Washington, D.C., Figure 43-4, p. 557.

best commercial capacitors today achieve specific power densities of 1 to 10 kWe/kg and specific energy densities of 1 to 300 J/kg. Repetition rates can be as high as several kilohertz, necessitating the development of capacitor technology integrated with that of negligible-loss switching topologies and voltage-multiplication pulse transformers. Most systems can be classified by voltage according to their average power and run-time characteristics. For future naval needs (30 years out), compact systems would be enabled by capacitor energy and power densities 10 to 100 times those available today. Development of new solid and liquid materials, in conjunction with advancing methods of manufacturing technology, will likely be feasible with the use of tools emerging from current technology programs. What will be required is a tightly integrated development program for materials and components, tailored to areas of need for the three main classes of capacitor technology: polymer films, ceramics, and electrolytic capacitors.

Figure 8.4 shows recent and projected progress in capacitor energy density. Industrial capacitor developers have taken a preeminent role in the integration of materials development with the practical realization of advanced capacitors. The Department of the Navy will need to support a system-responsive, technology-based development program in each of these capacitor areas. The program goals should be demonstration hardware that will operate at the required power and energy densities.

Capacitors with specific power densities of 200 to 1,000 kWe/kg and specific energy densities of up to 20 kJ/kg may be feasible in the future. Recent discoveries have resulted in capacitors whose performance degrades gracefully and does not result in single-point failure for the system. Now the potential exists for novel capacitor technologies that offer radically different availability over the system life. Indeed, throughout the system's normal life, even in adverse environments, the failure modes of new systems will provide graceful and predictable reduction in performance, so that total system operation can be retained, although at a reduced performance level.

Table 8.3 projects the performance trends for state-of-the-art capacitor technology. It lists selected examples of several classes of advanced capacitors that further research could develop as practical, highly compact units. For large production volumes, the unit cost of these advanced units is projected to be comparable to the cost of current technology.

Inverters

The naval forces' requirement for efficiently conditioned electrical power will continue to grow dramatically over the next 30 years. Part of this requirement will come from the conditioning needed to provide a low noise—hence dc—electrical bus with the power quality, reliability, and maintainability that will keep electric-powered systems operational during warfare. Eliminating the

TABLE 8.3 Performance of State-of-the-Art and Projected Advanced Capacitor Systems

	Energy Density (kJ/kg)		Power Density (kW/kg)			
Capacitor System	Now	Future (2020)	Now	Future (2020)	Repetition Rate (Hz)	Critical Research Areas[a]
Polymer film	1.5	10 to 20	60	2,000	<100	Film impregnants, foils, >150 °C, 1 MJ/unit
Ceramic	0.01	5	100	50,000	10,000	Ceramics, electrodes, >500 °C, 1 kJ/unit
Electrolytic	0.2	2	20	200	100	Electrolytes, separators, >200 °C, 10 kJ/unit

[a]Technology factors that must be addressed to achieve the projected future performance. SOURCE: Adapted from the Board on Army Science and Technology, 1993, "Electric Power Technology for Battle Zones," *STAR 21: Strategic Technologies for the Army of the Twenty-First Century,* National Academy Press, Washington, D.C., Table 43-9, p. 558.

present single-point failure topologies that result in catastrophic failure of electrically powered assets can be accomplished by developing zonal control architectures that will be enabled by advanced inverters.[7]

The objective is to provide the platform with a power source that degrades gracefully when damaged during war, continuing to operate, albeit at reduced performance, so that electric-powered assets that are not directly damaged remain available for their missions. In point-failure situations, bus conditioning via bidirectional inverters would continue to supply power from generator-to-generator feed lines redundantly, isolating damaged generator members of the system and permitting load sharing under inverter control. Because no other power-conditioning technology can provide this control, inverter dc bus power conditioning is vital to the electrically powered naval forces of the future, and the Department of the Navy must ensure the development of this technology. As illustrated in Figure 8.5, up and into the early 1990s little substantive progress in cost-performance ratio was made in the power-density scaling of bus power elements for main computer systems.

Integrated, Modular Power Electronics

The Department of the Navy in partnership with industry has pursued the development of standardized, solid-state, integrated power-conditioning modules and has succeeded in achieving reductions in size and increases in efficiency, if not reduction of cost. The requirements of future weapon systems could be met through a focused development program that increases the inverter subsystem power densities by 10 to 100 times over the 0.01 to 0.2 kWe/kg of current military systems. There is a need, for example, for modular, molecularly grown, block power conditioners that are fault tolerant and whose performance degrades gracefully and for voltage management modules that are interchangeable and stackable and can be connected in parallel to create the quality of ac/dc voltages needed in the field.

Interchangeable voltage management modules would provide universal, intelligent, dc bus power conditioning, with per-module, power-break sizes that are cost effective for the Navy. Modules would probably come in sizes of 1, 10, 100, 1,000 and 10,000 kW. Power quality and electromagnetic interference management functioning would be integrated into each module. Such modules will need to integrate the passive component technologies for capacitors, inductors, and the like, with advanced power switches and intelligent control capability. A program now under way within the Department of the Navy to develop a PEBB is a step in the right direction. This effort must incorporate the best available technology into successive generations of power-conditioning systems for naval applica-

[7]Board on Army Science and Technology. 1993. *STAR 21: Strategic Technologies for the Army of the Twenty-First Century*, National Academy Press, Washington, D.C.

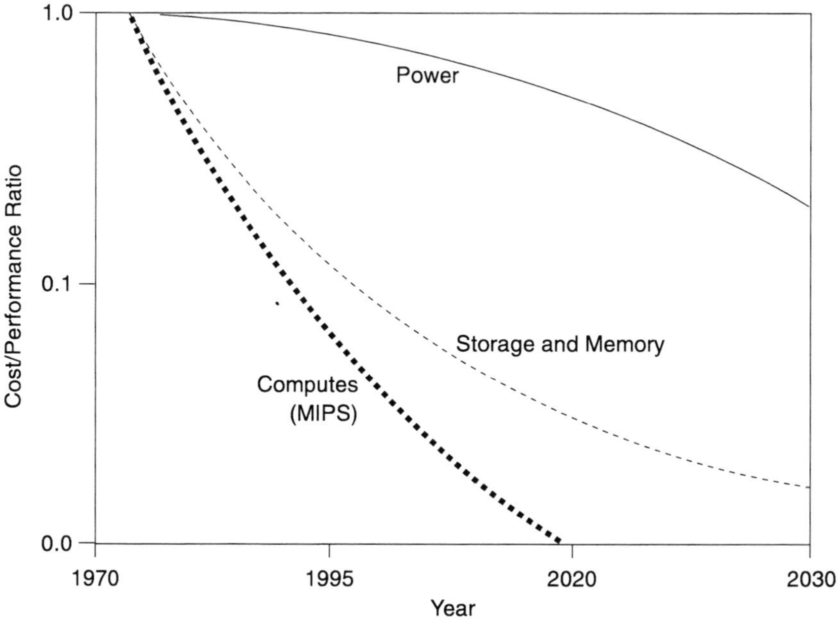

FIGURE 8.5 Technology advancement with time of major computer systems elements as a function of cost versus performance ratio. SOURCE: Adapted from the Board on Army Science and Technology, 1993, "Electric Power Technology for Battle Zones," *STAR 21: Strategic Technologies for the Army of the Twenty-First Century*, National Academy Press, Washington, D.C., Figure 43-5, p. 560.

tions. The modular approach to improved power electronics will leverage commercial products for military use while reducing the effort required to develop new applications. Detailed circuit design will be unnecessary, allowing designers to concentrate on optimizing system performance.

Energy Storage and Recovery

By the year 2035, there will be a greater dependence on electric power for vehicle drives as well as for a wide array of battlefield electronics and for new electrically driven weapon systems. The reduction of the signature of power generation units will become increasingly important in the battle zone of the future, and this will likely increase the dependence of the naval forces on compact energy storage systems, either chemical or mechanical, in which the stored energy can be rapidly recovered as electric output. Electric vehicle research in the automotive industry is an important contributor to developments in long-life, high-capacity, quick-response electric energy systems and currently sustains work in the development of reliable and highly efficient chemical and mechanical energy storage systems.

Five classes of energy storage and electric-power recovery technologies that could apply to the forces' future needs were considered by the panel, as follows:

- Batteries,
- Flywheels,
- Inductive energy storage,
- Capacitive energy storage, and
- Compressed gas or steam.

The panel concluded that of these five storage technology areas, batteries show the greatest promise for meeting near-term needs, although flywheels may provide some capability in hybrid vehicle propulsion systems. Inductive and capacitive energy storage media will require continued development to reach a level of performance to be useful in fielded systems. Compressed gas is currently used to start engines on surface ships and for ballast blow and weapons launch in submarines. Steam is used for aircraft launch on carriers. The compressed gas or steam approach is fundamentally limited and ripe for replacement by the other candidates listed above.

Batteries

The science of electrochemical energy storage is fundamentally mature. The pace of development is relatively slow, but the need for safe and reliable electric power is so pervasive that there is a constant high level of research and development. New commercial battery products find their way onto the market, as historical shortcomings in the safety and stability of materials and chemical reactions are resolved. This was the case, for instance, in the recent development of the nickel metal hydride (NiMH) battery as a replacement for the environmentally unsound nickel cadmium (NiCd) battery.

Technology development promises evolutionary progress in the realization and utility of the physical potential within the basic chemistries. Several long-term efforts in commercial and government establishments focus on improvement in shelf life, safety, volumetric and gravimetric energy values, and, in the case of secondary batteries, smart controls for charging efficiency and cycle lifetimes.

Nonrechargeable, or primary, batteries are currently used in a wide range of applications in the Navy where very low levels of electric energy (i.e., watt-hours) are required. Lithium cells with lives of 5 to 10 years are available and will undoubtedly be improved over the next 30 years. Tables 8.4, 8.5, and 8.6 summarize primary battery technology.

TABLE 8.4 Primary Cell Types With Little or No Chance of Further Development

Chemistry	Cell Volt	W-h/kg	Notes
Aluminum-Manganese Dioxide $Al\text{-}MnO_2$	1.6		Unpredictable storage life; anode corrosion problems; voltage delay
Magnesium-Air $Mg\text{-}O_2$	3.0		Anode corrosion problems
Magnesium-Bismuth Oxide $Mg\text{-}Bi_2O_3$	1.6		Voltage delay; poor, intermittent performance
Magnesium-Sulfur Mg-S	1.6		Voltage delay; high impedence; poor, intermittent performance; H_2S byproduct
Zinc-Potassium Iodate $Zn\text{-}KIO_3$	1.9	31	High rate; low shelf life; high reaction temperature; poor low-temperature performance
Zinc-Sodium Dichromate $Zn\text{-}Na_2Cr_2O_7$	2.1	0.25	Anode corrosion problems; H_2 products

SOURCE: Compilation of data from Bis, R.F., J.A. Barnes, W.V. Zajac, P.B. Davis, and R.M. Murphy, 1986, *Safety Characteristics of Lithium Primary and Secondary Battery Systems,* NSWC TR 86-296, Naval Surface Weapons Center, Silver Spring, Md., July; and Bis, R.F., and R.M. Murphy, 1986, *Safety Characteristics of Non-lithium Battery Systems,* NSWC TR 86-302 Rev. 1, Naval Surface Weapons Center, Silver Spring, Md., July.

TABLE 8.5 Primary Cell Types with Promise as Prospective Power Sources

Chemistry	Cell Volt	W-h/kg	Notes
Aluminum-Silver Dioxide $Al\text{-}Ag_2O_2$	1.57		Pumped electrolyte; heat exchanger required
Calcium-Thionyl Chloride $Ca\text{-}SOCl_2$	3.0		Increased tolerance to abuse over Li
Calcium-Sulfuryl Chloride $Ca\text{-}SO_2Cl_2$	3.2		Increased tolerance to abuse over Li

SOURCE: Compilation of data from Bis, R.F., J.A. Barnes, W.V. Zajac, P.B. Davis, and R.M. Murphy, 1986, *Safety Characteristics of Lithium Primary and Secondary Battery Systems*, NSWC TR 86-296, Naval Surface Weapons Center, Silver Spring, Md., July; and Bis, R.F., and R.M. Murphy, 1986, *Safety Characteristics of Non-lithium Battery Systems*, NSWC TR 86-302 Rev. 1, Naval Surface Weapons Center, Silver Spring, Md., July.

TABLE 8.6 Primary Cell Types Currently Available

Chemistry	Cell Volt	W-h/kg	Notes
Cadmium-Mercuric Oxide Cd-HgO	0.9		Long shelf life; sealable
Calcium-Calcium Chromate (thermal) $Ca-CaCrO_4$	3		Long shelf life; high voltage; 1 s to 1 h operating life; tolerates extreme environments
Magnesium-Lead Chloride $Mg-PbCl_2$	1.1		Seawater activated: torpedoes; emergency signaling; sonobuoys
Magnesium-Manganese Dioxide $Mg-MnO_2$	1.8		Transceivers; sonobuoys; beacons; good temperature performance
Magnesium-Silver Chloride Mg-AgCl	1.7		Seawater activated
Zinc-Air $Zn-O_2$	1.4	440	High capacity
Zinc Manganese Dioxide (Leclanché) $Zn-MnO_2$	1.5		
Zinc-Carbon Zn-C	3	105	Everyday flashlight cell
Alkaline $Zn-MnO_2$	1.25	100	
Mercury Zn-HgO	1.3	105	
Zinc-Silver Chloride Zn-AgCl	1.0	2.1	Extremely long shelf storage; low rate supply
Zinc-Silver Oxide (Silver Zinc) $Zn-Ag_2O_{(2)}$	1.6/1.8	50 to 200	High energy/high rate: torpedo; space; submarine; ignition; rocket

SOURCE: Compilation of data from Bis, R.F., J.A. Barnes, W.V. Zajac, P.B. Davis, and R.M. Murphy, 1986, *Safety Characteristics of Lithium Primary and Secondary Battery Systems*, NSWC TR 86-296, Naval Surface Weapons Center, Silver Spring, Md., July; and Bis, R.F., and R.M. Murphy, 1986, *Safety Characteristics of Non-lithium Battery Systems,* NSWC TR 86-302 Rev. 1, Naval Surface Weapons Center, Silver Spring, Md., July.

Rechargeable battery systems capable of a large number of charging and discharging cycles will play a much greater role in the Navy and Marine Corps of the future. A significant part of future electric power requirements during combat engagements can come from advanced battery systems. Tables 8.7, 8.8, and 8.9 summarize secondary battery technology.

TABLE 8.7 Secondary Cell Types with Little or No Chance of Further Development

Chemistry	Cell Volt	W-h/kg	Notes
Cadmium-Silver Chloride Cd-AgCl	1.0		Very fast charging; fast storage loss
Zinc-Lead Dioxide Zn-PbO_2	2.5	108	Anode corrosion problems; poor charging performance

SOURCE: Compilation of data from Bis, R.F., J.A. Barnes, W.V. Zajac, P.B. Davis, and R.M. Murphy, 1986, *Safety Characteristics of Lithium Primary and Secondary Battery Systems*, NSWC TR 86-296, Naval Surface Weapons Center, Silver Spring, Md., July; and Bis, R.F., and R.M. Murphy, 1986, *Safety Characteristics of Non-lithium Battery Systems*, NSWC TR 86-302 Rev. 1, Naval Surface Weapons Center, Silver Spring, Md., July.

As the ships and aircraft of the future incorporate more electronics, they will require lightweight, reliable primary power sources and backup systems. This need has spurred interest in further development of thermal batteries that are designed to use the high thermal energy of combustion or an explosion to generate electricity for a very short period of time. Current thermal systems are characterized by nearly unlimited shelf life because of their nonvolatile chemistries and short-term, high-energy power output. They are currently in use in tactical missiles and short-term power backup systems. Longer operating life will enable their use in cruise missiles and sonobuoys.

Where weight is a major concern or the application calls for both long shelf and operating lives, lithium-based chemistries are a clear leader.[8] Primary lithium batteries with 20-year shelf lives are in current use in smart mines and kinetic kill weapons. The U.S. Army has been a leader in the development of high-density, high-power, 6- to 12-V battery power systems for vehicles, pulsed-power weapons, remote power backup, and missile systems.[9] The utility of high-power systems that are capable of operating in severe environments is driving further research on lithium battery safety issues such as handling and mitigation of volatile chemistry systems.

Nickel-hydrogen has long been the preferred spaceborne secondary battery technology. Capable of high discharge rates and nearly infinite charge-discharge cycling, nothing currently matches its reliability. Underwater applications have been dominated by silver zinc (AgZn) systems. Some silver aluminum units are in use for lower power applications such as small torpedoes and the smaller

[8] "The Military Runs on Lithium Batteries," *Battery and EV Technology*, 18(10):6, February, 1994.

[9] Shephard, Jeffrey. 1994. "Military Leads in Battery Development," *Military & Aerospace Electronics*, 5(2):27-29, February.

TABLE 8.8 Secondary Cell Types with Promise as Prospective Power Sources

Chemistry	Cell Volt	W-h/Kg	Safety
Aluminum-Air Al-O_2	1.5 to 2.1	80	Seawater activated
Cadmium-Air Cd-O_2	1.9	< 440	High cost
Hydrogen-Air $LaNi_5H_6$-O_2	1.2	3,650 theoretical; 380 actual	Gas cross-leakage; short cycling life; low rate
Iron-Air Fe-O_2	1.2	715 theoretical; 90 actual	Electric vehicle (EV); H_2, O_2 production, poor thermal operation
Iron-Chromium (redox) Fe-Cr	1.2	120 theoretical; 30 actual	Reactant cross-diffusion problems
Nickel-Iron Fe-NiOOH	1.3	263 theoretical	EV; long life; high H_2
Iron-Silver Oxide Fe-$Ag_2O_{(2)}$	1.2/1.5	106	To improve Zn-$Ag_2O_{(2)}$ life
Sodium-Sulfur (glass) Na-S	1.8	760 theoretical	High temperature (350 °C)
Zinc-Bromine Zn-Br_2	1.8	80	EV; electric load leveling; Br release danger
Zinc-Chlorine Zn-Cl_2	2.1	826 theoretical; 200+ actual	EV; electric load leveling; chilled recharge required
Zinc-Nickel Zn-NiOOH	1.5	345 theoretical; 81 actual	EV; H_2, O_2 production
Zinc-Silver Oxide Zn-$Ag_2O_{(2)}$	1.8/1.6	130	Torpedoes, submarines, and so on
Zinc-Air Zn-O_2	1.2	185	High-capacity replacement for Ni-MH

SOURCE: Compilation of data from Bis, R.F., J.A. Barnes, W.V. Zajac, P.B. Davis, and R.M. Murphy, 1986, *Safety Characteristics of Lithium Primary and Secondary Battery Systems*, NSWC TR 86-296, Naval Surface Weapons Center, Silver Spring, Md., July; and Bis, R.F., and R.M. Murphy, 1986, *Safety Characteristics of Non-lithium Battery Systems*, NSWC TR 86-302 Rev. 1, Naval Surface Weapons Center, Silver Spring, Md., July.

TABLE 8.9 Secondary Cell Types Currently Available

Chemistry	Cell Volt	W-hr/kg	Notes
Nickel-Cadmium (NiCd) Cd-NiOOH	1.2	25/35/210	Cyl/button/vented; long life; memory problems
Nickel-Metal Hydride Ni-MH	1.2	65	Replaces Cd in NiCd
Cadmium-Silver Oxide $Cd\text{-}Ag_2O_{(2)}$	1.4	75	Like NiCd but lower rate
Hydrogen-Nickel $H_2\text{-}NiOOH$	2.5	50	Pressure vessel; extremely long life; satellites
Hydrogen-Silver $H_2\text{-}Ag_2O_{(2)}$	1.4/1.1	80	Pressure vessel; extremely long life; deep discharge
Lithium-Thionyl Chloride $Li\text{-}SOCl_2$	3.4	700	Primaries available as buttons; very wide temperature range; 20-year life
Lithium Ion Family			
Lithium-Sulfur Dioxide $Li\text{-}SO_2$	2.8	275	Wide temperature range; 20-year life
Lithium-Manganese Dioxide $Li\text{-}MnO_2$	3.1	300	
Lithium-Titanium Sulfide $Li\text{-}TiS_2$	3.6		
Lithium-Manganese Disulfide $Li\text{-}MnS_2$	3.6		
Lithium-Sulfur Dioxide $Li\text{-}SO_2$	2.8	275	Wide temperature range; 20-year life
Lithium Carbon Monofluoride $Li\text{-}(CF)_n$	2.8	310	Long shelf life
Lithium-Iodine $Li\text{-}I_2$	2.7	230	Long service life (>10 years)
Lithium-Copper Sufide Li-CuS			
Lead-Acid $Pb\text{-}PbO_2$		100	Fuses; torpedoes; radio sonobuoys
(sulfuric acid)	2.0		
(fluoroboric)	1.8		
(perchloric)	1.9		
Zinc-Silver Oxide $Zn\text{-}Ag_2O_{(2)}$			
Zinc-Manganese Dioxide $Zn\text{-}MnO_2$	1.2		"Renewal" brand; needs smart charger

SOURCE: Compilation of data from Bis, R.F., J.A. Barnes, W.V. Zajac, P.B. Davis, and R.M. Murphy, 1986, *Safety Characteristics of Lithium Primary and Secondary Battery Systems*, NSWC TR 86-296, Naval Surface Weapons Center, Silver Spring, Md., July; and Bis, R.F., and R.M. Murphy, 1986, *Safety Characteristics of Non-lithium Battery Systems*, NSWC TR 86-302 Rev. 1, Naval Surface Weapons Center, Silver Spring, Md., July.

offboard vehicles, but AgZn batteries are widely used for high-power applications. These can provide up to 100 kW over 15 minutes and, together with advanced electric motors, can propel a torpedo at high speed.

Anticipated Technology Advances

The maturity of research in electrochemical power systems provides the benefit of some measure of predictability of future developments. There is nothing that currently matches the high power densities of vented nickel-cadmium batteries to provide the starting power for an aircraft engine in –40 °C weather, and there will not likely be a shift from conventional lead acid and NiCd batteries in this application for the next 10 or 15 years. New aircraft designs, including the F-22, are incorporating existing battery systems. Sealed NiCds are slowly being developed for these applications and are in limited flight use in the commercial sector, but they have yet to demonstrate the high power densities and low maintenance levels of the mainline systems. And, only slightly further in the future, NiMH developments show great promise for increased power densities and lower cost.

Most battery chemical reactions are bidirectional and can be implemented in either primary or secondary cell designs. Those chosen for today's array of secondary battery types are those for which the characteristics of the charge/discharge cycle are the most benign. Whether for these more natural secondary batteries or for tomorrow's more critical designs, positive control over the battery's operational status and charging parameters is becoming a critical enabling technology. The recent commercial availability of rechargeable alkaline batteries is a case in point.

Battery charger technology has advanced markedly in recent years.[10] Propelled by the adoption of the more exotic commercial batteries such as lithium cells, the smart charger is becoming ubiquitous in laptop computers, for instance. Most of the problems associated with electrochemical cell development have to do with their susceptibility to anode corrosion, the production of explosive or noxious gases during charging, and the many deleterious effects of overcharging. Built-in sensors for determining temperature, pressure, along with state of charge, and microcontrollers built into the battery effectively control most of these effects and consequently improve battery life, performance, reliability, and safety. Cooperative strategies between battery controllers and the host system controller can optimize selected system/battery performance parameters such as individual cell control, load shedding, or temporary capacity reduction in all or part of the host.

For advanced rechargeable battery systems, significant improvements in mobility and supportability will derive from improvements in the energy density of new batteries by as much as fivefold over current state-of-the-art lead-acid

[10]"Special Report: Part 2 of Power Technologies Comes with Batteries Included," *Electronic Engineering Times*, Issue 848, May 15, 1995.

systems. Most batteries today operate at specific gravimetric capacities well below their theoretical limits. Significant improvements in battery operation can be afforded by active management techniques such as charge and discharge rate management within instantaneous chemistry, temperature, and load conditions. In addition to these operational improvements, fundamental research continues in the materials and structure of alloys, separator materials, and seals that will inevitably produce tomorrow's breakthroughs in battery size, output, and efficiency. A factor of two to three improvement over lead-acid batteries is predicted for power density. To be applicable to the needs of the naval forces, devices in research today must evolve into rugged, fully reliable batteries. Table 8.10 presents some parameters of interest for several rechargeable battery concepts.

It will be important that the Department of the Navy focus its development efforts with the objective of developing, at most, a few battery systems that will cover the broad range of future Navy and Marine Corps needs. Attention to the standardization of form factors, interfaces, and charging and handling systems will be important for reducing the logistics burden associated with widespread battery usage. For example, of the 500 battery types available to U.S. Army quartermasters, probably the world's largest battery user, 10 types represent about 90 percent of usage. Consolidation and rationalization of the mix of battery types and applications are ongoing. Military expenditures on batteries have grown from approximately $50 million is 1981 to over $190 million today.

Special Application of Batteries: Power for Remote Sensors

Future naval forces will depend on remote sensors in a variety of situations: Unmanned vehicles will patrol the skies; microbuoys will monitor the seas; target-sensitive triggers will proliferate throughout the arsenal; randomly deployable sensors will locate and monitor enemy command and control; and the marines will cast off miniature, self-forming communications networks as they advance through the littoral. These capabilities will result from the size reduction and increased functionality of microelectromechanical systems and small, long-lived, power sources. The implications for the direction of battery power evolution are clear: higher density; higher efficiency, and intelligent power management.[11]

Sensor Types

Airborne Sensors

Much battle space awareness information will be derived from sensors mounted on unmanned aerial vehicles (UAVs). Increased capability aboard these

[11] Powers, R. 1995. "Batteries for Low Power Electronics," *Proceedings of the IEEE,* 83(4):11, April.

TABLE 8.10 Performance of State-of-the-Art and Advanced Battery Systems

Battery System	W-h/kg		W/kg		W-h/liter		$/kW-h		Problems
	Now (to 2000)	Future (2020)	Now (to 2000)	Future (2020)	Now (to 2000)	Future (2020)	Now (to 2000)	Future (2020)	
State-of-art lead-acid	35	40	110	150	80	100	150	100	Low energy density, fairly high self-discharge
Bipolar lead acid	38	50	—	79	—	85	—	100	Costs more than conventional lead-acid, low specific power
Na/S	90	120	110	140	140	180	1,000	100	Operates at 350 °C
Ni/Cd	35	40 to 50	50 to 60	80 to 90	80	90 to 100	250 to 300	200	Sensitivity to overcharge
Zn/Air	90	140	100	150	—	—	—	35 to 45	System weight dominated by electrolyte weight Low energy efficiency
AL-Li/FeS	80	150 to 200	95	180 to 220	—	—	High	100	Operates at 400 °C

SOURCE: Board on Army Science and Technology. 1993. "Electric Power Technology for Battle Zones," *STAR 21: Strategic Technologies for the Army of the Twenty-First Century*, National Academy Press, Washington, D.C., Table 43-5, p. 571.

space-limited vehicles will pay dividends in the reach and efficiency of the fleet sensor suite. Miniaturization and modularity will allow rapid tailoring of these vehicles for a variety of jobs at hand, from microclimatology to target detection and location, to battle damage assessment.

Microbuoys and Unmanned Undersea Vehicles

The same technology that will empower the UAV as an inexpensive and versatile sensor platform will also enable the creation of seaborne platforms with specialized capabilities for use in the future. Deployed fleets of these robots will be capable of underwater communications, underwater intelligence actions, antisubmarine warfare, mine detection and destruction, oceanography, and seabed mapping, particularly in the littoral waters where operation with conventional submarines is problematic.

Smart Triggers

As smart weapons gain autonomy the on-board sensor and control electronics requirements will grow. As autonomous capability becomes more robust and reliable, smart triggers will be deployed on a wider variety of weapons including mines and antimine devices, long-dwell ASW robots, and long-range bombardment projectiles.

Sensor Deployments

Autonomous Sensors

High-energy, low discharge rate batteries extend available operating times, thus enabling more autonomous operation of sensor systems. Current systems are occasionally fielded for special-purpose remote sensing with electronics packages that are as large as a suitcase and with a battery suite that may be two or three times larger than that. Although the number, capability, and support requirements for smaller, more portable designs will enable a new range of remote sensor possibilities, these new sensors will also require the enhanced computational capabilities of microelectronic processors and validated, robust control programs.

Networked Sensors

A clear enabler of networked cooperative sensors is the development of reliable, very low power command and control and data exfiltration mechanisms. Microelectronic transceivers will be the key to the development of self-forming communications networks. Current commercial technology has deployed medium-bandwidth, shoebox-sized, self-forming, and self-managing RF communications networks. These digital networks provide convenient access to mobile users. Each network component knows or can calculate its own location and locate its nearest neighbors dynamically. It is an easy extrapolation of this concept to the military battle space in the form of randomly deployable sensors or network nodes. Microelectronics and efficient power will enable miniaturization and disposability.

Stationary and Vehicle Flywheels

Flywheels are covered briefly in a previous section as energy-generation devices. Flywheels coupled directly to a variable-speed motor-generator have the potential of compactly storing large amounts of energy and rapidly delivering electric power. The motor-generator would function as a generator when the system was drawing on the energy stored in the flywheel and as a motor when energy was being stored in a flywheel by spinning it up. For many configurations, flywheels could also be spun up mechanically through the use of reciprocating or turboshaft engines. Through the use of advanced composite materials, flywheels could store as much as an order of magnitude more energy per unit of flywheel mass than can be obtained with flywheels of the best alloy steel.

The government-industry partnerships addressing hybrid electric drive for wheeled or tracked vehicles have produced innovative flywheel and motor-generator technology at the few tens of kilowatt level. U.S. Army-supported programs have developed technology for rotating inertial energy for tracked-vehicle, pulsed-power weapons applications that, although not successful to date, may have future applications to both ships and amphibious or land vehicles.

Inductive Energy Storage

Inductive energy storage, generally for use with pulsed-duty applications, has been evolving over the last decade. Cryogenically cooled aluminum inductors have been developed for low-loss, very-short-term energy storage in pulse-forming networks. SMES as described above was developed for use in high-power, directed-energy weapons applications but has evolved into applications for long-term energy storage for uninterruptable power sources. The key technologies for SMES development in the future are superconducting materials and high-strength composite materials for containment of the large forces associated with magnetic energy storage. Trends in these areas would indicate that practical superconductors operating above 100 K and composite materials with yield strengths approaching 10^6 psi should be available 20 to 30 years hence.

THE ELECTRIC SHIP

The Navy is engaged in an ongoing effort to develop an electric ship. When all the key technologies are sufficiently developed to allow the electric power generation and distribution system to be integrated with the electric propulsor, the resulting capability will represent a major change in the performance features and operating characteristics of shipboard power and propulsion systems.

The Electric Ship Concept

The electric ship approach is enabled by the so-called second electronic revolution where the rapid advancements in solid-state semiconductor technology that led to integrated circuits, microprocessors, and computers are now being applied to power semiconductor devices. Advances in the materials used in electromagnetic machinery such as permanent magnets and high-temperature superconductors also present opportunities for substantial increases in power density and efficiency. And the emergence of direct electric conversion technologies such as fuel cells provides the potential for fuel-efficient, low-emission, and low-noise sources of electrical power.

The Navy's IPS program utilizes an open-architecture, common-module approach that builds on existing commercial efforts. Within the IPS program, a family of modules is being defined that will serve as the building blocks for designing, procuring, and supporting marine power systems applicable across a broad range of ship types. To this end, shipboard power systems are divided into the following seven elements or module types:

- Power generation,
- Energy storage,
- Power conversion,
- Power distribution,
- Propulsion motors,
- Platform loads, and
- Power control.

IPS is a flexible and scalable power system by virtue of its open architecture and use of common machinery modules. It has the potential to revolutionize the design, construction, and operation of U.S. naval ships by using electricity as the primary shipboard energy medium. IPS minimizes the number of prime movers by permitting any generating unit to supply either propulsion or ship service power to support ship operational priorities at any given time. The flexibility of electric power transmission allows power generation modules of up to 20 MW to be located independently from the propulsion motors and ship service power converters in whatever arrangement supports the ship's mission at the lowest overall cost. The ability to independently position the minimum amount of machinery components in small, unmanned modules avoids the need for large engine rooms, which in turn permits greater separation and compartmentation in the ship, accruing significant benefits in manning, safety, and ship survivability. The use of small, unmanned machinery modules also permits the use of non-chlorofluorocarbon (CFC)-based fire suppression systems similar to those currently used in fleet propulsion gas turbine enclosures.

The Department of the Navy's full-scale advanced development (FSAD) system prototype incorporates a commercial marine approach to electric power generation and propulsion distribution in combination with a zonal dc electrical distribution system for ship service power. The generator is a conventional two-pole 60-Hz machine that produces the required 22-MW output power at 4,160 volts alternating current (Vac). The power is distributed in this form via conventional medium-voltage cable and switchgear to the propulsion converter that produces the voltage and frequency required by the propulsion motor. The converter consists of a diode-rectifier front end and multiple pulse-width-modulated (PWM) inverter sections connected in parallel to achieve the desired power levels. The inverter section development extends the state of the art in insulated gate bipolar transistor (IGBT)-based PWM technology to the medium-voltage range. The motor is a squirrel cage induction motor based on severe-duty commercial marine and industrial designs. The ship service dc power is produced by a conventional transformer and rectifier and distributed via dc cable and converted by ship service inverter modules (SSIMs) to 60 or 400 Hz, or variable frequency and voltage as required.

The IPS architecture will allow incorporation of developing technologies such as PM electric machines, fuel cells, and PEBB into future ship designs as programmed, preplanned replacements for the core technology in the first-generation IPS modules. Use of technology-independent module interface standards will facilitate technology insertion with minimum impact on the ship design and construction process. Figure 8.6 is a diagram of the basic IPS architecture.

Power Generation for Surface Ships

Intercooled Recuperated Gas Turbine

The intercooled recuperated (ICR) gas turbine currently under development is a derivative of the Rolls-Royce RB211 aeroengine used in air fleets around the world. Three elements have been added to the core engine to produce its 29,000 brake-horsepower (bhp) rating and 30 percent total mission fuel savings (relative to comparable existing power plants). First, an intercooler cools the gas after low-pressure compression and before second-stage compression. This reduces compressor work, allowing more power from a given engine core. This is the key to attaining the 29,000 bhp from the RB211, and it also adds to overall engine

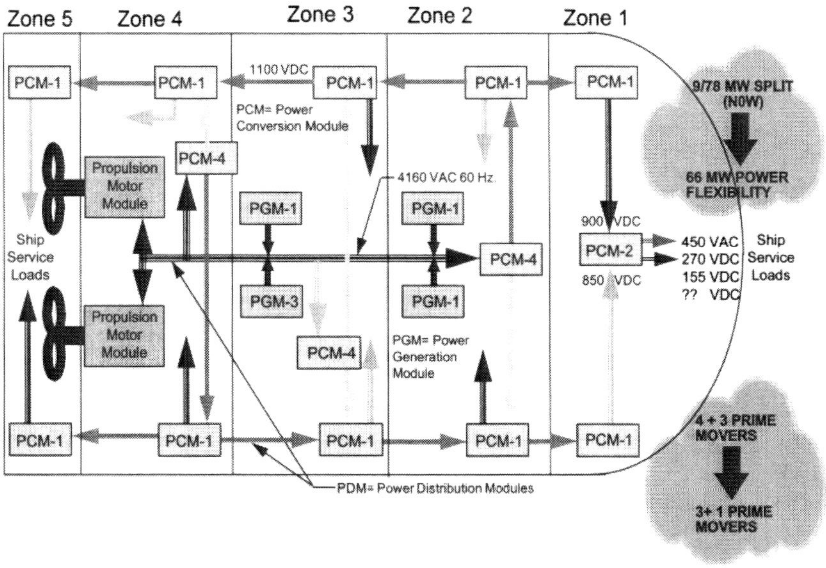

FIGURE 8.6 Integrated power system basic architecture. SOURCE: Adapted from (1) Advanced Surface Machinery Programs Office, 1996, "A Strategy Paper on Power for U.S. Navy Surface Ships," draft, Naval Sea Systems Command, Arlington, Va., Figure A-1, April 8, and (2) Krolick, C.F., 1996, "Technology Insertion," in the briefing "Advanced Surface Machinery Programs (ASMP)," presented to the Panel on Technology by the Naval Sea Systems Command, Arlington, Va., May 2.

efficiency. Next, a recuperator recovers exhaust heat wasted by simple-cycle engines to preheat combustion air before ignition. Finally, variable area nozzles on the first stage of the power turbine restrict flow when operated at less than full power, raising exhaust temperature and thereby increasing recuperator effectiveness at low power. The ICR has a lower exhaust temperature than comparable engines, resulting in reduced IR signature.

Although the ICR maintains the footprint of the engines it is designed to replace, it weighs roughly twice as much and costs about $1.5 million more per engine. Recent studies on fleet introduction in the DDG-51 (forward-fit) indicate an average follow-on ship acquisition increase of $10 million per ship (including engine costs) while saving $1.5 million per ship per year in fuel costs. Studies for a two-engine surface combatant indicate that, in a new design, the fuel load and intake/exhaust reductions compared with those in simple-cycle engines more than offset the increased engine weight. The net result is that even with an increase in engine cost the overall ship acquisition cost is the same, with fuel cost savings comparable to those for the DDG-51.

The current development schedule (incorporating the recuperator recovery plan) supports introduction on the DDG-51 during FY 2001, with 3,000-hour qualification and shock testing in FY 1998-1999. The ICR is the engine incorporated into the first IPS power generation module (PGM-1), which is envisioned to be the main power unit of the future fleet.

Advanced-cycle Gas Turbines

In addition to the 29,000 bhp ICR, advanced-cycle gas turbines that are derivatives of existing commercial engines can be incorporated into turbine-generator sets in the 1- to 4-MW power range. Examples of these are the Solar Taurus-R and Allison 501 K34 ICR. The alternatives employ various combinations of recuperation, intercooling, and variable geometry to achieve substantial improvements in fuel efficiency over currently available engines—which includes the Navy's Allison 501 K34. It is anticipated that these developmental engines would cost more than their simple-cycle counterparts, and feasibility would have to be based on comparison of fuel savings versus development, fleet-introduction, and acquisition costs. For some of these engines, the improved fuel efficiency comes at the expense of response time. Current ship service generator set voltage and frequency specifications require transient-response characteristics that these advanced turbine generator sets may not be able to meet without the addition of solid-state power conversion equipment on the output side. The IPS electrical distribution interface specification has the potential for relaxing engine transient response requirements, allowing realization of fuel savings without additional equipment size or cost. As with the large ICR, recuperation will reduce IR signature, and the higher efficiency should lead to emissions reductions.

Since each core engine is unique, these engines would not constitute a modular family. Development and qualification costs on the order of $100 million dictate selection of the single rating best matched to the range of fleet requirements. Typical development times for these engines is about 5 years from contract award to delivery of production engines.

Medium-/High-speed Diesels

There are a number of existing families of medium- to high-speed diesel generator sets that cover the 1- to 4-MW power range. Typically, these generator sets are substantially more efficient than simple-cycle gas turbines and often less expensive. Although they take up comparable space as gas turbines and have much lower intake/exhaust air requirements, they are much heavier and require more acoustic treatment (mufflers, isolation mounts), which adds to ship size and cost. Although poor emissions have been attributed to the diesel in the past, microprocessor-based control and monitoring systems are being employed in modern diesel engines to improve performance and reduce emissions.

Within a given family, diesels are somewhat modular in that they are based on multiples of a particular cylinder size. Since these are typically existing commercial engines, members of a family could be qualified as needed to support the range of fleet requirements. For example, the CAT 3512 diesel generator set is being considered for qualification for submarine emergency generator applications. The qualification effort is estimated to cost approximately $5 million and should take about 2 1/2 years.

Fuel Cells

Fuel cells convert chemical energy directly into electrical energy and as a power generation module can be viewed as a continuously fueled battery. They take in fuel and oxidant and produce electricity, water, and heat. In addition to the fuel cell stacks, the power generation module includes heat exchangers, regulators, controls, and fuel-processing and other auxiliary equipment. Besides enhanced energy density and efficiency, fuel cells have the potential for providing power with little or no emission of regulated pollutants. Other attributes include low-noise, reduced-exhaust IR signature, improved reliability and maintainability, and a high degree of modularity. Table 8.11 provides a summary of characteristics for the alternative fuel cell technologies being considered. It is anticipated that when fully matured, fuel cell costs could be as low as $500/kW.

Most of the fuel cell development to date has been focused on electric utility applications that use natural gas for fuel and for which compactness is not required. Recent developments include transportation applications using methanol as fuel and where power density is an issue. For shipboard use, a diesel fuel reformer and desulfurizer would be required, and, although power density is an issue, it must be weighed against impact on acquisition cost—the further removed from commercial practice, the more expensive it will become. Existing Navy exploratory development efforts are focused on the fuel-processing issues of reforming and desulfurization with a 10-kW fuel cell demonstration planned. DARPA is supporting development of fuel cells that operate on logistic fuels (JP8 or DF2).

Although the fuel cell stacks are very modular, the fuel processor and auxiliary equipment will benefit from economics of scale. Therefore, a family of ratings is likely (e.g., 500 kW, 1 MW, 2 MW, 4 MW) that uses multiples of a common stack element with appropriately rated fuel processor and auxiliaries. Commercial stacks and auxiliaries might be used with a fuel processor designed to meet ship requirements. This might take about 3 1/2 years and cost roughly $11 million to develop.

Integrated Power System Family of Modules

At present the IPS notional family of modules includes the following power generation modules (PGMs):

TABLE 8.11 Available Fuel Cell Systems

Type	Electrolyte	Operating Temperature (°F)	Efficiency	lb/kW	ft^3/kW
Phosphoric acid	Phosphoric acid (liquid)	350 to 450	38 to 40	30 to 46	0.93 to 1.5
Molten carbonate	Carbonate salts (liquid)	1,200 to 1,400	40 to 55	40 to 60	0.98 to 2.1
Solid oxide (tubular)	Zirconium oxide (solid)	1,700 to 1,900	45 to 60	20 to 30	0.6 to 1.2
Solid oxide (planar)	Zirconium oxide (solid)	1,700 to 1,900	45 to 60	8 to 13.5	0.3 to 0.81
Proton exchange membrane	Sulfonated polymer (solid)	180 to 250	39 to 42	6 to 11.9	0.19 to 0.3

SOURCE: Adapted from the Advanced Surface Machinery Programs Office, 1996, "A Strategy Paper on Power for U.S. Navy Surface Ships," draft, Naval Sea Systems Command, Arlington, Va., Table A-1, p. A-7, April 8.

- PGM-1—21 MW (29 KHP) 4,160 Vac, three-phase, 60-Hz ICR gas turbine-driven generator;
- PGM-2—3.75 MW 4,160 Vac, three-phase, 60-Hz diesel-driven generator;
- PGM-3—3 MW 4,160 Vac, three-phase, 60-Hz DDA501-K34 gas turbine-driven generator;
- PGM-4—TBD MW efficient auxiliary gas turbine-driven generator;
- PGM-5—TBD MW fuel cell;
- PGM-6—8 MW 4,160 Vac, three-phase, 60-Hz diesel-driven generator; and
- PGM-7—12 MW 4,160 Vac, three-phase, 60-Hz diesel-driven generator.

The IPS architecture does not preclude configurations that use combinations of these PGMs for providing electrical power, but not propulsion, e.g., gas-turbine or steam-turbine (fossil or nuclear fueled) driving propellers via reduction gears. Much of the benefit of IPS is in power integration and optimization of installed and operating power sources. Determination of the set of future PGMs must take into account all of the attributes of the alternatives and how those attributes match up with the overall fleet requirements, as well as the power system architecture. In an ICR/IPS-based fleet, fuel efficiency of the lower rated PGMs may not have as much leverage as in today's ships. Acquisition cost and the maintenance and manning aspects of operations and support costs may be the big drivers. Emissions quality will become increasingly important, as will arrangement flexibility to facilitate integration into low-observable platforms. Also, logistics support considerations may dictate selection of the module(s) that covers the widest range of requirements versus those that optimize for each requirement.

Power Distribution and Conversion

Distribution

As indicated in the IPS discussion, a dc zonal electrical distribution system (dc ZEDS) is envisioned as the ship service power architecture for future surface combatants. It employs the dual port/starboard bus configuration used in the ac ZEDS introduced on the DDG-51 in FY 1994, but with dc buses replacing the ac buses, and integrated power conversion centers (IPCCs) replacing the load centers in each zone. The IPCCs contain power conversion modules that convert dc power to the type required by the loads in the respective zones. On existing ships this is 450 or 120 Vac, at 60 or 400 Hz, but in the future there is opportunity to provide dc directly to the end users. For many users, particularly the combat system loads, the various forms of ac power are immediately rectified to dc for internal distribution within the cabinets, and the availability of a dc feed would eliminate the internal conversion equipment and also simplify the power interface. Many of the auxiliary systems are connected to the electrical distribution system via electric motors and their associated controllers. As the size and cost of solid-state motor controllers continue to fall, they will begin to replace the existing electromechanical devices in ship applications. It will be possible to provide a dc feed to these solid-state controllers and to any power supplies on the auxiliary skid for monitoring and control systems or sensors.

A study on dc power for combat systems on the DDG-51 indicated potential for reductions of internal power supplies totaling 17 ton and 400 ft3. A comparison of dc zonal systems with partial load integration (dc feeds to some auxiliary equipment only) to the ac zonal systems on the DDG-51 showed a potential savings of 35 tons and $1.1 million. In addition to these cost and weight savings, dc zonal systems provide enhanced survivability in that the IPCCs prevent damage or faults occurring in one zone from propagating electrically to other zones.

Power Electronics

A major factor in pursuing the more electric architecture for Navy ships is the anticipation of major advances in the state of the art of power electronics much as has occurred in microelectronics. Power electronic devices, like microelectronic devices, are semiconductor devices that switch electrical currents on and off. Microelectronic devices switch very small currents at low power levels to control the flow of information. Power electronic devices switch large currents at high power levels to control, distribute, and process electrical power. Compared with traditional electromechanical switches, they are faster, smaller, more precise, easier to control, less expensive, and more efficient. Improved performance will be spurred by development of the n-type metal oxide semiconductor (MOS)-controlled thyristor (n-MCT), an advanced power semiconductor device, and the PEBB, a multichip microsystem enabled by the n-MCT.

The PEBB will integrate within a single unit all the elements required for generalized power processing and will replace many single-application, multicomponent power control circuits with a single device that delivers digitally synthesized power. This advanced power transformer will be a three-port device—two ports for power and one port for control—with bidirectional power flow converting power from one type to another based on the control algorithm being executed within the PEBB and the information available via the control port. The PEBB family will constitute a standard set of snap-together parts that start at the semiconductor-chip level and build up to the system level while integrating intelligence at various levels for custom performance—a power electronic analog of a microprocessor.

A PEBB consists of an array of power electronics switches, gate drivers, sensors, a microcontroller, filters, a fiber-optic control interface, and advanced packaging. Although PEBBs could be constructed using virtually any semiconductor device, the n-MCT meets the broadest range of requirements for military and commercial users. Its principal advantage is its capability for simultaneous high-speed switching (> 70 kHz), high withstand voltage (> 1,400 V), low on-state voltage (< 2 V), and high currents (up to 1,000 amps). High-speed switching reduces the size of the input and output filters by greatly increasing the ripple frequency. For many applications, the size of the filters can be reduced to such an extent that they can be integrated directly into the PEBB package. Thus, by reducing the size of the components with high-speed switching, reducing conduction losses with low on-state voltage n-MCTs, and reducing switching losses with soft-switching topologies, it is possible to design compact and highly efficient power conditioning equipment. For an IPS ship, which employs a significant amount of power electronics, the PEBB can reduce ship acquisition cost by $4.4 million, not taking into account any effect on the ship resulting from reduced equipment size and weight.

The ONR-sponsored PEBB program uses hybrid, integrated circuit, power-semiconductor, and power-electronic technologies to design, construct, and demonstrate modular power electronics architectures for intelligent, distributed power processing. The PEBBs will be manufactured on high-volume production lines to demonstrate reliability and techniques for low-cost production. Affordability and commercial viability will be critical to the success of the PEBB. Therefore, concurrent engineering that takes manufacturing issues and customer feedback into account early in the design process and fast introduction of commercial products are integral parts of the development program. ONR's long-range plans, supported by a science program, include the insertion of advanced controls such as neural networks and artificial intelligence and the replacement of silicon with wide-bandgap semiconductor materials such as silicon carbide to increase the operating temperature and voltage of the PEBB.

Several generations of PEBBs are envisioned with each proceeding through the definition, development, and demonstration cycle. Three iterations, or generations, are planned in the 5-year program, with each generation building upon lessons learned in the previous cycles. First-and second-generation PEBBs could be employed in preprototypes of the dc ZEDS power converters and, if cost effective, could be considered for forward-fit into the ac ZEDS on the DDG-51. Third-generation devices would be available in time for production power conversion equipment for the SC-21. When available, the higher-voltage devices would be considered for propulsion power converters and for medium-voltage solid-state switchgear.

Pulse-power Distribution

This distribution provides for the generation, conversion, and transmission of prime power for high-energy pulsed loads such as electrothermal chemical (ETC) guns, electromagnetic (EM) launchers, and EM armor. In general, prime power refers to the first-step energy conversion from fuel to a form of electrical energy that is best for transmission to the point of use. At the user end, this electrical energy is converted as required to a pulse or series of pulses tailored to the application. The most likely application for surface combatants is the ETC gun that employs a pulse-forming network (PFN) at the user end, near the gun mount, to minimize impedance in the very high current pulse cable. The PFN stores electrical energy before each shot and then discharges a pulse with characteristics specific to the gun type. Assuming capacitor-based PFNs, the various combinations of muzzle energy and repetition rate used with the ETC gun alternatives of interest all translate to roughly the same prime power requirement of 2 MW. Thus a nominal 2-MW prime power module or set of modules could be developed independent of which gun alternatives are selected.

A notional system would consist of a 13.8-kVac generator, a rectifier, and a 15-kilovolts direct current (kVdc) power cable. If the main engines are available as the prime power source, the generator could be attached to the output shaft or main reduction gear via a power-takeoff (PTO) gearbox. For IPS, the PTO, generator, and rectifier would probably be submodules within a PGM. If for some reason the main engines are not available, a stand-alone pulse prime power PGM could be developed that would include the engine, gearbox, generator, and rectifier and would add a flywheel off the PTO gear for energy storage. The IPS architecture also supports the alternative of a transformer/rectifier either as a submodule within the PGM or as a separate power conversion module. Fuel cell-based PGMs could eliminate the engine, generator, gearbox and rectifier but would require a step-up converter to provide 15 kVdc.

The IPS program is addressing the pulse power load requirement under its systems engineering efforts by using computer simulations to estimate the capacity that can be derived from the 4160 Vac distribution subsystem (PDM-1).

Propulsion Motors

The squirrel cage induction motor being produced under the IPS FSAD contract represents the state of the art in commercial ac-drive motor technology for power levels associated with ship propulsion. PM motor technology is currently competing with the commercial induction machine for lower-power-level, adjustable speed-drive applications based on higher efficiency or greater power density, depending on the needs of the application. The potential for increased power density without significant increase in cost makes PM motors attractive for ship applications where there may be some size constraints, such as small waterplane area twin hull (SWATH) or trimaran hullforms or podded propulsors. Even in-hull arrangements can benefit from reduced size for smaller-displacement ships as indicated by the projected $5.6 million acquisition cost savings for substituting the PM motor for the induction motor in the Advanced Surface Machinery Program (ASMP) surface combatant studies. In parallel with the FSAD contract, the IPS program is developing a PM motor module in the power range suitable for ship propulsion. This propulsion motor module will conform to the IPS interface specifications and standards and will serve as a demonstration of technology insertion in an open-architecture system.

Other machinery technology under development both by the military and commercial sectors includes homopolar motors and superconducting machines. Although the two concepts are usually linked together, homopolar motors can be excited by either conventional or superconducting magnets, and superconducting magnets can be used in both homopolar and ac motors. As with the PM motors, homopolar and superconducting motors would ideally be developed to conform to the IPS interface specifications and standards. The IPS architecture allows for extension of the specifications and standards if warranted either by commercial demand or by critical Navy-unique requirements (e.g., acoustics).

IPS with electric propulsion motors allows consideration of integrated motor/propulsor concepts. The advanced modular propulsor, for example, incorporates steering into a propulsion module that replaces the existing appendage arrangement of exposed shafting, struts, and rudder. The goal of the modular propulsor program is to define a module or family of modules based on a common set of parts that can provide the propulsion requirements for Navy and commercial ships. Affordability, which is the key to commercial viability, must include equipment acquisition and operating costs, propulsor hydrodynamic efficiency, and ship effects such as producibility enhancement and revenue-earning capacity of any space freed up by moving the motor and shaftline outside the hull. The hydrodynamic efficiency issue is complicated because Navy surface combatants must be designed for 30-plus knots maximum speed, whereas they spend most of their time operating at cruise speeds of one-half to two-thirds of maximum. For Navy applications, there is the additional consideration of hydroacoustic performance—for this effort the requirement was to be at least as good as today's surface combatants (e.g., DDG-51).

The steerable podded propulsor provides the greatest potential for meeting Navy hydrodynamic and hydroacoustic requirements while also being competitive with conventional propulsors on commercial ships. Past Department of the Navy R&D efforts on podded propulsors have adopted both improved efficiency and noise reduction as design objectives. Model tests on a 1/20th-scale DD-963 resulted in 10 to 20 percent reduction in fuel consumption. Later efforts focused on significant reductions in both propulsor and propulsion machinery noise and developed full-scale pod conceptual designs based on liquid-cooled synchronous motors driving counterrotating and postswirl propellers via epicyclic gears. Scale-model tests were conducted with geometrically similar pods on a scale model of the PG-100 (an intermediate-scale demonstration using the PG-100 was planned, but not completed). Results confirmed hydrodynamic performance projections. Also, as part of the same overall program, scale counterrotating propellers were tested in open water in the Navy's large cavitation channel, with results confirming the projections for significant increase in cavitation inception speed over fleet propellers. Most recently, ASMP has been tracking the progress of Kvaerner Masa/ABB on their Azipod joint commercial venture. The Azipod is a steerable pod housing an electric motor driving an external propeller and is being marketed on the basis of reduced ship construction costs and improved ship maneuvering and has been demonstrated up to the 15-W power level for an icebreaker application. Counterrotating versions of the Azipod design are planned. This is a prime example of an opportunity for the Navy to leverage a commercial development.

The objective of the current modular propulsor effort was to identify technology options for providing a family of cost-effective modular propulsors for military and commercial use. The approach was to conduct a survey of future commercial and military ships and then develop a strategy for producing a common family of standardized propulsor modules. Based on the performance and affordability criteria provided by the Navy, the steerable pod propulsor emerged as the preferred concept. It was noted, however, that if reduction in propulsor noise under all operating conditions were required, then ducted propulsors (both in-hull and external configurations) should be considered. It is anticipated that motor developments under the IPS program will support pod-mounted configurations; therefore, the modular propulsor effort will concentrate on the rest of the equipment needed to produce the steerable pod module. This would include a full-scale, at-sea demonstration of propulsor performance.

Power Control

Standard Monitoring Control System

The standard monitoring and control system (SMCS) will integrate the sensing, transmission, interpretation, and display of hull, mechanical, and electric (HM&E) parameters necessary for machinery control, condition monitoring and assessment, signature management, and damage control. The system design is consistent with the total-ship integrated command and control (IC^2) architecture and supports and enhances the proposed integrated survivability management system (ISMS) and the integrated condition assessment system (ICAS).

SMCS offers the potential to reduce acquisition costs and to introduce a standard system for application across multiple platforms, taking advantage of open systems architectures and commercial industry standards. It will also provide the necessary architecture to support critical imperatives for embedded readiness assessment, mission planning and training, and condition-based maintenance.

The IPS system monitoring and control subsystem consists of the software necessary to implement power management, fault response, and system human-computer interface. This subsystem is composed of system-level control software (PCON-1) and of zonal-level control software (PCON-2). The system monitoring and control subsystem is assumed to reside on a computational and networking infrastructure that is external to IPS. Currently, this external infrastructure is SMCS compliant. Designing the IPS control software to be independent of the host hardware as much as possible should enable the exploitation of future advances in computers and networks.

Intelligent Ship Control

Support of reduced manning on future ships requires safety critical automation systems that can be trusted to operate reliably under all conditions, including battle damage. They must possess sufficient system integrity and fault tolerance to enable significant reduction in the ship's crew. The SMCS provides an affordable, modular, distributed, open-architecture, machinery-monitoring and control system based on commercial and dual-use standards and technology, but requires enhanced automation, integration, system integrity, and fault tolerance to satisfy future affordability and manning goals.

ONR has an ongoing program to develop and demonstrate the required integrated ship control (ISC) automation technology base. This automation will support both low manned and advanced machinery ship concepts. ISC will fully support highly integrated ship operation and vital combat system support within emerging U.S. Navy dual-use and open-architecture standards for communications and computer resources.

Future Impact on the Navy

The Advanced Surface Machinery Program employs a systems engineering approach that maintains flexibility and minimizes investment until technologies are demonstrated, evaluated, and brought together for optimum total-ship cost-effectiveness. Affordability assessments have been made of individual program elements, but it is in the context of the full set of ASMP elements, integrated first as a total machinery system and further as part of the total-ship concept, that the most significant gains are achieved. Figure 8.7 indicates the system improvements and expected fuel savings.

These fuel savings are based on equal shaft power requirements and do not take into account any change in hull drag that may result from resizing the ship. ASMP has conducted affordability assessments that take into account ship impact costs as well as component cost and ship producibility effects as part of overall ship acquisition cost. Operating and support costs included manning, maintenance, and training in addition to fuel.

ASMP systems engineering has followed a design policy of common modules and standard components and interfaces and is developing modules to perform each of the critical functions of naval ship machinery systems. A family of modules is being defined that meets the range of projected fleet requirements at minimum fleet life-cycle cost. It employs a common module interface standard

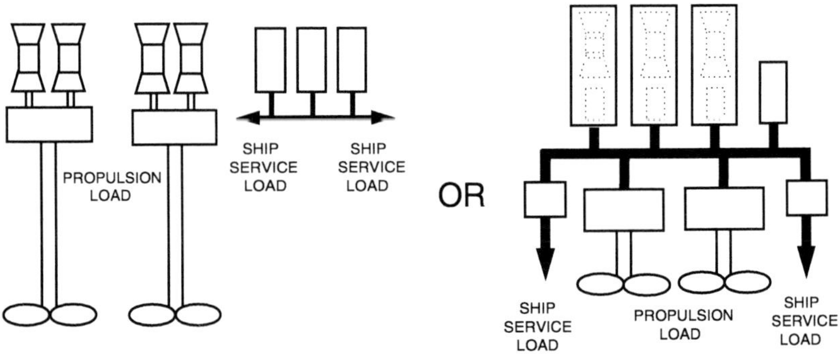

FIGURE 8.7 Integrated power system (right) advantage: Operators can determine how best to use the power, and the system saves 15 to 19 percent fuel in gas turbine surface combatants. SOURCE: Adapted from Krolick, C.F., 1996, "Advanced Surface Machinery Programs (ASMP)," presented to the Panel on Technology by the Naval Sea Systems Command, Arlington, Va., May 2.

to facilitate rapid and affordable introduction of new technology, as driven by industrial market forces and future warfighting requirements.

Although the current emphasis is on reducing the cost of meeting existing levels of performance, it is also necessary to assess the ability to deliver enhanced performance as required to meet new threats. For example, the fuel efficiency advantages of ICR/IPS can provide operational flexibility for a notional eastern Mediterranean transit and station-keeping mission by increasing transit speed by 7 knots, increasing days on station by 9 days, or increasing gauge by 1,100 nautical miles, while keeping fuel load constant compared with the baseline.

Developments Needed

A concurrent engineering approach is necessary that identifies modular-technology building blocks that can be used in various combinations to cover the range of possible requirement sets and then engages a development process with three overlapping phases for those modules to minimize risk and financial exposure. Development of test hardware and systems and component technologies at less than full scale would constitute the first phase. This would provide the technical basis for definition of the complete set of modules needed to support the range of projected fleet requirements, including the many alternatives being considered for the next-generation surface combatant. Application of the systems engineering process would lead to identification of a small subset that would actually be developed and demonstrated in the second phase. These would be selected on the basis of minimizing the risk of producing ship equipment that would meet some as yet unspecified requirements during the full-scale engineered development phase.

Time Scale for Development/Deployment

ASMP has recognized the need to get products into the fleet at the earliest opportunity. The goal of rapid deployment can sometimes conflict with the systems engineering approach of optimizing the total-ship design. There are two ways to address this dilemma. Designers can anticipate where technology is headed and plan for the phased introduction of advanced components as they become available. In addition, the use of open architecture and modular designs will facilitate the insertion of technology upgrades. ASMP has followed this approach with introduction of the zonal electrical distribution and SMCS architectures in DDG-51 class ships. The zonal electrical system required no technology development, and the SMCS may wind up using only the architecture, software, and local level hardware from the advanced development model (ADM) by taking advantage of its openness in choice of supervisory-level host hardware. Figure 8.8 provides a schedule of the ASMP products and their fleet transition possibilities.

IPS Architecture and Core Technology

The IPS architecture and core technology have their maximum impact on new-design ships and would be hampered in forward-fit applications by the extensive nonrecurring engineering costs and limited opportunity for taking advantage of arrangement flexibility. The first surface combatant new design opportunity is the SC-21 (FY 2003 leadership award). Completion of FSAD will enable

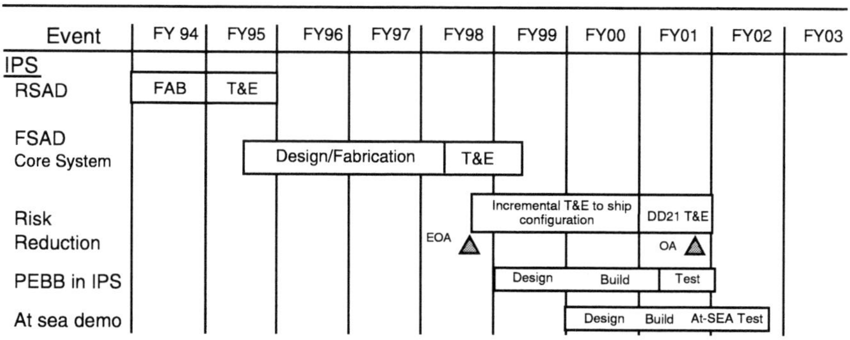

FIGURE 8.8 Integrated propulsion system. SOURCE: Adapted from (1) Advanced Surface Machinery Programs Office, 1996, "A Strategy Paper on Power for U.S. Navy Surface Ships," draft, Naval Sea Systems Command, Arlington, Va., Figure A-1, April 8, and (2) Krolick, C.F., 1996, "Advanced Surface Machinery Programs (ASMP)," presented to the Panel on Technology by the Naval Sea Systems Command, Arlington, Va., May 2.

introduction of IPS in noncombatant applications in FY 1998, which would lessen the fleet introduction cost and logistics support burden on the SC-21 class.

Introduction of IPS on aviation platforms will follow a two-step approach that straddles the SC-21 time frame. Although the ultimate objective is to provide a full-up IPS configuration for the CVX (FY 2006 award) that addresses all potential electrical loads including electric propulsion and electromagnetic catapult, the near-term CVN-77 (FY 2002 award) presents a transitional platform for introduction of a partial IPS configuration that addresses ship-service electrical loads only. The upfront systems engineering will define the CVX IPS configuration and identify those features that can be cost effectively demonstrated on CVN-77. Maximum utilization will be made of IPS SC-21 modules that will minimize development and fleet introduction and support costs and lead to identification of any additional modules needed to meet specific CVX requirements. These new modules would then be developed and qualified under the appropriate R&D program. For the large-deck amphibious ship (LHX) application, IPS offers arrangement flexibility for locating machinery relative to the well deck and engine operational flexibility, important when ship-service load is a large fraction of total installed power. Again, maximum reuse will be made of modules from previous ship applications.

Power Generation

The ICR gas turbine and its derivative, power generation module-1 (PGM-1) is envisioned as the main power unit for the IPS fleet starting with the SC-21. Selection of technology for any lower-power-rated PGMs for application to SC-21 will be driven by the requirements that emerge from the SC-21 cost and operational-effectiveness analysis (COEA) process. While these requirements are being finalized and the selection criteria developed and prioritized, it will be necessary to establish a technology base for the competitive selection process.

Commercial medium- to high-speed diesels are readily available in the power range of interest and could be deployed in the FY 2003 time frame. In any case, IPS diesel-based PGMs are likely to be identified to support the commercial end of the applications spectrum.

Advanced cycle gas turbines in the 1- to 4-MW range with fuel efficiency appreciably better than the existing DDA-501 are not commercially available at this time, although some commercial and DOE funding is going into their development. The technology exists and adequate time may be available to develop and qualify a new Navy engine from a commercial core in time for SC-21, assuming that a requirement emerges in the next year or so. The present ASMP strategy is to initiate militarization of a commercially available engine starting in FY 2000, based on the overall fleet requirements at that time.

Fuel cell developments are being supported with commercial and government funding, but largely for utility or land transportation applications that do not use Navy distillate as fuel. The Navy should continue to ensure demonstration of the fuel-processing system at a scale adequate to level the playing field with the alternative technologies in time of the SC-21 selection process. Also, if commercial or other government-funded efforts will not provide a full-scale stack with the characteristics required by naval force applications in time for demonstration with the fuel processor, then the Department of the Navy should consider investing in accelerating development of the stack. As fuel cell technology matures, it may achieve power density and costs that make it competitive as a larger-power-rated PGM as early as the 2010 time frame.

Other direct electrical conversion technologies such as thermionics and thermophotovoltaics will begin to emerge in the 2010 time frame as part of a cogeneration scheme using conventional engines.

Power Distribution/Conversion

Direct current zonal electric distribution systems, as with IPS in general, will have the greatest impact in a new design, but the power-conversion equipment can be approached in a way that allows their introduction in the ac ZEDS on new-construction DDG-51 and/or as part of the partial IPS configuration demonstrated in CVN-77. As part of this early introduction of power conversion equipment, first- or second-generation PEBB devices may be used. This will be driven by commercial viability of these early PEBB products that will encourage investment in production capacity. The dc ZEDS for SC-21 and CVX will be fully integrated with the auxiliary and combat system equipment, and generation-three or -four PEBBs will likely be employed in the power conversion equipment. By 2010, the PEBB will be fully utilized in the implementation of an autonomic dc ZEDS. Generation-five to generation-seven PEBBs will include high-voltage applications such as solid-state switchgear for the IPS 4160V bus or for the rectifiers and inverters for a higher voltage dc bus.

Pulse-power Systems

These systems could support back-fit, forward-fit, and new design, since the ETC gun concepts conform to existing weapons modules. However, gun and associated prime power system development has been delayed such that the SC-21 now appears to be the earliest target for these systems. Pulse loads are more likely for the CVX. By 2010, high-voltage, pulse-duty PEBBs could be employed in the pulse-power rectifiers.

Propulsion Motor

Although IPS core technology can support existing ships, the PM motor and a PEBB-based propulsion power converter are candidates for the SC-21, particularly in the lower displacement variants where power density has a greater value. Steerable pod propulsors are also viable for SC 21, most likely in conjunction with PM motor technology. Homopolar and superconducting motor technology will be available by 2010, if commercial development of high-T_c superconducting machinery is successful or if Navy-unique critical requirements (e.g., acoustics) drive their development.

Power and Propulsion

For Submarines

Future submarine power and propulsion technology will inevitably move toward increased reliance on electricity. Advanced dc power distribution using high-speed, turbine-driven rectified alternators feeding solid-state inverters through a dc distribution bus is already targeted to the new nuclear-powered submarine (SSN). Electric propulsion offers an affordable opportunity to enable new propulsor concepts that can improve signature. Studies over the past few years indicate that integration of the electric-propulsion and ship-service power-generation functions at the prime mover can result in reduced ship size and displacement because of fewer power plant components and improvements in their arrangement. Studies have also shown that eliminating centralized hydraulic systems and transporting energy electrically to the point of use will result in up to 70 tons of weight savings and potential cost reduction. In the future, this concept might be extended to other systems, such as heating, ventilation, and air conditioning.

In general, power and propulsion technologies for surface ships are also applicable to submarines. However, the optimization of submarine systems will be driven by stealth, safety, power density, and other requirements that differentiate surface and undersea vehicles.

The technologies that may be required for submarine power and propulsion systems of the future will include the following:

- Very low harmonic motor controllers;
- High-power-density, high-performance, solid-state inverters and converters and the components thereof;
- Motors and generators with very low acoustic and magnetic signature, including versions that can operate submerged in seawater at high pressure;
- New technologies for motors and generators such as superconducting magnets, cryogenic coolers, current collectors, high-field permanent magnets, liquid cooling, and active noise control; and
- Electrical substitutes for systems and components that now rely on fluid transport for energy and actuation, including electric actuators, electromagnetic launchers, and thermoelectric coolers.

For Amphibious and Land Vehicles

The Marine Corps relies heavily on power and propulsion technology for its vehicle power systems. The mandates for higher-water-speed amphibious vehicles, signature reduction, adaptability to electric power, future weapons, and reduced vehicle weight all lead to the need for an integrated electric power system for future land and amphibious vehicles.

Key drivers will be the following:

- Electric propulsion to allow the prime mover/generator to supply power to both water and land propulsion trains;
- Electric power distribution and control to direct and condition power for propulsion, weapons, and sensors as required; and
- Electric power generation and energy storage to allow operation of critical systems with reduced IR and other signatures.

Key technologies, as listed below, mirror those described above in this report for the surface-ship systems but are driven by the land and amphibious requirements:

- Pulsed-power systems including energy storage pulse-forming networks and very high power solid-state switching;
- Flywheel energy storage;
- High-power density engines;
- Electrical steering, suspension, and actuators; and
- High power density electric drives at the unit size of several hundred horsepower.

For Air Vehicles

The trend toward increased reliance on electricity is also evident in recent technical developments directed toward future aircraft.[12] The drivers toward a more electric aircraft include the following:

- Reduction in aircraft weight;
- Improved survivability and reliability;
- Reduced maintenance;
- Reduced cost;
- Higher fuel efficiency; and
- Reduced need for ground-support equipment.

---

[12]Office of Navel Research. 1994. *Power by Wire Technology Assessment,* Arlington, Va., May.

One solution to these drivers is to replace hydraulic, pneumatic, and mechanical aircraft systems with electrical functional equivalents and to provide the means to supply the increased electrical load (while reducing other engine loads). The key technologies that support the development of more electric aircraft mirror those that enable the electric ship. In most cases, however, the electrical power levels in aircraft systems are an order of magnitude less than those in a typical ship. Key technical developments for more electric aircraft power and propulsion include the following:

- Combined electric starter/generator for direct engine mounting;
- Smart, fault-tolerant electric power distribution and control;
- High-temperature-capable (~500 °C), solid-state power switches for engine-mounted electronics and power distribution control elements; and
- High-temperature-capable electrically powered actuators.

RECOMMENDATIONS

The panel examined a wide range of technologies, trends, and the Department of the Navy drivers that will push the development of new technologies in the future. It concluded that a focus on technologies that enable higher power density and more efficient and more effective propulsion, power generation, and distribution in all naval vehicles will be necessary to ensure future performance and affordability. The general areas of electrical power science and technology, and technology that will enable the efficient conversion of fuel to electrical power, hold great promise for major technical progress and payoff in the future. The Department of the Navy should place a high priority on the development of new all-electric ships with the associated drive, power-conditioning, and distribution systems. Investment in the following areas is warranted:

- Increasing switching speed, current, voltage, and operating temperature of solid-state power switching devices;
- Solid-state power controllers that are adaptable to multiple uses, modular, and compatible with the commercial market;
- Electrically driven substitutes for mechanically driven, hydraulic, pneumatic, or steam-driven actuators;
- Means for rapidly detecting and controlling the state of complex power distribution systems;
- Electric motor and generator technologies that increase power and torque density, reduce acoustic and electromagnetic signature, operate at elevated temperatures, have higher efficiency, and/or allow operation submerged in seawater;
- High-energy-density means for both long- and short-term electrical energy storage;

- Means for high-efficiency direct conversion of fuel to electric power;
- Higher-speed, more efficient engines that increase power density while reducing fuel consumption; and
- Higher-efficiency devices at the user end of the power system where the impact of efficiency is greatest (e.g., counterrotating propulsors).

9

Future Naval Forces and the Operating Environment

INTRODUCTION

Effective Navy and Marine Corps operations of all types require a comprehensive knowledge of the operating environment and, in turn, an understanding of the impact of those operations on the environment. The tools for characterizing the operating environment include long- and short-term weather forecasting and mapping and modeling of ocean and littoral waters, including positions of submarines, ships, and mines.

Readily available supercomputing-scale computational power (see Chapter 2, Computation), combined with high-resolution, pervasive sensor information (see Chapter 4, Sensors), increasingly sophisticated sensor fusion and filtering of data, and improved data display and assimilation tools (see Chapter 3, Information and Communications), will provide Navy and Marine Corps decision makers with access to accurate and predictive battle space environment information.

Public awareness of the impact of commercial and military maritime operations on the ocean environment is growing, and new restrictions on ocean dumping have been placed on ships. Recent Marine Pollution (MARPOL) Annex V regulations include a total ban on plastics discharge anywhere at sea and restrictions on the discharge of other solid waste materials in specially designated areas. Restrictions will likely become more pervasive in the future with the international regulation of atmospheric emissions from ships scheduled to begin in 1999. Tomorrow's naval forces will have to be able to reduce the sources of waste and to safely dispose of waste in a more efficient manner than currently fielded technology now allows.

TERRESTRIAL WEATHER AND CLIMATE PREDICTION

Recent advances in remotely acquired data, mainly from satellites, are providing a wealth of information about the ocean and atmospheric environment not previously available. The Navy complements such remote sensing with ship-based measurements of ocean depths, temperatures, salinities, and other parameters and has good historical data sets derived from more than 100 ship-years of dedicated time. New distributed sensors will provide real-time data at high resolution representing large areas, with deployment possible in remote or otherwise inaccessible regions as needed.

A major thrust for the future will be the enhancement of environmental data through the use of increasingly sophisticated models of the ocean/atmosphere system. Assimilation of these data into the Coupled Ocean-Atmosphere Dynamic System (COADS) model will be enabled by the rapid advances in computational power and modeling and simulation technology. The real-time weather prediction made possible by this combination of massive database modeling and computational power will allow tactical users to anticipate events in real time and strategic planners to more accurately predict seasonal weather.

Meteorological Thrusts

On the meteorological side, the current major emphasis of Department of the Navy research conducted at the Naval Research Laboratory, Marine Meteorology Division, at Monterey, California, is on data assimilation for the Navy Operational Global Atmospheric Prediction System (NOGAPS) and the Navy Operational Regional Atmospheric Prediction System (NORAPS) models. The data being assimilated includes those from both surface- and space-based platforms, data from the latter including the remotely sensed measurements of the ocean-surface wind speed, precipitation rates, and total vertical column moisture.

A new NASA scatterometer (NSCAT) was successfully launched on a Japanese satellite in August 1996 and is delivering accurate wind speed and directional data spanning the world's oceans. Future instruments, called SEAWINDS, are scheduled for launch on Japanese platforms in 2002 and 2007. These and other satellites will provide a major new critical data set for naval open-ocean operations. They are expected to revolutionize boundary-layer modeling for upper-ocean environmental changes as well as provide a complete array of electromagnetic surveillance of the battle-group environment. The wind is the basic parameter that determines the vertical structure of all atmospheric variables below the cloud base. Currently, these data are available only in a delayed mode; future naval operations will require real-time data acquisition from U.S. spacecraft.

In all likelihood, planners soon can expect to see significant development of very-high-resolution weather prediction models, such as the Coupled Ocean-Atmosphere Mesoscale Prediction System (COAMPS), which is the planned successor to NOGAPS and NORAPS. COAMPS will provide data on a 5-km grid. Higher-resolution models will be necessary, however, for precision tactical warfare purposes. Expected increases in computing power, high-density data storage, and real-time sensor data will allow the necessary increases in both spatial and temporal resolution, with a concomitant increase in prediction reliability for such phenomena as localized rainfall, cyclone tracking, ocean wave height and direction, and coastal mesoscale issues.

Forecasting ability has improved continuously since 1970, and expectations are that this trend will continue beyond the year 2000.[1,2] These expectations are based on data that do not take into account the high-resolution satellite data now available, which should significantly increase cyclone predictability. Local forecasting of rainfall conditions was performed by a special group of the National Weather Service using IBM and other supercomputers for the 1996 Atlanta Olympics.[3] The availability of this type of information allowed the organizers of the Olympics to manage their multiple widely separated venues effectively. The application of similar real-time or near-real-time information to battle-group operations should be straightforward.

Data and Data Assimilation Thrusts

The Department of the Navy has provided a great service by making available large amounts of extremely valuable data, such as the global digital bathymetric database (DBDB). A gridded 5-arc-minute product from these data is publicly available and widely used. In the aftermath of the Cold War, efforts to declassify data that are most useful for understanding emerging environmental issues (e.g., global warming, fisheries, pollution) are increasing, along with the development of technology to assist the user with assimilating large amounts of data through the use of graphical representations and other visualization means. The focus is on how such data may be enhanced by combination with mathematical models, for both reporting current conditions (best estimate of current state of the environment) and forecasting. Such a combination has been accomplished successfully in meteorology, which has been driven by commercial interest and which is currently more advanced than ocean modeling. This combination of models and data will be a major area of research in the atmospheric as well as

[1]Anthes, Richard, A. 1995. "Tropical Cyclones: Their Evolution, Structure, and Effects," *Journal of the American Meteorological Society*, 19(41):1-203.

[2]Foley, G.R., H.E. Willoughby, J.L. McBride, R.L. Elsberry, I. Ginis, and L. Chen. 1995. "Global Perspectives on Tropical Cyclones," *WMO Report,* TCP-38, R.L. Elsberry, ed., Secretariat for the World Meteorological Organization, Geneva, Switzerland.

[3]Skinrud, E. 1996. "Georgia on Their Minds: Olympic Weather Team Pushes the Limits of Forecasting," *Science News*, 150:29, July.

ocean sciences in the near future, especially with the greatly expanded current databases and new types of data available from remote sensing.

Computing and Oceanography Thrusts

Modeling of such complex systems as the oceans, the atmosphere, and their interactions requires massive computational power. The world oceans have major circulation features such as the Gulf Stream and associated eddies that require horizontal grid resolution of a few kilometers and many vertical levels to adequately resolve important features. For example, to address the entire global ocean, including the effects of these small-scale but energetic features, which may be critically important, requires a computational grid of at least 100 million points. Assuming a typical time-step size of about 5 minutes and about 250 operations per grid point per time step, this translates to about 10 million-billion operations to calculate 1 year of activity. This does not include coupling of models for the ocean and the atmosphere, which is even more computationally intensive. Thus, the High Performance Computing Initiative's (HPCI) minimum computing allotment to the DOD would not be enough for serious climate studies.

For regions of special naval force or other environmental interest, it may be desirable to use nested grid approaches in which ultrahigh-resolution regional grids are nested within coarser grids to reduce the computational load. Nesting has the major advantage of providing physically consistent open boundary conditions. This technology is particularly promising for shipboard computing applications. For example, it allows a ship-based computer to assimilate local observed data into a resident regional model in which the boundary conditions are provided via communications from a large-scale model center, as depicted in Figure 9.1. This approach avoids the transmission of information that may compromise strategic advantage and decreases the need for large-bandwidth shipboard communications. Improved nesting approaches will be a major area of technology development for the foreseeable future.

Goals of the Ocean-Atmosphere Weather Modeling Effort

In the near term (3 to 5 years), planners can expect to achieve high-resolution weather modeling and prediction using HPCI facilities, including 3- to 5-day forecasts of mesoconvective rainfall, cyclone tracks and intensity, ocean waves, and coastal mesoscale phenomena.

In the medium term (5 to 10 years), it should be possible to provide meso- and microscale weather information to users in the field, improved weather models with ocean/atmosphere interactions, high-resolution forecasts for battle-group and field users supported by high-resolution local observations via deployable sensors, correlation of local sensor data with atmospheric and ocean models to

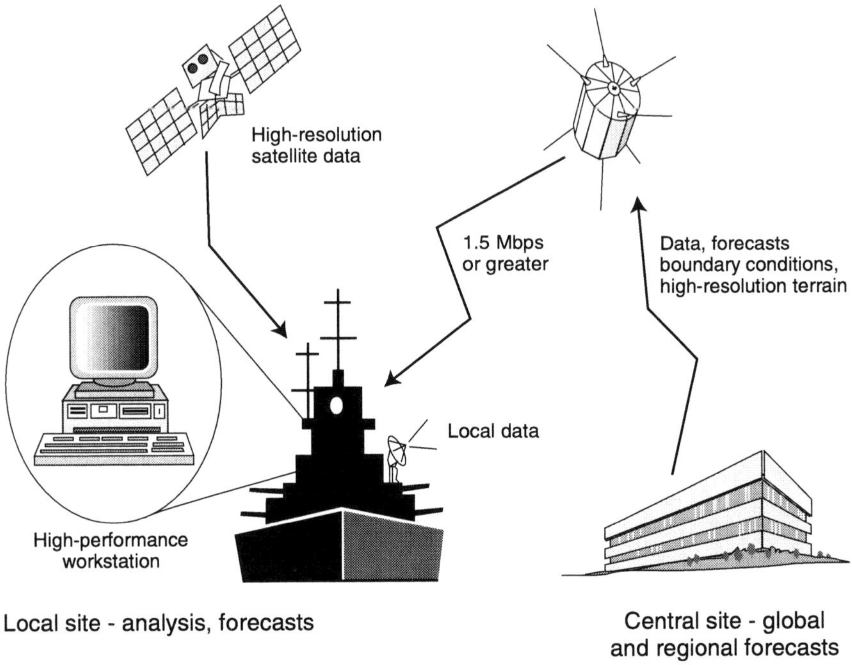

FIGURE 9.1 Schematic representation: A future battle-group weather forecast.

provide accurate automatic target recognition and acquisition, and 6- to 7-day forecasts of large-scale weather systems such as cyclones.

In the long term (10 to 20 years), ships should be provided with local high-performance computing with massive stored databases for autonomous high-resolution, small-scale weather reporting of current environmental conditions and lower-resolution, larger-scale 30-day weather forecasts displayed multi-dimensionally and multisensorially.

SPACE WEATHER PREDICTION

The space environment around Earth, known as the magnetosphere/ionosphere system, can have adverse effects on naval communications, navigation, and weapons systems. The ONR and NRL conduct research on space phenomena, but the Department of the Navy generally does not field operational sensors that collect environmental data that can be used in space weather forecasting. The U.S. Air Force 50th Weather Squadron at Offutt Air Force Base, Nebraska, has responsibility for monitoring space weather for all of DOD. The Air Force employs a full suite of modeling and prediction tools, but because the space environment is so large and complex, these models and tools are necessarily limited. In terms of forecasting, the most important variables are the interplanetary magnetic field orientation and the density and speed of the solar wind. NRL is now intimately involved in the Solar Heliospheric Observer (SOHO) mission that provides a point measurement of the solar wind density and speed. Observations from more than one platform are crucial to providing adequate temporal and spatial characterization of the parameters that ultimately determine the state of the space environment.

With respect to satellite operations that support naval interests, satellites in geosynchronous and semisynchronous orbits are exposed to a flux of high-energy particles. Electronics on orbiting space platforms are affected by the radiation dose and the naturally occurring variations in these energetic particles. Charging of dielectrics to several thousands of volts under certain space environment conditions can occur with subsequent discharges and permanent damage to sensitive electronics. Energetic particles may emanate from regions that involve huge volumes of space that cannot reasonably be sampled in situ. Recent efforts supported by a consortium of civilian and governmental institutions have shown some promise in remotely imaging this vast magnetosphere using space platforms. Continuing efforts in magnetospheric remote sensing is needed if the DOD and the Department of the Navy are to have reasonable lead time in forecasting disturbed conditions in space.

The value of specifying the state of the on-orbit space environment in real time cannot be overstated. Every operational spacecraft should carry basic particle- and field-sensing equipment to characterize the local radiation environment. Equipping new spacecraft with compact environment sensor packages would be the space equivalent of collecting atmospheric and sea surface measurements by the surface fleet.

Scintillation

The ionosphere is often viewed as the cooperative medium that makes over-the-horizon communications possible, but the ionosphere is also the largest source of error in civilian GPS[4] navigation and can cause communication problems, notably those arising from the phenomenon of scintillation. At sunset, the ion and electron content of the ionosphere near the equator becomes irregular. These irregularities cause significant rapidly varying phase and amplitude distortions in communication signals. Radio waves appear to scintillate, causing uncertainty in geo-locating the satellite transmissions. At high latitudes the auroras cause similar effects. Scintillation is more widespread and intense during the maximum portion of the 11-year solar cycle, such as the upcoming period from 1998 to 2003.

---

[4]Ionospheric effects on military GPS systems are partially mitigated through the use of two frequencies.

Mid-latitude areas, where most weapons systems are deployed, are not normally affected by large regions of ionospheric scintillation, but the low- and high-latitude regions that are affected may extend to 40 percent of Earth's surface. The occurrence, extent, and impact of ionospheric scintillation needs to be better understood, along with the physics of scintillation forecasting, if increasing geo-positioning accuracy with the GPS satellite system is to be achieved.

To understand the extent of scintillation, surveys must be made worldwide. Inexpensive sensors need to be fielded at both shore locations and on ships that collect data in HF, UHF, and GPS frequencies. After data collection and analysis to correlate the frequency of occurrence of scintillation with weapons-system interference, the Department of the Navy should examine its options for implementing scintillation mitigation measures. Finally, 24- to 72-hour forecasts of ionospheric scintillation could be generated if the DOD developed an operational computer model of the sun and interplanetary space, with a corresponding sensor suite to drive the model in real time. The output of the solar/interplanetary model could be used as input to radiation models that forecast the condition of the magnetosphere and to those models that produce background and scintillation forecasts of the ionosphere.

DEEP-WATER MODELING

NRL is a world leader in modeling the world oceans for aspects critical to naval operations. It routinely calculates ocean currents, fronts, and eddy locations (in hindcast mode) using daily surface weather fields including those obtained from the European Forecasting Center. Despite this excellent capability, naval forces in the future must have real-time access to the output of a global model driven by current input data on (1) surface wind fields, (2) surface sea level, and (3) the internal variability of thermal fields.

Wind data may be obtained from scatterometers on polar-orbiting satellites, such as NSCAT. Two satellites are required to give the Navy the coverage needed to monitor all remote areas. Altimeters mounted on the Navy GEOSAT and NASA TOPEX-POSEIDON satellite have demonstrated the ability to measure the shape of the ocean surface. Given an accurate geoid and tidal model, such data can be assimilated into ocean models to provide an estimate of ocean currents, front location, and eddy location. Measuring the internal variability of thermal fields is more difficult because the information comes from the mid-water (on the order of 500- to 2,000-m) ocean thermal structure. The two technical systems that are candidates for future development are acoustic tomography and drifting smart floats that measure current and temperature and maintain position in a constant water density. Both systems have been tested by the academic community, and both show promise. The global acoustic monitoring of ocean thermometry (GAMOT) project, for example, has demonstrated that the feasibility of deploying very inexpensive drifting passive sonar buoys capable of measuring deep-water travel time using fixed-sound-source acoustic transmission and data uplink to satellites. Such measurements allow the development of models of the upper 1,000 m of a 4- to 5-km-deep ocean. The other 80 percent is below the level of most conceivable naval operations. Because the ocean is stratified, slow moving, and, consequently, hydrostatic, almost all of the fundamental environmental information that is time dependent is transferred to the deep ocean through conservation of mass and momentum.

Since the atmosphere cannot be accurately forecast more than 1 to 2 weeks in advance, accurate upper-ocean forecasts cannot be obtained because the ocean is forced by the wind. Fortunately, the ocean structure changes relatively slowly, so that current ocean reports will be viable for a week or more, which is long enough for most critical naval operations.

LITTORAL-WATER MODELING

The ocean's response to the atmosphere, radiant energy, and tides may be characterized according to its local depth. The domains are the deep ocean (depths greater than 1,000 m), the continental slopes (depths from 2,000 m to 200 m), the continental shelf (depths from 200 m to 20 m), and the near-shore littoral zone (depths shallower than 50 m). The overlap in depths accounts for different configurations of shelf-slope geometry around the world. For example, off Peru, one might find water depths of 100 m within 100 m offshore, whereas near deltas off the southern United States, depths of only 10 to 20 m might be found 10 km from shore.

A thorough understanding and modeling of near-shore littoral waters is critically important for future naval forces. A physical understanding is reasonably well established, but this zone also encompasses much greater variability in current, sediment transport, visibility, salinity, and so on, than the deeper regions of the ocean. The deep ocean is quite variable in many regions, but the time scale of variation is generally longer than that for shore waters, and, as discussed above, high-resolution real-time data are not as critical for many naval operations.

The physical variability, the acoustic environment, and the visibility of the littoral zones are not well modeled. Each region on Earth is affected differently primarily because of variations in bottom topography, water runoff, climate, and bottom composition. There are, however, well-understood (except for turbulence) physical principles to assist in characterizing each regime. If appropriate data are acquired and retrospective studies performed, it is possible to anticipate the physical state of the littoral zone that might be encountered in naval operations. Multimodality sensor systems that provide real-time data are being developed for this highly variable environment.

Physical Modeling of the Littoral Zone

Near-shore waters are driven by the sun, tides, winds, and land runoff of fresh water and sediments and by off-shore currents. The actual shape of the bottom and its capacity to reconfigure are also significant. Tomorrow's naval forces will have access to large-scale ocean models capable of accurately predicting off-shore currents needed for boundary conditions and will have accurate large-scale weather modeling for atmospheric boundary values of such phenomena as wind stress, heat flux, and solar radiation. For each threat region, a local, portable physical model could be created that could accurately simulate the physical state of a littoral zone under various scenarios of weather, both typical and extreme (typhoons, tsunamis, climate variability, etc.).

Model scenarios for each potential area could be developed and validated. Codes, climatologies, topographic configurations, and the like, can be stored in a modern high-speed, large-memory workstation. Modern data assimilation protocols (either variational adjoint or Kalman-Bucy filtering) can be overlaid on the basic model database. Either in anticipation of actions or during an event, all data gathered in the region can be assimilated to produce the most probable physical environment needed for surface, underwater, acoustic, and countermeasure operations.

The data output from these physical models is time dependent and three-dimensional. Shipboard personnel generally do not have the technical background to interpret the complex environmental fields. A solution is to develop four-dimensional graphical visualization systems to be used to identify patterns of such phenomena as currents, temperature fronts, and low-visibility regions. Such software is not currently available except in primitive form but is under development.

Mapping the shape of the bottom is currently straightforward, but visualization of the flow in the water volume is difficult because of the enormous databases involved and the inherent problem of mapping a four-dimensional picture onto a two-dimensional computer screen. Workstations of the future will have enough cycle power, memory, and storage to handle the computations, but new visualization techniques have to be developed, such as four-dimensional virtual reality systems. Within 10 years, it should be possible to provide on shipboard modest-sized virtual-reality sites for naval personnel to see the present and evolving underwater physical environment.

Shallow-water Acoustics

The physics of acoustic phenomena in the open ocean is well understood. Given a source location, a receiver, and information on such things as water density and the shape of the bottom, the behavior of sound can be calculated. But in most littoral zones, this is almost impossible because the temperature and salinity of the water change with season and weather conditions. Moreover, surface sea waves change quickly and frequently and the behavior of the bottom reflectivity and absorption changes within waters. Thus, reliable interpretation of acoustic transmission in near-shore waters is technologically challenging.

Unfortunately, foreign navies are investing in electric submarines that are relatively small and quiet. These, along with inexpensive and plentiful mines, pose serious threats to naval forces in shallow water. Special attention needs to be directed toward pattern recognition and signal recognition of moving and stationary objects in shallow water. Extensive simulation of acoustic systems must be carried out with the various scenarios predicted by the physical modeling system. For example, high-frequency active sonar may be effective for mine detection in shallow water but not have enough range for ASW in deeper water. Modeling will provide the information necessary to make informed choices of sensor-system deployment and enable the development of protocols to identify interdicted structures not encountered in tests of the anticipated acoustic environments.

Visibility

Various manned and unmanned operations in the littoral zone are affected by changing visibility conditions. Intense storms can reduce underwater visibility by moving silt, and the typical clear waters seen by divers on tropical shelves during pleasant weather do not characterize the visibility on most continental shelves. Fog, rain, and sandstorms can affect atmospheric visibility. To offset these problems, physical models of the visible environment under a variety of conditions should be developed. Such simulations can be used for training personnel on the conditions and hazards to be experienced in naval operations.

Near-shore Currents

In many littoral zones, strong unpredictable currents, such as storm surges, rip tides, turbulence, and breaking waves, can occur. These shallow energetic currents can create havoc for surface or underwater transportation of people and equipment. Databases need to be developed for climatologies and probabilities of strong near-shore currents. Beaches must be characterized by their physical properties and ability to support energetic events. It is currently technically feasible to predict storm surges up to a day or two in advance, since strong atmospheric storms are very well predicted on that time scale.[5]

In the United States, a priori storm-surge prediction is typically used to determine coastal building codes, for example. Hurricane storm surges predicted when

[5]O'Brien, James J. 1996. "Changes to the Operational Early ETA-Analysis/Forecast System at the National Centers for Environment Protection," *Weather and Forecasting*, 11(3):391-413.

a storm arises are based on empirical information, not accurate modeling. The Navy Department could benefit from the models being developed by the academic community, not only for naval bases but also for potential action locations.

SHIPBOARD WASTE AND POLLUTION MANAGEMENT

Cost-effective, environmentally sound, and technically feasible methods for handling solid and hazardous waste will continue to be a vital concern for the Navy, particularly with regard to the unique logistical constraints associated with shipboard operations. Increasing restrictions placed on ship discharge by the international community also drives the need for efficiency improvements on board Navy ships. Although the Navy is currently exempt from the MARPOL 73/78 Annex V Regulations for the Prevention of Pollution by Garbage from Ships, these guidelines are applicable to all classes of ocean-going ships and the U.S. Congress has indicated that the Navy will have to comply with Annex V at some point in the future. In this document, only those waste and pollution management problems unique to ship operation are considered.

Waste management alternatives such as waste minimization, pollution prevention, recycling, and waste use and reuse, when used in conjunction with innovative technological advances, can offer a variety of options to ensure the effective management of the Navy's shipboard solid, liquid, gaseous, and hazardous waste. Several recent reports by the National Research Council[6,7] on shipboard waste management summarize currently available and near-term waste management technologies. As pointed out in the Guidelines for Implementation of Annex V of MARPOL 73/78, education of personnel is an important component of any waste management system.

Adapting existing technologies or developing new technologies to help manage shipboard waste requires knowledge of the types and quantities of waste produced during naval operations. The types of waste streams typically generated on U.S. Navy vessels can be categorized as (1) solid (e.g., metal, glass, plastics, food waste, paper, cardboard, medical waste); (2) liquid (nonoil waste such as blackwater [sewage] and graywater [shower/laundry/dishwashing discharges] and oily waste such as bilge cleaning and derusting fluids, antifouling paints, and waste oils); or (3) gas (air emissions from marine vessel diesel or gas turbine engines and on-board incinerators).

The ideal waste management system would be completely closed where wastes are used as feedstocks for useful products or energy, preferably on board.

[6]Naval Studies Board. 1996. *Shipboard Pollution Control; U.S. Navy Compliance with MARPOL Annex V*, National Academy Press, Washington, D.C.

[7]Marine Board. 1995. *Clean Ships, Clean Ports, Clean Oceans*, National Academy Press, Washington, D.C.

Closed-cycle systems would be necessarily complex, heavy, and bulky and therefore not suitable for shipboard use. Partial systems for recycling some materials such as metals, plastics, and water and converting waste paper into energy are in use on shore.

Waste Minimization

A primary goal of any waste management system that must operate in a closed and contained environment, such as in Navy vessel operations, is to have a logistics system in place that minimizes the production of waste. To help facilitate this goal, all shipboard supplies and maintenance procedures should be evaluated in terms of the types and quantities of wastes generated, and a thorough study should be conducted to see if other options such as process changes, engineering design changes, product or packaging specification, or substitution can accomplish the same goal while simultaneously reducing the waste stream. For example, a significant portion of the space on a logistics support ship is occupied by soda cans. Substitution of the cans with bulk soda syrup and water carbonation systems would increase logistical-support ship capacity and decrease waste while not significantly affecting quality of life aboard ship.

A similar product and process design philosophy is being used by the automobile[8] and personal computer[9] industries in their design-for-recycling programs. Because of serious space limitations, the absence of at-sea replenishment, and limited air supply to support incineration, submarines have developed effective (albeit labor-intensive) waste minimization practices.

An example of waste minimization is the Plastic Removal in Marine Environments (PRIME) program that focuses on source reduction of plastic materials aboard Navy ships. By changing the packaging specifications for supplies, PRIME has reduced the annual amount of plastics brought on board by 475,000 pounds.[10] This is approximately 10 percent of the total 4.5 million pounds of plastic estimated to have been discharged each year until 1988. PRIME eliminated another 62,000 pounds of waste (per ship per year) by using paper instead of polystyrene drinking cups. This amounts to about 8 million large, 14-oz. styrofoam cups or over 10 million small cups. Such programs can be expanded to produce significant additional savings at little additional cost.

Managing potentially hazardous substances can best be accomplished by ensuring that the materials are consumed. A current example of hazardous waste

---

[8]Ashley, S. 1993. "Designing for the Environment," *Journal of Mechanical Engineering*, 115(3):53-55.

[9]Sheng, P., and J. Sutanto. 1995. "Environmental Factors in Parametric Design of Computer Chassis," *Journal of Electronics Manufacturing*, 5(3):199-216.

[10]Marine Board. 1995. *Clean Ships, Clean Ports, Clean Oceans*, National Academy Press, Washington, D.C.

minimization is the Consolidated Hazardous Material Re-utilization and Inventory Management Program (CHRIMP), a computer-based inventory system based on the concept of centralized control and cradle-to-grave management of hazardous materials and waste such as paint, solvents, and cooling agents.

Nonminimizable hazardous materials such as those generated by various cleaning and industrial operations are stored on board in drums until docking, where they are transported to landfills or land-based incinerators. Identifying and/or developing technologies that could minimize, clean and reuse, or detoxify these waste materials on board the vessel while at sea should be an important research priority. Another approach is the development of nontoxic, environmentally benign alternatives, particularly for cleaning operations. The use of citrus-based cleaners instead of chlorofluorocarbons for degreasing in the semiconductor industry is a good example of the latter.

Waste Treatment

A major roadblock in developing effective waste treatments is generally a separation process. For example, recycling of metal cans has been proven to be economically feasible because there are cost-effective methods of separating steel and aluminum cans with the resulting materials easily recycled. Similar arguments can be made for plastics, paper, solvents, and waste water of all kinds. Ultrafiltration is being used in the food-processing and paint industries but needs further development for the complex mixtures encountered on Navy vessels. In current use are ceramic membranes for concentrating oily waste and polymeric membranes for blackwater and graywater. The discharged filtered water has waste concentrations below 15 ppm, within the current acceptable discharge limit. The removed dried solid material may be incinerated, eliminating the ocean discharge of the food waste, graywater, and sewage streams.

Plastics, a major waste management concern on ships, are currently treated for disposal by shredding the material and then thermally compressing it into flat disks for storage until off-loading at port. Shredding plus thermal compression yields a volume reduction of almost 50 percent. The resultant disks are composed of several types of plastic, usually contaminated by food and other container contents, thereby significantly limiting their ability to be recycled.

More than 200 types of plastic now exist in waste streams, and these should be separated for effective recycling. Commercial recyclers are developing infrared techniques for identifying different types of plastic, which can then be sorted for recycling. Currently, seven types can be distinguished in this manner. Such technology may not be feasible for shipboard use, however. A more practical solution may be to limit the number of plastic types on board and sort by use. The use of biodegradable plastics would not eliminate the disposal problem but would allow such plastics to be treated as food or paper waste. Biodegradable plastics would be particularly useful for forms such as plastic film that are readily contaminated.

Shredding and pulping are efficient methods for the treatment of food and paper waste. Pulpers are able to treat up to 600 lb/h of waste, and the resulting slurry can be mixed with seawater and dumped overboard in permitted regions. The slurry presumably does not adversely affect plant and fish life. This technology can have significant benefits as food and paper waste constitutes approximately 60 percent of shipboard solid waste by weight. Additional reduction in weight can be achieved by biological treatment, but more compact systems need to be developed for Navy shipboard use. To meet future, more stringent, discharge requirements, the biologically treated slurry would have to be dewatered and stored or oxidized.

After all possibilities for minimization, recycling, and reduction are exhausted, residual concentrated carbon-based waste may be pyrolyzed and oxidized to produce volatile oxides with minimal solid residue. Compact, efficient incineration or alternative pyrolysis/oxidation processes specifically adapted for use aboard U.S. Navy ships should be developed. Some of the methods now in various stages of development are supercritical water oxidation, high-pressure hydrothermal processing, electrochemical oxidation, semiconductor photocatalysis, ozonation, electrohydraulic cavitation, plasma arc thermal conversion, pulsed-power cold plasma reactors, and advanced incinerator designs. In addition to factors such as process efficiency and throughput and emissions that must be considered for any disposal technique, other factors unique to Navy vessels that must be considered include power requirements (a net power producer, if possible), size (small), infrared signature (low), EMI (little or none), reliability (under normal and battlefield conditions), and the need for operator expertise.

Although the Navy is exempt from MARPOL regulations for the time being, it is still required to achieve zero plastic discharge. Because restrictions on ocean dumping will only increase, it is prudent for the Department of the Navy to continue development of adequate waste disposal systems. The two alternatives appear to be (1) storage of compacted separated waste for port recycling and disposal, and (2) adaptation of current cruise ship or similar systems involving biological pretreatment and incineration of wastes except for plastics, glass, and metal, which are stored and recycled on land. In either case, properly packaged waste could be retrograded to combat logistics support ships after offloading supplies. Both of these options require significant on-board storage space. For waste destruction, near-term development should concentrate on compact, low-power, low-signature, high-reliability incineration systems.

Air Pollution

The main source of air pollution in any navy are propulsion and auxiliary power systems burning fossil fuels. In fact, restrictions on nitrogen oxides (NO_x), unburned hydrocarbons, carbon monoxide, and particulate matter will be imposed on new marine diesel-powered vessels by 1999. The most direct method, therefore, of reducing air pollution from naval operations is to develop alternative energy sources (see Chapter 8, "Electric Power and Propulsion") for most, if not all, shipboard energy needs. At present, nuclear power is the cleanest power source available for ship use with regard to air pollution. Although disposal of spent nuclear power plant waste is a serious problem, it is not one that is unique to ship operations.

Fuel cells for clean oxidation with few environmental byproducts and solar cells for direct electrical energy production are two alternative energy sources that have been under development for several decades. Compact, high-density fuel cells, based on immobilized enzymatic redox reactions for power storage, should be available in the near term. Similar advances in solar cells have occurred regularly over the past decade.

Certainly, increasing the efficiency of ships' engines and their distribution systems will reduce reliance on nonrenewable and potentially embargoed oil resources. The centralization of power sources possible with the electric ship would enhance the development of more efficient engines.

For the near term, however, air pollution from chlorofluorocarbon (CFC) refrigerants is a much more serious problem because of the deleterious effect of CFCs on atmospheric ozone. Under the auspices of the Copenhagen Amendments to the Montreal Protocol, the production of ozone-depleting fluorocarbon refrigerants has been banned since December 31, 1995. The Navy currently has approximately 850 CFC-114 chilled water air-conditioning plants that provide mission-critical electronics cooling on surface ships and submarines. It has therefore been necessary for the Navy to identify a suitable refrigerant alternative to CFC-114 and to develop and qualify suitable backfit modifications for shipboard CFC-114 air-conditioning plants. Currently, 12 different Navy shipboard CFC-114 air-conditioning plant designs are in effect—ranging from 125 to 363 tons in cooling capacity—all of which require backfit modifications.

Noise Pollution

An issue of critical importance to today's naval forces, and almost certainly to be important to 21st-century forces, is the growing concern about the possible harmful effects of certain underwater sounds on marine life, and the restrictions this might place on the use of sonar systems and other naval activities that generate underwater sounds, including weapons and hull-integrity tests.

In point of fact, the Navy Department has had to seek permits to conduct ship-shock tests (using explosive sources) and tests of prototype sonar systems, as well as for exercises to develop tactics and strategies for employing low-frequency active sonars. Weapons-testing in the Gulf of Mexico has been constrained. The Navy has to prepare extensive environmental impact studies for these operations and in some instances has altered venues and test plans. In all instances, delays have been incurred, and the expense in dollars and manpower has been significant.

Largely as a result of an increasingly aware, vocal, and powerful environmental movement that is demanding compliance with marine mammal and endangered-species protective laws, recent naval practices have been called into question—specifically, the low-frequency active acoustic ASW development programs, which are based on very-high-intensity sonars operating in the vocalization frequency bands of many marine mammals, especially whales.

The steady, historical reduction in submarine-radiated noise in all submarine variants—nuclear, diesel, and air-independent—has reduced the effectiveness of passive ASW and necessitated the development of active methods. Over the next several decades the proliferation of quiet, capable, and effective submarines through foreign sales and indigenous manufacture may result in even more reliance on active acoustics. As a result, the issue of compliance with environmental laws will almost certainly be a major problem for the future naval forces unless mitigation measures are undertaken.

Therefore, it is entirely possible that the Department of the Navy will be prevented from developing and fielding the systems it requires to execute its future missions, unless a rational basis for mitigating risk to marine populations is developed. The Navy Department must pursue the R&D necessary to develop sufficient understanding of the issues to enable rational, informed decisions.

RECOMMENDATIONS

Battle-space awareness, communications, target identification, navigation, weapon guidance, and tactical planning all require real-time understanding and forecasting of the atmospheric, space, and sea environments of operation. Global weather models with improved satellite data on winds, temperature, solar inputs, and so on will permit the generation of accurate weather forecasts. Space weather forecasting of solar disturbances, scintillation phenomena, and other disturbances will be modeled based on real-time satellite data. The Department of the Navy must support the development of this modeling capability.

With respect to terrestrial weather and climate prediction, the panel presents the following recommendations:

- To take advantage of the vast quantities of new environmental data from an increasing array of remote-sensing devices—in space, in the upper atmosphere, on land, and on and under the sea—the Department of the Navy should pursue the development of shipboard computational and data communication systems coupled to land-based high-performance computers.
- Emphasis should be placed on developing the higher-resolution climate prediction models that are necessary for tactical warfare operations.

- R&D emphasis should be placed on the high-performance hardware, algorithms, and memory storage required to enable real-time applications under realistic battle-space conditions.

With respect to space weather prediction, the panel presents the following recommendations:

- Greater emphasis should be placed on educating naval personnel about the effects on weapon systems of interference caused by electromagnetic disturbances and energetic particles in the space environment.
- Greater R&D emphasis should be placed on technologies for mitigating or eliminating adverse effects owing to space weather, such as the scintillation phenomena, on communication and weapons systems.
- The Department of the Navy should monitor and contribute to the development of space-based remote sensors, such as magnetospheric imagers, and Earth-based sensors that monitor the sun and the ionosphere to provide reasonable lead times in forecasting disturbances in space weather.
- Development of a robust computer model of the sun and interplanetary space, driven by real-time sensor data, should be a high priority within the DOD, and should be encouraged by the Navy Department, to permit reasonably accurate forecasts of ionospheric scintillation.

With respect to deep-water modeling, the panel presents the following recommendations:

- To take advantage of the vast quantities of new data from an increasing array of remote-sensing devices and thus be able to accurately forecast ocean currents, fronts, and eddy locations and behaviors, the Department of the Navy should pursue the development of multidimensional, on-board data retrieval and analysis systems involving high-performance models.
- Deployment of at least two additional remote-sensing satellites may be required to permit complete monitoring of the ocean's surface.
- Greater R&D emphasis should be applied to obtaining thermal profiles of depths ranging to 1,000 m, a critically important element in ocean forecasting.

With respect to littoral-water modeling, the panel presents the following recommendations:

- Greater attention should be paid to the dynamics of near-shore environments, including such phenomena as shallow-water acoustics, variable visibility, and strong, near-shore currents, which at present are not well understood.
- Databases should be developed that sufficiently describe near-shore phe-

nomena, including beach characteristics, that can lead to accurate forecasting of littoral conditions.

- To enable naval personnel to use such databases, graphical interface technologies need to be developed, including on-board, multidimensional, multi-sensorial virtual-reality systems capable of representing real-time and evolving underwater physical environments near shore.

With respect to shipboard waste and pollution management, and management of noise pollution, the panel presents the following recommendations:

- To mitigate risks to the environment and to shipboard personnel, greater emphasis in reducing and handling a variety of waste streams associated with naval operations is a pressing need. Where possible, successful land-based applications of waste minimization and treatment technologies should be adapted for shipboard operations.
- Emphasis should be on curbing waste at its source and treating waste as a renewable resource or useful energy source.
- Increased R&D on subsurface noise propagation and its effects on marine populations must be conducted. There are few data currently available on an issue that is likely to pose serious impediments to a host of naval operations.

10

Technologies for Enterprise Processes

INTRODUCTION

The Department of the Navy is a huge large business enterprise that manages multiple large-scale processes including platform and weapons acquisitions, supplier management, logistics management, resource planning, and personnel management and training. Powerful new information technologies are becoming available that can be applied to these enterprise processes to significantly improve overall efficiency and effectiveness. The panel believes that these technologies will enable new ways of conducting business that will lead to faster, lower-cost development and acquisition programs; more efficient logistics systems; reduced manpower requirements; faster and more nimble planning capability; and greatly improved warfighting effectiveness.

The complexity of the Navy Department enterprise is growing, while at the same time funding constraints are forcing the organization to become more efficient. The imperative of "faster, better, cheaper" is now being applied to every phase of the acquisition process, as well as to operations. The combination of increasing complexity and constrained resources will drive the development of new operational constructs, or processes, to achieve mission-required capabilities. Building the capabilities necessary to meet Navy and Marine Corps requirements demands highly specialized analyses and understanding and the integration of multiple disciplines into a set of enterprise processes that will extend across the entire spectrum of naval activities.

The panel has identified categories of Navy Department activity as being particularly amenable to the application of information and management technology and has defined the following set of model enterprise processes: simulation-based acquisition, agile commerce, real-time logistics management, resource planning, dynamic mission planning, and system of systems.

SIMULATION-BASED ACQUISITION

Description of the Technology

Simulation-based acquisition (SBA) is a potentially revolutionary process for the acquisition of complex platforms or systems. It integrates all acquisition activities starting with requirements definition through production, deployment, and operational support into a computational network with a common database. This integrated approach enables the optimization of design, manufacturing, and life-cycle support functions associated with complex systems. At the initiation of a proposed SBA procurement, government and industry teams will create virtual prototypes of candidate systems that could include warships, aircraft, communications systems, logistic systems, information systems, or other complex systems. These virtual prototypes will be fully linked to an integrated product database and are used by all participants during acquisition, development, and operation throughout the life cycle of the products. SBA is based on a computer environment that integrates existing models and simulations (and new ones if required), engineering and physics, and operations and doctrine to evaluate overall product value and cost, to guide the development process, and to support operation of the product during its service life. Figure 10.1 shows a schematic representation of the SBA process.

SBA comprises an advanced scalable architecture, an integrated product database that includes product attributes; an interactive, immersive synthetic environment; integrated collaboration tools; analysis and modeling tools packaged with standard, interoperable interfaces; smart agents to assist in information collection and integration; and advanced product and process-optimization tools.

Relevance to the Naval Forces

SBA allows decisionmakers the unprecedented ability to examine design and operational issues for proposed systems by interacting with the virtual prototype of the product. High-fidelity simulations allow acquisition authorities to visualize and analyze the virtual prototype in distinct phases of its life cycle. Virtual prototypes allow officials to "fly before you build" to examine alternative designs, performance, and cost. They are thus able to conduct studies that include mission value, affordability, performance tradeoffs, manning, maintenance, and the like. Government and industry personnel work together to build the right product, the right way, at the right price, on schedule.

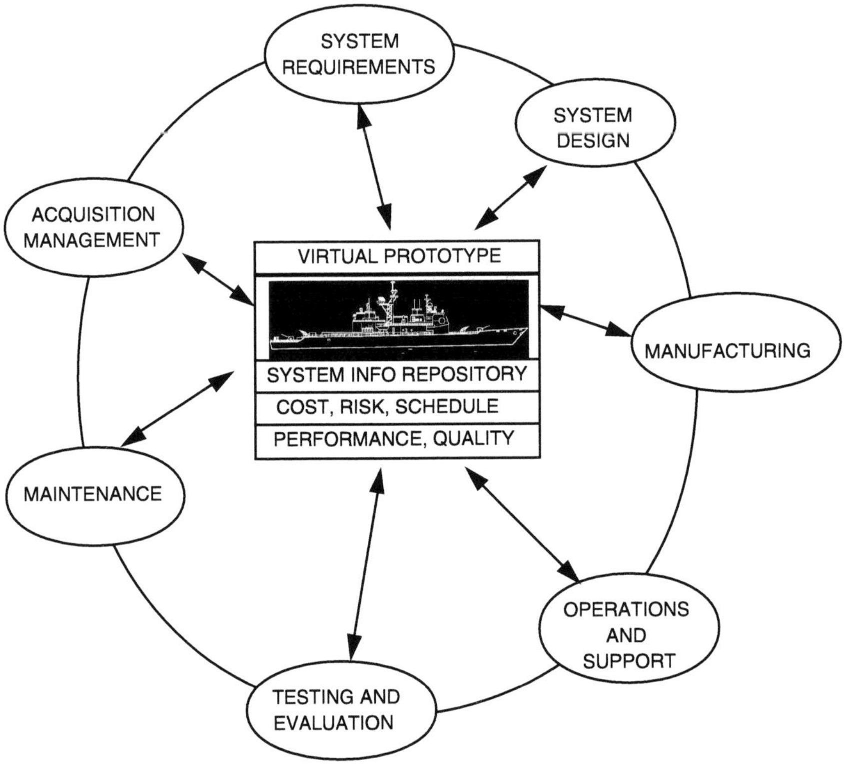

FIGURE 10.1 Schematic representation of the SBA process.

Technology Status and Trends

A key technology for SBA is the smart product model (SPM), which is a software product consisting of a set of standards and protocols that allow legacy databases—object oriented, relational, and/or flat file—to be integrated into a persistent distributed repository for all information related to a design/development project. Tools and users can access any piece of information and/or any method in the SPM without having to consider which database or databases contain the data. Users can avoid large monolithic and unmanageable data storage approaches, and their attendant problems, through the use of agile distributed and modular information management solutions.

The SPM product provides users with standard base classes that are usable by any selected instance. Users have complete flexibility in defining or recycling the schema they use for storing information. Similarly, users have complete flexibility in determining which databases to use and how to physically and logically partition data so that information relevant to a particular context can be stored locally for fast access. A context is a group of tools joined together in a federation to perform some analysis or design task. Every context has a run-time-interface (RTI), which provides the infrastructure for information exchange, and a context manager, which controls and manages the tools. For example, a design context may consist of a computer-aided design (CAD) tool, a database, and the RTI. The database would be the local SPM and could contain pointers to data in other local SPMs.

The SPM's distributed heterogeneous architecture allows users to create logical groupings of information germane to a specific analysis task without recompiling or relinking tools or databases. These capabilities allow companies to form virtual enterprises and share design information without having to redo their internal data-management approaches.

The product SPM provides this functionality via a globally unique naming convention, the minimal set of base classes mentioned above, and an integrating SPM (ISPM) that provides a card-catalog functionality for all persistent data stored in any database associated with a selected instance. By keeping the infrastructure simple, the amount of work that users have to do to transition their legacy data into the SBA environment is minimized. SBA users can use the SPM product to produce a global SPM that is the virtual integration of all databases in a selected instance as well as local SPMs that are groupings of one or more databases. Local SPMs are usually part of a context, a high-level architecture federation of tools established to work selected aspects of the design, analysis, or manufacturing processes. Local SPMs are formed to improve efficiency, but they also allow members of a virtual enterprise to recycle their SPMs from project to project. As long as the databases in the local SPMs conform to the SPM product standards, they can seamlessly integrate simultaneously into both local and global SPMs.

The SPM architecture does the following:

- Supports legacy databases with minimal changes.
- Allows organizations to integrate their current data repositories with minimal changes so that the transition to SPM is smooth and low cost.
- Offers extensibility and scalability so that it can accommodate large designs as well as the additions of new databases as the design progresses.
- Creates virtual enterprises by integrating logically and physically distributed COTS databases and custom databases. Much of this functionality is provided by COTS databases, but the integrating functionality must be provided by the SPM product.
- Provides seamless access to all data for SBA tools. Tools must be able to locate any information in the SPM and, once it is located, work with it without requiring database-specific interfaces.

The SPM architecture supports the implementation of intelligent information that is capable of responding to changes, such as the relocation of a component, and to user inputs for data. The current paradigm to implement intelligent information lets users and tools invoke object behaviors/methods as needed, but the overall SPM architecture supports other approaches to facilitate the integration of legacy data and future developments in data-storage technology.

Much of the requisite technologies for SBA exist or are being developed under multiple efforts in the information technology area. In particular, a DARPA program focused on simulation-based design (SBD) is developing a prototype system that is a collaborative, multidisciplinary environment for developing and using virtual/real prototypes. A SBD system is configured for a specific product application, linking copies of the SBD core-processing system together with application-specific software tools selected for the particular functions being addressed.

Engineers are able to quickly generate a virtual prototype for an initial conceptual design to validate feasibility and provide a basis for cost estimation by reusing data from previous systems and by importing data from external sources. The virtual products can be analyzed with existing or legacy analysis tools and can be operated in virtual environments combining real and simulated products.

With SBD, engineers are able to capture design processes, such as the steps in designing a power subsystem for a ship, as mega-programs—programs made up of several programs—that are manipulated and operated exactly like the virtual prototypes. Development proceeds within SBD by establishing product constraints and requirements and by constructing increasingly detailed virtual prototypes of the product. The virtual prototypes (software models) can be viewed, interacted with, analyzed, and operated like real prototypes. As design changes are made, the prototypes are analyzed and evaluated by operation within synthetic physics-based environments. SBD not only manages these design artifacts, with built-in configuration management tools, but also allows engineers to incorporate components of different levels of fidelity within a virtual prototype.

Complex product development typically involves multiple heterogeneous organizations. The interconnectivity needed for product development requires support for defining, managing, and enforcing development processes and the resulting work flow. In a large-scale product development enterprise, each team has its own data, product, and process models. A collaborative SBD system is configured as a collection of copies of the common software. The common software is called a core processing system. The software units that make up the core processing system work together to provide seamless access to all public resources in the entire SBD system so that tools do not need to deal with resource location and allocation issues. Each user interface provides access to the rest of the SBD system as mediated by the core processing system units.

SBD allows users to maintain the product data in one or more databases—legacy or new, flat file, relational, or object oriented—that can be either centralized or distributed. Each core processing system maintains an object model that is accessed from its user interface for visualization and interaction.

To keep the data consistent across all participating sites in a collaborative environment, SBD uses a publish-and-subscribe protocol to communicate between core processing system sites. A core processing system manages the communication and coordination by publishing classes of objects that it owns (i.e., that it can create and manage) and subscribing to the classes that it is interested in as well as through messages (remote method invocation). Each core processing system publishes the kinds of data and services that it can provide. It will respond to requests for data, services, or publication of certain data. In this sense the core processing system provides a gateway for all communication with other SBD systems. Tools subscribe to the data that it is using so that it can be automatically informed of changes. The publish-and-subscribe protocol ensures that relevant attribute information is propagated to all interested users.

The publish-and-subscribe interface can be used to protect proprietary data and enforce security policies. It is up to the overall system management to define minimum information interchange requirements. This design gives tools the ability to access any public data in the product model via standard procedures without needing to consider the physical or logical location of the data.

The primary technology areas involved in SBA and SBD are multidisciplinary analysis and optimization, advanced modeling and simulation, data visualization and interaction technologies, collaboration technologies, software interface and standards technologies, and advanced computing technologies.

Future Impact on Naval Force Capabilities and Missions

SBA will have a major impact on the Department of the Navy because it will reduce both resource requirements and risk associated with the acquisition process. In addition, SBA will substantially increase the quality of the systems being acquired. As a result, the Navy Department will be able to more efficiently procure systems that meet their validated requirements. Platforms will be able to perform a wider variety of missions and interoperate more efficiently with other platforms and services because of refinements designed into platforms during the concept-development phase. Life-cycle costs can be determined early in the design phase so that appropriate trades can be made with real-time review of the cost and benefits. By using SBA, the Navy Department should be able to improve the value-price ratio for future systems by a factor of at least two.

Technology Developments Needed

Recent advances in computer and networking technology are bringing SBA closer to reality. Achieving the full capability promised by SBA, however, presents a number of technology challenges. One of the most important needs is the establishment of standards for software interfaces. Object-oriented technology offers promise toward these needs but will not be adequate without some significant extensions. Also, it does not address the use of existing legacy software and databases. One method of dealing with these needs is the use of wrappers, which are interfaces added to existing code to make it compliant with the common object request broker architecture (CORBA). Because these methods are not optimized for data exchange and generally are relatively slow, they are only a temporary solution.

A second area that must be improved to meet SBA needs is that of database management. Complex systems often contain tens of millions of parts. Each part can have tens of thousands of bytes of data associated with it. Accessing, interacting, and managing this massive database is a substantial challenge. Human-computer interaction technologies that use immersive visualization to view data in a hierarchical manner and parse the data consistent with viewer needs are an important part of developing database management capabilities.

The ability to perform SBA in a rapid manner will be limited by computation rates for at least a decade. Design and analysis codes, such as computational fluid dynamics (CFD) codes to deal with fluid flow around a ship or structural-loading codes, frequently take hours to run today. If these codes are to be run interactively, improved efficiencies and faster computers are needed. The panel predicts that computer power will continue to grow at about the same pace it has over the past 40 years (see Chapter 2 of this report), so that this area should not be a fundamental limitation to SBA growth but rather will support evolutionary growth over the next few decades.

Product life-cycle development already makes use of extensive computer-based analysis and simulation models. However, the current tools and products cannot be integrated to form a unified SBD system. Today there are no requisite sets of tools that are directly suitable for defining and developing virtual prototypes. The tools used in the product-development process do not integrate well with each other or with other types of tools and do not always have compatible data types. Although there are tools to support collaboration between people engaged in collaborative design, there are no software tools to coordinate the collaboration between tools within concurrent engineering processes.

Commercial technology and standards that address tool-to-tool communication (e.g., CORBA and/or Internet protocols) are beginning to become available. Also, standards are beginning to emerge for representing virtual prototypes. The virtual-reality modeling language (VRML) has been explicitly designed to produce virtual prototypes that can be placed in synthetic worlds and interact with other objects in these worlds. Leveraging these commercial developments and Navy development focused on integration and mission dependent software will enable rapid advances in overall SBA capability.

Development Sources

Most of the individual technologies in this area are commercially driven. However, the development of a high-level architecture that allows Navy, Marine

Corps, and DOD models to interoperate with other simulations, the integration of the software from different sources, and the modeling and simulation of systems and mission performance should be supported by the U.S. government. It is anticipated that government and commercial development will continue along these lines.

Foreign Technology Status and Trends

The United States leads the technology is this area and is expected to maintain a lead as long as sufficient investment continues. The lead may narrow if other countries increase their spending is this area.

Time Scale for Development and Insertion

DARPA is currently running a demonstration project in SBA. SBA technology appears to be ripe for adoption by the military Services, although the simulation tools will have to be tailored for specific acquisition applications. The capability to perform multidisciplinary optimization for complex product acquisition should be readily available by the year 2000, with ever increasing capability available thereafter.

AGILE COMMERCE

The Cold War has left the United States with two parallel industrial infrastructures—one for defense and the other for commerce. Each sector relies on distinct manufacturing technologies and business practices. This legacy has not only made many defense products unaffordable but also has encumbered our industrial competitiveness in the global manufacturing scene. Today's marketplace is increasingly characterized by fierce global competition and complex customer behavior. Success in the 21st century will depend on the ability to respond quickly to a niche market demanding high-quality, customized products at the lowest possible cost. This drives the need for a unified, dual-purpose industrial base that can cater to both defense and commercial needs and has the ability to succeed in an environment dominated by continuous and unpredictable changes.

The dual-use strategy can be realized by deploying various technologies and business practices. A flexible infrastructure allows a reconfigurable supplier network that is capable of producing a wide range of products. Lean manufacturing eliminates unnecessary inventory and product defects and minimizes the use of resources. Agile commerce is a new paradigm that encompasses and enhances the above strategies and reflects the ability to quickly adapt in a continuously changing requirements environment and to link those requirements to industrial suppliers and their manufacturing capabilities.

Description of the Technology

Agile commerce is the integration of innovative information technologies and business practices to implement an agile electronic network of customers, suppliers, and other service brokers to provide rapid procurement and production of supplies and materials. It is based on agent-based architecture and offers a set of services to enable the Navy Department and companies tied to the network to form and operate as a virtual enterprise.

The Navy Department and industrial suppliers will be able to easily access and use agile commerce for efficient procurement and production of high-quality, customized products at substantially lower cycle time and costs. Agile commerce is envisioned to provide to the United States an integrated industrial base equipped to rapidly respond with highly customized solutions to customer requirements of any magnitude, thus reinstating the United States as the world leader in manufacturing.

Relevance to the Naval Forces

With increasing reliance on commercial development and suppliers, agile commerce will enable the Navy Department to accomplish the following:

- Rapidly respond to changing requirements;
- Optimize the use of human and machine resources;
- Form virtual enterprises with suppliers for efficient acquisition; and
- Perform customization of high-quality products at competitive prices.

These capabilities are necessary to enable acquisition planners to acquire the right platforms and systems at the right cost and rapidly enough to meet operational requirements. To meet evolving operational needs, agile commerce capabilities allow a manufacturer-in-the-loop process for dynamic planning and conduct of operations.

Technology Status, Trends, and Developments

The agile commerce strategy is driven by matching business needs with advancements in information technology and electronic commerce. Advanced networking capabilities will link the information infrastructure of manufacturing enterprises to support agile business transactions. These advanced networks will connect and seamlessly integrate information about customers, multitiered suppliers, third-party brokers, and other relevant users. Accepted and emerging standards will be used to ensure the widest utility and to avoid duplication of existing hardware and software technologies. The agile commerce net will consist of three major technological elements as follows:

- Communication infrastructure;
- Intelligent agent-based services; and
- Information-formatting and delivery tool kits.

Networking technology will continue to advance rapidly, driven by commercial interests and needs. To provide adequate security and interfaces to DOD databases, the Department of the Navy must support complementary standards, interfaces, and selected network technology.

A DARPA-sponsored project, agile infrastructure for manufacturing (AIMS), is developing a pilot that is currently being applied to a limited set of major, second-tier, and small-business suppliers. Once the information infrastructure, technologies, and business practices have been validated, it is anticipated that AIMS or a similar rapid commerce capability would be ready for application to a large defense/commercial supplier base, spanning the U.S. industry. This should begin in 1998. Capability is expected to grow rapidly with time. Leveraging developments from other related areas, capability should double every 12 months for the next several years.

REAL-TIME LOGISTICS MANAGEMENT

Description of the Technology

Real-time logistics management (RTLM) is the near-real-time ability to plan, execute, monitor, and replan the availability of people, equipment, units, and supplies to support mission operations. To facilitate the complex interactions required for successful multidisciplinary coordination, all elements in the logistics chain should have access to the most current information related to the supplies and operations under consideration. That information must be presented in a form that can be understood by various disciplines and in a manner that highlights the impact of recent information on their own sphere of influence. The explosion in information technology makes possible revolutionary improvements in RTLM, and detailed knowledge of logistics information will be available instantly to all personnel from logistics managers to operators.

Relevance to the Naval Forces

The naval forces rely on the timely availability of assets and supplies as it supports a variety of missions at forward locations throughout the world. The Navy also transports the majority of supplies to support conflicts wherever they occur. Advanced RTLM technologies will provide the Department of the Navy with detailed planning, optimized scheduling, coordinated execution, and accurate visibility into assets, all of which are necessary in future operations. Recent experiences have often shown that needed supplies are not available because of incorrect requisitions, misshipments, or even incorrect labeling or storage locations. Currently, a work-around for these problems results in excessive ordering of needed supplies, frequently resulting in an overage of a factor of two. RTLM should eliminate these problems and reduce supply substantially.

Technology Status, Trends, and Developments

The United States is currently on the leading edge of an explosion in information-management technology that is making possible revolutionary improvements in RTLM. Among the most important technologies relevant to logistics management are autonomous smart agents for information collection and integration; interactive, integrated simulation; and process optimization. Recent developments in these areas and others are providing initial capabilities in several applications. In 2 years, there should be notable progress in functionality. In 10 years, a high degree of automation should be available for many of the logistics-management functions. By 2035, most of the RTLM information functions will be highly automated.

RESOURCE PLANNING

Description of the Technology

Resource planning is the decisionmaking process of acquiring necessary resources to meet organizational requirements. It requires that end-to-end thinking and systems engineering be applied to the planning process. It must extend beyond the individual system or platform and consider the joint mission requirements with other Services and nations and the concomitant capability. Resource planning must include the capability to compare alternative systems or platforms and to consider acquisition options. Full and accurate accountability must be employed to determine life-cycle costs. These capabilities are necessary to enable efficient resource planning.

Relevance to the Naval Forces

Because of the complexity of many systems, processes, or operations, resource planning becomes a highly nonlinear process. In these circumstances, integrated simulations for careful development and review of alternative solutions and informed decisionmaking are required. Employed in this manner, resource planning allows operators to optimize performance, schedule, and costs for selected plans.

Technology Status, Trends, and Developments

Resource-planning capabilities will be integrated into the simulation-based acquisition capability described above. This integration will allow the resource issues to be coupled with other mission plans and operations to construct alternative mission solutions and review in detail the value and cost of each option as it affects the total mission. By optimizing resource planning across the total mission, the naval forces will achieve significant improvements in mission effectiveness, efficient resource utilization, and large overall cost reductions.

DYNAMIC MISSION PLANNING

Description of the Technology

Dynamic mission planning (DMP) is the rapid, intelligent assembly and reassembly of necessary mission information to construct a dynamic representation and context for past, current, and projected mission operations. DMP will support simulations of plans and missions, using advanced distributed simulation technologies developed through DARPA and other agencies for programs such as the synthetic theater of war (STOW) and simulation-based design. DMP will incorporate coastal systems and fixed elements, as well as ships, weapons systems, environmental effects, and other dynamic elements. Behaviors of ships and other platforms and groups represented by higher-level commands will also be included. A major element of DMP is the analysis, evaluation, and integration of data that contain uncertainty or that may be totally incorrect.

Relevance to the Naval Forces

Dynamic mission planning will create a virtual battle space that will allow the Navy and Marine Corps to visually interact with all elements of a selected scenario. It will contain multiple levels of fidelity and enable operators to rapidly develop and evaluate multiple options. DMP will support three integrated functions:

- Strategy development,
- Plan generation, and
- Operations assessment.

Because of the escalating complexity of these functions, development of the most effective plan requires complex analysis and understanding of multiple disciplines. DMP and the virtual battle space will provide multiple, interactive interfaces to the mission planning process and be usable by multiple users at all levels of activity. DMP, when coupled with the technologies in simulation-based acquisition, will provide a capability to create a much more optimized mission plan and to dynamically maintain that plan.

Technology Status, Trends, and Developments

The capability to incorporate data that contain uncertainty is one of the greatest challenges in constructing and maintaining an accurate plan. Multiple approaches are being developed to deal with uncertainty and should lead to substantial capability by the year 2000. The status and trends of these technologies are discussed in the simulation-based acquisition section of this chapter.

SYSTEM OF SYSTEMS

Description of the Technology

A system of systems is an integration of a number of elements to form a larger, more complex system to accomplish a desired objective. Future missions will require coordinated action from integrated sea, air, and land forces that cannot be achieved by a single system or platform. A system of systems would include constellations of multiple system types, such as satellites, aircraft, ships, expeditionary units, and control centers. These individual elements, complex in and of themselves, together would constitute a complex system of systems that, when operated as an integrated entity, can achieve warfighting effectiveness greater than the sum of the individual elements. Efficient information exchange and communication links will be essential to the system of systems approach.

Relevance to the Naval Forces

The naval forces operate in a highly complex environment that contains many system components that change dynamically. Each of the Navy and Marine Corps platforms or weapon systems has the potential to be more effective in its mission objectives if it is interconnected with complementary systems in an appropriate manner. This approach will enable the Navy and Marine Corps to deploy its resources and achieve mission success at lower cost and risk.

Technology Status, Trends, and Developments

The technology and capabilities described in the section titled "Simulation-based Acquisition" in this chapter provide the overall capability to model and evaluate the system-of-systems approach. These simulations allow the construct of a plan that uses the same capabilities as described in the section on dynamic mission planning. The status and trends of the technologies involved are as described in both of these sections.

RECOMMENDATION

Large-scale processes within the Department of the Navy, such as platform acquisition, logistics management, resource planning, mission planning, and personnel management, are major cost drivers of naval operations. Information technologies are becoming available that can revolutionize the execution of these enterprise processes with a resultant substantial reduction in manpower, cycle time, risk, and cost. Simulation-based acquisition is an example of a revolutionary approach for the acquisition of complex platforms and systems. The Department of the Navy should strongly embrace and support these information technologies for enterprise-wide processes.

11

Other Technology Application Areas

Four other application areas in which many of the above technologies have important implications for the Department of the Navy and which are important to the future dominance of the battlefield are (1) space, (2) signature management, (3) chemical and biological warfare, and (4) combat identification.

SPACE

For military use, the four major applications of space are navigation of platforms and weapons, communications, environmental monitoring, and surveillance. It is clear to the panel that for the Department of the Navy, space offers innumerable advantages for rapid and secure global communications, for monitoring of the air and sea environment, for locating and targeting adversary naval and air forces, and potentially for soft and hard kills of targets. Space will continue to provide the high ground for line-of-sight access to the ocean areas of the globe and is integral to the mission responsibility of the naval forces.

The panel recognizes that the Department of the Navy, while a major user of space, is not a developer of major space systems. However, the panel believes that to remain a "smart buyer and user" of national space assets, the Navy Department must maintain a knowledgeable cadre of professional personnel versed in the latest space technology, equipment, and systems. The Department of the Navy space cadre must participate in an active manner with the other major organizations constituting the National programs through joint training and job rotation. This cadre must have opportunities for professional growth and recognition. It must also have access to the highest decision levels of the Department of the Navy so that space requirements, alternatives, and options can be considered and acted on in the best overall interests of the naval forces.

SIGNATURE MANAGEMENT

Signature management is the application of many of the technologies described in this report to permit the purposeful reduction of the observability of a platform, vehicle, facility, person, and so on in order to accomplish one or more of the following objectives: (1) hide its existence from the enemy, (2) confuse the enemy's ability to locate it, (3) confuse the enemy's ability to identify it, or (4) reduce its vulnerability to attack. Over the past decade, the DOD has made remarkable use of signature management to enhance the probability of success on the battlefield. The F-117 stealth fighter clearly demonstrated the advantage of surprise in Desert Storm in terms of inflicting initial devastating destruction on the adversary while remaining virtually invulnerable. U.S. Navy submarines through their shape, material coatings, and operational practices have been the silent service for decades. The B-2 stealth bomber and the Sea Shadow stealth ship developments, while not yet deployed in warfare, are other examples.

In the past decade or so, materials to control the absorption, reflectance, and emissions of electromagnetic and optical signals, the muffling of acoustic emissions, and the shaping of objects have been the principal methods of controlling and reducing the signature of platforms below detectable levels. Electromagnetic analysis codes based on fundamental physical principles, such as Maxwell's equations, attempt to determine the signature response of a platform to incident electromagnetic energy. The sophistication and fidelity of these codes have evolved at the rate of progress in computers. Progress in the future should track the computer technology trends described in Chapter 2, provided that the Department of the Navy invests in these software code developments. This is important because high-fidelity computer modeling of electromagnetic wave interactions with objects is a powerful design tool for future stealthy platforms. Confidence in the accuracy and fidelity of such modeling codes can greatly reduce expensive testing time on outdoor ranges.

Materials for these applications have been complex in many cases and expensive. To make signature management available to a wider variety of applications, the Department of the Navy should invest in appropriate material technologies with a goal of significantly reducing the cost and improving the ease of application and maintainability.

With the advent of MEMS technology and the availability of systems-on-a-chip and with advances in smart materials, computing power, and intelligent software systems, the possibility exists in the future to actively control the signature of important assets such as aircraft, UAVs, and UUVs. This could lead to signature warfare, the ability to present in real time a different signature to an adversary as the warfighting scenario evolves. Signature warfare could change the outcome of a battle by using the unique knowledge of how signatures can change the perception of the adversaries' visualization of the battle space. Signature management remains critically important to the naval forces.

CHEMICAL AND BIOLOGICAL WARFARE

Chemical and biological warfare (CBW) is an area that has become a credible threat, both tactical and strategic, without having been demonstrated effectively.[1] The use of mustard gas in World War I was sobering and led to international agreements to ban further use of chemical weapons. Developments in recent decades have altered this situation in a number of respects. The proliferation and advances of technology have brought CBW within the reach of almost any nation or well-organized terrorist group. People of all nations are being trained in techniques that can produce biological weapons of mass destruction. Facilities need not be vast and expensive and may be easily disguised. These weapons have been employed in the Iraq-Iran war and in the release of nerve gas in the Tokyo subway in 1995.

Chemical warfare and biological warfare are often lumped together, but in fact they are distinct. Biological warfare, like nuclear warfare, involves weapons of mass destruction. These are strategic weapons. Chemical warfare is more localized. Estimates of the extent of the danger of biological weapons have been made by various agencies. Biological weapons are predicted to be effective over several hundred square kilometers, causing hundreds of thousands to millions of deaths. Chemical weapons are estimated to be effective over a few square kilometers, with deaths in the hundreds to thousands. Toxins form a third class of agent that is often included with chemical and biological warfare agents. Toxins are associated with biological warfare because they have their origins in biological media. Toxins, however, are substances that can be synthesized outside of biological media, making them more appropriately an agent of chemical warfare. In this regime where chemistry and biology are merging, synthesis of toxins will often be accomplished by biochemical or genetic engineering techniques.

Detection of facilities for the production and delivery of chemical and biological weapons is very difficult and often may not be possible. Unlike the apparatus of nuclear programs, that for biological warfare is relatively small, inexpensive, and indistinguishable from facilities used in peaceful pursuits. For example, development of insecticides by genetic-engineering methods and adapting methods for their distribution closely resemble procedures required for producing offensive biological weapons. With the proliferation of techniques and the availability of equipment, it is virtually certain that many nations will acquire

[1]Dando, Malcolm. 1994. *Biological Warfare in the 21st Century*, Brasseys, London.

the capability for manufacturing CBW agents. Clandestine operations should be relatively easy to implement.

Offensive Aspects

The United States is a party to both the Chemical Weapons Convention and the Biological Weapons Convention, international bodies that are working toward the elimination of these weapons worldwide. Accordingly, the United States does not now have programs for the production of offensive chemical and biological weapons, and it is unlikely that this position will change. Following World War II, the United States and other nations had chemical weapons programs, but the United States terminated its activities in 1969.

Defensive Aspects

Defensive chemical and biological weapons programs are acceptable, and the United States is active in the area. The Army is the lead agency, and a substantial program is in place. Development of vaccines for immunization is an example of defensive CBW. Imaginative approaches are being pursued. For example, vaccines that will respond to a range of disease threats are being devised. The defensive technologist will, however, always be at a disadvantage in view of the wide range of options open to the offensive technologist.

Antidotes for nerve gases (e.g., sarin, tabun, and VX) are unlikely, owing to the rapid action and high lethality of these agents. These gases at milligram doses are deadly and are readily absorbed through the skin. Fortunately, these compounds hydrolyze to less lethal products fairly rapidly, but local use is a real threat. Of course, new agents could be (and may have been) designed that are longer lived in normal ambient conditions. The compact and crowded nature of Navy ships, for example, could be a problem in defending against a chemical weapons attack. The use of sarin in a Japanese subway station in 1995 did not result in mass death, but it might have in the hands of more competent terrorists.

Detection, Tracking, and Prediction of Dispersal

Dispersal of CBW agents is accomplished predominantly by use of aerosols. There may or may not be a visible cloud. Dispersal of aerosols over wide areas is dependent on wind patterns. From the Navy's point of view, delivery by up-wind release could be a problem, but the mobility of the target vessel would complicate the attacker's mission. Even so, the advantage would seem to be on the side of the CBW perpetrator. Evasion by a battle group, especially when a cloud of harmful substances is not visible, would be very difficult.

Protection Techniques

Although vaccines are known for most of the likely biological warfare agents (e.g., anthrax), there are many agents available (at least a dozen), and vaccination of a general population for the full range of agents might not be feasible. The naval forces could presumably be immunized for the obvious agents, but this strategy might not work against a sophisticated foe.

The methodology of self-protection from CBW agents involves detection and identification of the agents and evasion and/or provision of a safe haven until the threat has abated. The U.S. military, and the Department of the Navy specifically, is investing heavily in the science and technology of detection and identification of CBW agents. Building on the rapidly advancing field of MEMS, sophisticated instruments are being developed that give promise of being small, rugged, broad gauge in coverage, and reasonable in cost. MEMS employs fabrication methods developed in the semiconductor industry to manufacture mass spectrometers, chromatographs, and miniature chemical laboratories on a wafer. Fiber-optic sensors have been fabricated that can detect toxins and bacteria at levels relevant to CBW. Objectives include complete analysis systems that operate at less than a watt of power, are less than a pound in weight, are an inch or two across, and can be produced at a unit cost of tens of dollars. This kind of instrumentation is well adapted to transport by unmanned vehicles, and control and readout should be readily integrated into conventional electronics. The magnitudes of progress reported and promised in the MEMS-sensor-instrumentation arena are sufficiently out of the ordinary as to suggest hyperbole, but there is substance here and the progress is real. The investment implied is considerable (on the order of $100 million per year), but the capability offered is essential. Continued support of this exciting technology offers an effective way to stay ahead of the CBW aggressor.

The other aspect of the defensive problem is the safe haven to protect military personnel from CBW agents that are present. The Navy has been active in this area through the development of the Collective Protection System (CPS). CPS provides filters for CBW agents. High-efficiency air particulate (HEPA) filters, also employed in semiconductor clean rooms, are used to remove particulate material (e.g., spores), and charcoal filters are used to adsorb gaseous molecules, as in nerve gases. The protected zone is supplied with air drawn through the filters, and the atmosphere inside is operated at an overpressure to prevent inward leakage. A four-stage decontamination station allows entry of contaminated personnel to the protected zone. Although there are many areas for improvement, the CPS is already operational. Filter life is an issue that will require further development. Operation in heavy smoke must be demonstrated. Future enhancements may include destruction of CBW agents by catalytic oxidation units.

The CPS has been installed on the USS *Belleau Wood,* LHA-3, and about 30 ships are currently being or have recently been equipped to some extent. CPS can be designed into a ship's air-conditioning system to provide constant protection, or it can be backfitted into critical compartments (e.g., radar rooms) on existing ships. System cost is said to be about 2 percent of ship cost (excluding weapons systems). Overall, CPS represents a well-thought-out response to the novel threats of CBW.

Protective suits are available and provide good protection while they are being worn. Suits can be effective in a chemical weapons attack in which the agent can be neutralized or eliminated in some way. For CBW agents, one must consider what happens when the suits are removed. A further problem with suits is that they reduce the effectiveness of those wearing them. Problems aside, special protective clothing will play an important role in defense against CBW.

Effectiveness

The effectiveness of CBW against ships is largely untested. Biological warfare agents have an incubation period of at least a day and thus would probably not be the weapon of choice for battle situations. Chemical warfare agents, on the other hand, are fast-acting and can be lethal (as is the case for nerve gases) or incapacitating (e.g., U.S. advanced riot control agent), and could be used to advantage in a naval battle. The ship could be defeated without itself being damaged. CBW agents could also be used in a sabotage mode or in a port attack. If a substantial portion of the ship personnel were dead or incapacitated, the fleet would be of little use. Vulnerability of the naval forces to attacks of the nonbattle kind could be considerable in view of the naval forces' mission as the nation's advanced force in regional conflicts. CBW agents are required only in small quantities, and delivery could be made by swimmers, small boats, or unmanned aircraft. The manner in which this kind of warfare will be played out is yet to be determined. The Department of the Navy is supporting the right kinds of technology to face the issue strongly.

COMBAT IDENTIFICATION

In nearly every military operational scenario, the ability to identify friendly assets from enemy assets is extremely important. This capability has for many years been described as identification, friend or foe (IFF) but has recently become known as combat identification (CID). The capability to perform effective CID is a DOD-wide responsibility, since multi-Service involvement in battle requires uniform equipment, policies, and procedures. In the past, ineffective CID practices have caused many friendly-fire casualties as well as the destruction of friendly assets. As the tempo of warfare increases in the future, the incidence of friendly-fire accidents will increase correspondingly.

Navy and Marine Corps forces require effective CID in the following operations:

- Ground fire against ground targets,
- Ground fire against air targets,
- Air attacks against ground targets,
- Air attacks against air targets,
- Missile attacks against surface naval targets, and
- Ability to distinguish enemy and friendly assets in surveillance data.

Over the years, a number of systems have been developed primarily for the identification of friendly aircraft. Most early IFF systems took advantage of the radar sensor being used to detect and track aircraft. The ability to perform effective IFF required some form of answering transponder to identify an aircraft as friendly. Because most early radar systems used multiple high frequency (HF) and very high frequency (VHF), it was necessary for a friendly aircraft to carry an IFF set for each possible ground-radar frequency in use. These systems, such as the Mark XII and the Mark XV, had many deficiencies, including (1) the ability to be exploited by the enemy, (2) vulnerability to jamming, (3) reliability problems, (4) lack of uniform implementation by friendly nations, such as those nations belonging to NATO, (5) the inability to identify enemy aircraft, and (6) lack of covertness or low probability of intercept. These systems were also not tailored for naval or ground forces and assets. The Mark XV system was canceled in 1991, leaving only the older Mark II available to U.S. and allied forces.

In 1992, the Joint Combat Identification Program Office was established in response to congressional concern over friendly-fire casualties. Following the Gulf War, a number of "quick fixes" for friendly ground-target identification involving infrared sources were implemented. The Navy Department is currently developing a CID system that transmits GPS coordinates to a fleet satellite communications satellite and then back to a central command-and-control site. The system is encrypted but is neither covert nor resistant to jamming of the fleet satellite communications or GPS signals (see the discussion on GPS jamming in Chapter 5). The system may also be exploitable by an adversary. The U.S. Army currently has under development a millimeter-wave CID system for use on ground vehicles and perhaps helicopters. The Army has also developed a CID system using a special LPI waveform.

To date there has not been a multipurpose CID concept identified that uses common hardware for all of the various applications cited above. This serious deficiency has led to recent friendly-fire incidents, including the shoot-down of friendly helicopters in Iraq by Airborne Warning and Control System (AWACS)-controlled U.S. fighter aircraft. The panel strongly suggests that any effective

CID system must be interoperable across U.S. Service units, much like electronic warfare equipment and practices. Effective CID systems should be developed at the DOD level, and the current practice of assigning responsibility for certain CID elements to each of the services should be reconsidered.

The panel believes that there are sufficient technologies (as described in this report) in the areas of communications (especially LPI), signal processing, embedded computing, information, sensors, and signature management to develop an effective, multimission, interoperable CID system.

RECOMMENDATIONS

The following are the panel's recommendations:

- The Department of the Navy should maintain a knowledgeable cadre of professional personnel versed in the latest space technology, equipment, and systems; should ensure the professional growth, development, and recognition of this cadre; and should ensure that the cadre effectively enables the naval forces to satisfy their requirements for space-based information.
- The Department of the Navy should invest in and exploit the technologies described in this report for applications to reducing and controlling the signature of important assets.
- The Department of the Navy, in cooperation with the Joint Combat Identification Program Office, should initiate a fresh look at technologies to accomplish a robust, multimission, interoperable combat identification system. Inputs from universities, industry, and government laboratories should be sought to develop the best system that current technologies will permit.
- New combat identification technologies should be considered in concert with EW and signature management to optimize performance.

APPENDIXES

A

Terms of Reference

CHIEF OF NAVAL OPERATIONS

28 November 1995

Dear Dr. Alberts,

In 1986, at the request of this office, the Academy's Naval Studies Board undertook a study entitled "Implications of Advancing Technology for Naval Warfare in the Twenty-First Century." The Navy-21 report, as it came to be called, projected the impact of evolving technologies on naval warfare out to the year 2035, and has been of significant value to naval planning over the intervening years. However, as was generally agreed at the time, the Navy and Marine Corps would derive maximum benefit from a periodic comprehensive review of the implications of advancing technology on future Navy and Marine Corps capabilities. In other words, at intervals of about ten years, the findings should be adjusted for unanticipated changes in technology, naval strategy, or national security requirements. In view of the momentous changes that have since taken place, particularly with national security requirements in the aftermath of the Cold War, I request that the Naval Studies Board immediately undertake a major review and revision of the earlier Navy-21 findings.

The attached Terms of Reference, developed in consultation between my staff and the Chairman and Director of the Naval Studies Board, indicate those topics which I believe should receive special attention. If you agree to accept this request, I would appreciate the results of the effort in 18 months.

Sincerely,

J. M. BOORDA
Admiral, U.S. Navy

Dr. Bruce M. Alberts
President
National Academy of Sciences
2101 Constitution Avenue, N.W.
Washington, D.C. 20418

Enclosure

TERMS OF REFERENCE

TECHNOLOGY FOR THE FUTURE NAVY

The Navy-21 study (Implications of Advancing Technology for Naval Warfare in the Twenty-First Century), initiated in 1986 and published in 1988, projected the impact of technology on the form and capability of the Navy to the year 2035. In view of the fundamental national and international changes -- especially the Cold War's end -- that have occurred since 1988, it is timely to conduct a comprehensive review of the Navy-21 findings, and recast them, where needed, to reflect known and anticipated changes in the threat, naval missions, force levels, budget, manpower, as well as present or anticipated technical developments capable of providing cost effective leverage in an austere environment. Drawing upon its subsequent studies where appropriate, including the subpanel review in 1992 of the prior Navy-21 study, the Naval Studies Board is requested to undertake immediately a comprehensive review and update of its 1988 findings. In addition to identifying present and emerging technologies that relate to the full breadth of Navy and Marine Corps mission capabilities, specific attention also will be directed to reviewing and projecting developments and needs related to the following: (1) information warfare, electronic warfare, and the use of surveillance assets; (2) mine warfare and submarine warfare; (3) Navy and Marine Corps weaponry in the context of effectiveness on target; (4) issues in caring for and maximizing effectiveness of Navy and Marine Corps human resources. Specific attention should be directed, but not confined to, the following issues:

1. Recognizing the need to obtain maximum leverage from Navy and Marine Corps capital assets within existing and planned budgets, the review should place emphasis on surveying present and emerging technical opportunities to advance Navy and Marine Corps capabilities within these constraints. The review should include key military and civilian technologies that can affect Navy and Marine Corps future operations. This technical assessment should evaluate which science and technology research must be maintained in naval research laboratories as core requirements versus what research commercial industry can be relied upon to develop.

2. Information warfare, electronic warfare and the exploitation of surveillance assets, both through military and commercial developments, should receive special attention in the

review. The efforts should concentrate on information warfare, especially defensive measures that affordably provide the best capability.

3. Mine warfare and submarine warfare are two serious threats to future naval missions that can be anticipated with confidence, and should be treated accordingly in the review. This should include both new considerations, such as increased emphasis on shallow water operations, and current and future problems resident in projected worldwide undersea capability.

4. Technologies that may advance cruise and tactical ballistic missile defense and offensive capabilities beyond current system approaches should be examined. Counters to conventional, bacteriological, chemical and nuclear warheads should receive special attention.

5. The full range of Navy and Marine Corps weaponry should be reviewed in the light of new technologies to generate new and improved capabilities (for example, improved targeting and target recognition).

6. Navy and Marine Corps platforms, including propulsion systems, should be evaluated for suitability to future missions and operating environments. For example, compliance with environmental issues is becoming increasingly expensive for the naval service and affects operations. The review should take known issues into account, and anticipate those likely to affect the Navy and Marine Corps in the future.

7. In the future, Navy and Marine Corps personnel may be called upon to serve in non-traditional environments, and face new types of threats. Application of new technologies to the Navy's medical and health care delivery systems should be assessed with these factors, as well as joint and coalition operations, reduced force and manpower levels, and the adequacy of specialized training in mind.

8. Efficient and effective use of personnel will be of critical importance. The impact of new technologies on personnel issues, such as education and training, recruitment, retention and motivation, and the efficient marriage of personnel and machines should be addressed in the review. A review of past practices in education and training would provide a useful adjunct.

9. Housing, barracks, MWR facilities, commissaries, child care, etc. are all part of the Quality of Life (QOL) of naval personnel. The study should evaluate how technology can be used to enhance QOL and should define militarily meaningful measures of effectiveness (for example, the impact on Navy readiness).

10. The naval service is increasingly dependent upon modeling and simulation. The study should review the overall architecture of models and simulation in the DoD (DoN, JCS, and OSD), the ability of models to represent real world situations, and their merits as tools upon which to make technical and force composition decisions.

The study should take 18 months and produce a single-volume overview report supported by task group reports (published either separately or as a single volume). Task group reports should be published as soon as completed to facilitate incorporation into the DoN planning and programming process. An overview briefing also should be produced that summarizes the contents of the overview report, including the major findings, conclusions, and recommendations.

B

Acronyms and Abbreviations

ABM	Antiballistic missile
ac	Alternating current
ACTD	Advanced concept technology demonstration
A/D	Analog-to-digital
ADC	Analog-to-digital conversion (converter)
ADM	Advanced development model
AgZn	Silver zinc
AI	Artificial intelligence
AIMS	Agile infrastructure for manufacturing
AlGaAs	Aluminum gallium arsenide
AMRFS	Advanced multifunction radio-frequency system
ARG	Amphibious-ready group
ARM	Antiradar missile
ASAT	Antisatellite
ASIC	Application-specific integrated circuit
ASMP	Advanced Surface Machinery Program
ASW	Antisubmarine warfare
ATR	Automatic target recognition
AWACS	Airborne Warning and Control System
BASS	Bulk avalanche semiconductor switch
BAT	Brilliant antitank
bhp	Brake horsepower
C^3	Command, control, and communications
C/A	Coarse/acquisition (code)

CAD	Computer-aided design
CBW	Chemical and biological warfare
CCD	Charge-coupled device
CD	Compact disk
CEC	Cooperative engagement capability
CEP	Circular error of probability
CFAR	Constant false alarm
CFC	Chlorofluorocarbon
CFD	Computational fluid dynamics
CHRIMP	Consolidated Hazardous Material Re-utilization and Inventory Managment Program
CIC	Combat information center
CID	Combat identification
CINC	Commander in Chief
CINCPAC	Commander in Chief, Pacific
CISC	Complex instruction set computing (computer)
CMOS	Complementary metal oxide semiconductor
CMR	Colossal magnetoresistance
CNO	Chief of Naval Operations
CO_2	Carbon dioxide
COAMPS	Coupled Ocean-Atmosphere Mesoscale Prediction System
COADS	Coupled Ocean-Atmosphere Dynamic System
COEA	Cost and operational-effectiveness analysis
CONUS	Continental United States
CORBA	Common object request broker architecture
COTS	Commercial off-the-shelf
CPS	Collective Protection System
CPU	Central processing unit
CVD	Chemical vapor deposition
CVX	Next-generation aircraft carrier
CW	Continuous wave
D/A	Digital-to-analog
DAC	Digital-to-analog conversion
DBDB	Digital bathymetric database
dc	Direct current
DC	Digital circuit
DDS	Direct digital synthesis
DARPA	Defense Advanced Research Projects Agency
DEW	Directed-energy weapon
DFB	Distributed feedback
DMP	Dynamic mission planning
DOD	Department of Defense
DOE	Department of Energy

DRAM	Dynamic random access memory
DSB	Defense Science Board
DSP	Digital signal processing
ECAC	Electromagnetic Compatibility Analysis Center
ECL	Emitter control logic
EEG	Electroencephalogram
EHF	Extremely high frequency
EM	Electromagnetic
EMI	Electromagnetic interference
EMP	Electromagnetic pulse
ENIAC	Electronic Numerical Intergrator and Computer
EO	Electro-optic
ESM	Electronic support measure
ETC	Electrothermal chemical
EW	Electronic warfare
FET	Field-effect transitor
FFT	Fast Fourier transform
FLIR	Forward-looking infrared radar
FM	Frequency modulation
FP	Function point
FPA	Focal plane array
FSAD	Full-scale advanced development
GaAs	Gallium arsenide
GAMOT	Global acoustic monitoring of ocean thermometry
GaN	Gallium nitride
GCCS	Global Command and Control System
GLONASS	Global Navigation Satellite System
GMR	Giant magnetoresistance
GPS	Global Positioning System
HCS	Human-centered system
HDTV	High-definition television
HEMT	High-electron mobility transistor
HEPA	High-efficiency particulate air (filter)
HF	High frequency
HgCdTe	Mercury-cadmium-telluride
HM&E	Hull, mechanical, and electrical
HPCI	High performance computing initiative
HPM	High-power microwave
HSLA	High-strength, low-alloy
HTS	High-temperature superconductor
IC	Integrated circuit
IC^2	Integrated command and control
ICR	Intercoded recuperated (gas turbine)

IFF	Identification, friend or foe
IGBT	Insulated gate bipolar transistor
IHPTET	Integrated High-performance Turbine Engine Technology (program)
IMPATT	Impact avalanche transit time
InP	Indium phosphide
INS	Inertial navigation system
InSb	Indium antimonide
I/O	Input/output
IPCC	Integrated power conversion center
IPS	Integrated propulsion system
IR	Infrared
IMPATT	Impact avalanche transit-time (diode)
ISC	Integrated ship control
ISDN	Integrated services digital network
ISPM	Integrating smart product model
IW	Information warfare
JCS	Joint Chiefs of Staff
JJ	Josephson junction
JPEG	Joint photographic experts group
JPO	Joint Program Office
JSF	Joint Strike Fighter
JTF	Joint Task Force
ksi	Kips per square inch
kWe/kg	Kilowatts of electric power output per kilogram
LED	Light-emitting diode
LEO	Low Earth orbiting
LHX	Large-deck amphibious ship
LIDAR	Laser detection and ranging
LMRS	Long-term mine reconnaissance system
LNA	Low-noise amplification
LPI	Low probability of intercept
LWIR	Long-wavelength IR
Marisat	Maritime satellite
MARPOL	Marine Pollution (regulations)
MBE	Molecular beam epitaxy
MCM	Multichip module
MEMS	Microelectromechanical systems
MESFET	Metal-semiconductor field-effect transistor
MHD	Magnetohydrodynamics
MIMD	Multiple-instruction, multiple-data
MMIC	Monolithic microwave integrated circuit
MMW	Millimeter wave

MMWI	Millimeter-wave imaging
MOCVD	Metallo-organic chemical vapor deposition
MODFET	Modulation-doped field-effect transistor
MOS	Metal oxide semiconductor
MPP	Massively parallel processor
MQW	Multiple quantum well
MRI	Magnetic resonance imaging
MRLS	Multiple rocket launch system
MUD	Multiuser domain
MWIR	Medium-wavelength infrared
n-MCT	n-type MOS-controlled thyristor
NASA	National Aeronautics and Space Administration
NATO	North Atlantic Treaty Organization
NiCd	Nickel cadmium
NiMH	Nickel metal hydride
NiP	Nickel phosphide
NOGAPS	Navy Operational Global Atmospheric Prediction System
NORAPS	Navy Operational Regional Atmospheric Prediction System
NRAC	Naval Research Advisory Council
NRaD	Naval Research and Development
NRL	Naval Research Laboratory
NSB	Naval Studies Board
NSCAT	NASA scatterometer
NSF	National Science Foundation
OEIC	Optoelectronic integrated circuit
ONR	Office of Naval Research
OOP	Object-oriented programming
OOT	Object-oriented technology
OPIT	Oxide power in tube (process)
OR	Operations research
OSD	Office of the Secretary of Defense
OTH	Over-the-horizon
P(Y)	Encrypted precision code
PA	Power amplification
PC	Personal computer
PCR	Polymerase chain reaction
PCS	Personal communication system
PEBB	Power electronic building block
PET	Positron emission tomography
PFN	Pulse-forming network
PGM	Power generation module
PHEMT	Pseudomorphic HEMT
PM	Power magnet

PPS	Precise Positioning Service
PRIME	Plastic Removal in Marine Environments
PTO	Power takeoff
PtSi	Platinum silicide
PWM	Pulse-width modulated
R&D	Research and development
RAM	Random access memory
RF	Radio frequency
RSFQ	Rapid single-flux quantum (logic)
RISC	Reduced instruction set computing (computer)
ROM	Read-only memory
ROTHER	Relocatable over-the-horizon radar
ROV	Remotely operated vehicle
RTI	Run-time interface
RTLM	Real-time logistic management
S&T	Science and technology
SADARM	Search-and-destroy armor munition
SAR	Synthetic aperture radar
SAS	Synthetic aperture sonar
SATCOM	Satellite communications
SBA	Simulation-based acquisition
SBD	Simulation-based design
SEED	Self-electro-optic effect device
SEI	Software engineering initiative
SEP	Spherical error of position
SET	Single-electron transistor
SHF	Superhigh frequency
SiC	Silicon carbide
SiGe	Silicon germanium
SIMD	Single-instruction, multiple-data
SLOC	Source line of code
SMCS	Standard monitoring control system
SMES	Superconducting magnet energy storage (device)
SOA	State of art
SOHO	Solar Heliospheric Observer
SPM	Smart product model
SQUID	Superconductor quantum interference device
SQW	Single quantum well
SSGTG	Ship service gas turbine generator
SSIM	Ship service inverter module
SSN	Nuclear-powered submarine
STAP	Space-time adaptive processing
STAR	System Threat Assessment Report

STOL	Short takeoff and landing
STOW	Synthetic theater of war
SVD	Singular value decomposition
SWATH	Small waterplane area twin hull
TEWA	Threat Evaluation and Weapon Assessment
TMC	Titanium matrix composite
TOF	Time of flight
T/R	Transmitter/receiver
TRANSCOM	Transportation Command
UAV	Unmanned aerial vehicle
UHF	Ultrahigh frequency
ULCB	Ultralow-carbon bainite
UUV	Unmanned underwater vehicle
UV	Ultraviolet
UWB	Ultrawide bandwidth
Vac	Volts alternating current
Vdc	Volts direct current
VE	Virtual environment
VHF	Very high frequency
VLO	Very low observability
VLSI	Very large scale integrated
VOR	VHF omnidirectional range
VRML	Virtual-reality modeling language
VSTOL	Vertical short takeoff and landing
VTOL	Vertical takeoff and landing
WAAS	Wide area augmentation system
WBS	Wide-bandgap semiconductor
WIMP	Windows, icons, mouse, pointer
ZEDS	Zonal electrical distribution system